Lecture Notes in Computer Science 1554

Edited by G. Goos, J. Hartmanis and J. van Leeuwen

T0223230

Springer
Berlin
Heidelberg
New York
Barcelona
Hong Kong
London
Milan
Paris
Singapore
Tokyo

Shojiro Nishio   Fumio Kishino (Eds.)

Advanced Multimedia Content Processing

First International Conference, AMCP'98
Osaka, Japan, November 9-11, 1998
Proceedings

 Springer

Series Editors

Gerhard Goos, Karlsruhe University, Germany
Juris Hartmanis, Cornell University, NY, USA
Jan van Leeuwen, Utrecht University, The Netherlands

Volume Editors

Shojiro Nishio
Osaka University, Graduate School of Engineering
Department of Information Systems Engineering
2-1 Yamadaoka, Suita, Osaka 565-0871, Japan
E-mail: nishio@ise.eng.osaka-u.ac.jp

Fumio Kishino
Osaka University, Graduate School of Engineering
Department of Electronics and Information Systems
2-1 Yamadaoka, Suita, Osaka 565-0871, Japan
E-mail: kishino@eie.eng.osaka-u.ac.jp

Cataloging-in-Publication data applied for

Die Deutsche Bibliothek - CIP-Einheitsaufnahme

Advanced multimedia content processing : first international
conference ; proceedings / AMCP '98, Osaka, Japan, November 9 -
11, 1998. Shojiro Nishio ; Fumio Kishino (ed.). - Berlin ; Heidelberg
; New York ; Barcelona ; Hong Kong ; London ; Milan ; Paris ;
Singapore ; Tokyo : Springer, 1999
   (Lecture notes in computer science ; Vol. 1554)
   ISBN 3-540-65762-2

CR Subject Classification (1998): H.5, H.4, H.3, I.4, I.2.10, I.3.7, I.7.2

ISSN 0302-9743
ISBN 3-540-65762-2 Springer-Verlag Berlin Heidelberg New York

Typesetting: Camera-ready by author
SPIN: 10693211   06/3142 – 5 4 3 2 1 0   Printed on acid-free paper

Preface

This volume is the Proceedings of the First International Conference on Advanced Multimedia Content Processing (AMCP '98). With the remarkable advances made in computer and communication hardware/software system technologies, we can now easily obtain large volumes of multimedia data through advanced computer networks and store and handle them in our own personal hardware. Sophisticated and integrated multimedia content processing technologies, which are essential to building a highly advanced information-based society, are attracting ever-increasing attention in various service areas, including broadcasting, publishing, medical treatment, entertainment, and communications. The prime concerns of these technologies are how to acquire multimedia content data from the real world, how to automatically organize and store these obtained data in databases for sharing and reuse, and how to generate and create new, attractive multimedia content using the stored data.

This conference brings together researchers and practitioners from academia, industry, and public agencies to present and discuss recent advances in the acquisition, management, retrieval, creation, and utilization of large amounts of multimedia content. Artistic and innovative applications through the active use of multimedia content are also subjects of interest. The conference aims at covering the following particular areas: (1) Dynamic multimedia data modeling and intelligent structuring of content based on active, bottom-up, and self-organized strategies. (2) Access architecture, querying facilities, and distribution mechanisms for multimedia content. (3) Synthesis of virtual and augmented real environments using large amounts of multimedia data for the creation of multimedia content and the applications of such technology to mobile computing environments.

The Program Committee (PC) received 45 submissions from 12 different countries in Asia, Europe, North America, and South America. Each submission was reviewed by at least three PC members and in some cases by external reviewers. This volume contains 24 contributed papers and 6 invited papers presented at the conference. The contributed papers were selected by a highly competitive process, based on referee reports and painstaking deliberations by the PC members. We are very grateful to these members and the external reviewers, who are listed in the following pages, for their time-consuming and meticulous work in judging the papers. We also thank the invited speakers for sharing with us their vision of the future.

In addition to the conference, we also organized the Dynamic Media Contest, consisting of two tracks: the Network Media Content Track and the Exhibition Content Track. This volume contains the extended abstracts of two invited talks for the Dynamic Media Contest session, which was included in the conference as a special session.

Last, but not least, we thank all the attendees for contributing to the success of the conference. To you and to all those who did not have the opportunity to attend the conference, we are pleased to offer this volume as a snapshot of the state of advanced multimedia content processing research and practice today. We hope you will find it to be an invaluable reference and a source of inspiration for new ideas.

November 1998

Shojiro Nishio, Fumio Kishino
Co-Chairs, Program Committee

AMCP '98 Committees

Conference Chair:
Makoto Nagao — *Kyoto Univ., Japan*

Program Committee Co-Chairs:
Shojiro Nishio — *Osaka Univ., Japan*
Fumio Kishino — *Osaka Univ., Japan*

Dynamic Media Contest Co-Chairs:
Shinji Shimojo — *Osaka Univ., Japan*
Masatoshi Arikawa — *Hiroshima City Univ., Japan*
Yasuo Ariki — *Ryukoku Univ., Japan*
Kazutoshi Fujikawa — *Osaka City Univ., Japan*
Hiroyuki Tarumi — *Kyoto Univ., Japan*
Kuniaki Uehara — *Kobe Univ., Japan*

Finance Chair:
Masahiko Tsukamoto — *Osaka Univ., Japan*

Publicity Chair:
Yoshifumi Kitamura — *Osaka Univ., Japan*

Local Arrangement Co-Chairs:
Kaname Harumoto — *Osaka Univ., Japan*
Toshihiro Masaki — *Osaka Univ., Japan*

Sponsored by:
Japan Society of Software Science and Technology (JSSST)
Osaka University

Supported by:
Japan Society for the Promotion of Science (JSPS)

Program Committee Members

Masatoshi Arikawa	*Hiroshima City Univ., Japan*
Yasuo Ariki	*Ryukoku Univ., Japan*
Rudolf Bayer	*Tech. Univ. Muenchen, Germany*
Shih-Fu Chang	*Columbia Univ., USA*
Tat-Seng Chua	*NUS, Singapore*
Kazumasa Enami	*NHK, Japan*
Lennart E. Fahlén	*SICS, Sweden*
Henry Fuchs	*Univ. of North Carolina, USA*
Simon Gibbs	*Sony Dist. Sys. Lab., USA*
Yoshinori Hara	*NEC USA Inc., USA*
Michitaka Hirose	*Univ. of Tokyo, Japan*
Takeo Kanade	*Carnegie Mellon Univ., USA*
Yoshifumi Kitamura	*Osaka Univ., Japan*
Machiko Kusahara	*Kobe Univ., Japan*
Kin Fun Li	*Univ. of Victoria, Canada*
Zhi-Qiang Liu	*Univ. of Melbourne, Australia*
Kenji Mase	*ATR, Japan*
Steffen Meschkat	*ART⊕COM, Germany*
Shinichi Morishita	*Univ. of Tokyo, Japan*
Arcot Desai Narasimhalu	*ISS NUS, Singapore*
Ryuichi Oka	*RWC, Japan*
Igor Pandzic	*Univ. of Geneva, Switzerland*
Philippe Quéau	*UNESCO, France*
Jun Rekimoto	*Sony CSL, Japan*
Tetsuji Satoh	*NTT, Japan*
Shinji Shimojo	*Osaka Univ., Japan*
Keishi Tajima	*Kobe Univ., Japan*
Haruo Takemura	*NAIST, Japan*
Katsumi Tanaka	*Kobe Univ., Japan*
Demetri Terzopoulos	*Univ. of Toronto, Canada, & Intel, USA*
Yoshinobu Tonomura	*NTT, Japan*
Masahiko Tsukamoto	*Osaka Univ., Japan*
Kuniaki Uehara	*Kobe Univ., Japan*

External Reviewers

Hiroki Akama
Kenichi Arakawa
Kazutoshi Fujikawa
Junzo Kamahara
Toru Imai
Rieko Kadobayashi
Shahadat Khan
Volker Markl
Toshihiro Masaki
Yasuhiro Murasaki
Kazushi Nishimoto
Tomoyuki Nishita
Dirk Nitsche
Akira Sato
Guenther Specht
Yasuyuki Sumi
Hari Sundaram
Martine Wedlake

Table of Contents

Content Summarization

Video Summarization Based on Semantic Representation 1
  Rafael Paulin Carlos and Kuniaki Uehara (Kobe University)

Valbum: Album-Oriented Video Storyboard for Editing and Viewing Video 17
  *Yasuhiro Niikura, Yukinobu Taniguchi, Akihito Akutsu,
  and Yoshinobu Tonomura* (NTT)

Augmented Reality Technology and Applications

Beyond the Desktop Metaphor: Toward More Effective Display,
Interaction, and Telecollaboration in the Office of the Future via a
Multitude of Sensors and Displays (Invited Paper) 30
  Henry Fuchs (University of North Carolina at Chapel Hill)

A Method for Estimating Illumination Distribution of a Real Scene
Based on Soft Shadows 44
  Imari Sato, Yoichi Sato, and Katsushi Ikeuchi
  (University of Tokyo)

Integrating Real Space and Virtual Space in the 'Invisible Person'
Communication Support System 59
  Masahiko Tsukamoto (Osaka University)

Content-Based Video Indexing and Classification

News Dictation and Article Classification Using Automatically
Extracted Announcer Utterance 75
  Yasuo Ariki, Jun Ogata, and Masafumi Nishida
  (Ryukoku University)

Automatic Video Indexing Based on Shot Classification 87
  Ichiro Ide, Koji Yamamoto, and Hidehiko Tanaka
  (University of Tokyo)

Mutual Spotting Retrieval between Speech and Video Image Using
Self-Organized Network Databases 103
  *Takashi Endo, JianXin Zhang, Masakyuki Nakazawa, and
  Ryuichi Oka* (RWC)

Content-Based Retrieval

Content-Based Retrieval in Multimedia Databases Based on Feature
Models (Invited Paper) 119
    Peter Apers (University of Twente) *and Martin Kersten* (CWI)

An Efficient Index Structure for High Dimensional Image Data............................ 131
    Jae Soo Yoo (Chungbuk National University),
    Myung Keun Shin (KAIST), *Seok Hee Lee, Kil Seong Choi,*
    Ki Hyung Cho (Chungbuk National University), *and*
    Dae Young Hur (ETRI)

Color-Based Pseudo Object Model for Image Retrieval with Relevance
Feedback.. 145
    Tat-Seng Chua and Chun-Xin Chu
    (National University of Singapore)

System Environments for Virtual Reality

InvenTcl: A Fast Prototyping Environment for 3D Graphics and
Multimedia Applications 161
    Sidney Fels (University of British Colombia) *and Kenji Mase*
    (ATR)

The NAVL Distributed Virtual Reality System.. 177
    Martine Wedlake, Kin F. Li, and Fayez El Guibaly
    (University of Victoria)

Content Broadcast Systems and Applications

Research in Data Broadcast and Dissemination (Invited Paper) 194
    Demet Aksoy, Mehmet Altinel (University of Maryland),
    Rahul Bose (Brown University), *Ugur Cetintemel,*
    Michael Franklin (University of Maryland), *Jane Wang,*
    and Stanley Zdonik (Brown University)

Multimedia Database System for TV Newscasts and Newspapers........................... 208
    Yasuhiko Watanabe, Yoshihiro Okada, Kengo Kaneji, and
    Yoshitaka Sakamoto (Ryukoku University)

A TV News Recommendation System with Automatic Recomposition 221
    Junzo Kamahara (Kobe University of Mercantile Marine),
    Yuji Nomura (FFC Ltd.), *Kazunori Ueda* (Osaka University),
    Keishi Kandori (Asahi Broadcasting Corp.), *Shinji Shimojo,*
    and Hideo Miyahara (Osaka University)

Extended Digital Video Broadcasting with Time-Lined Hypermedia......................236
  Shinji Nabeshima, Kazuo Okamura, Takashi Kakiuchi,
  Kazutoshi Sumiya, Naoya Takao, and Yoshiyuki Miyabe
  (Matsushita Electric Industries, Co., Ltd.)

Video Images and Virtual Space

Active Image Capturing and Dynamic Scene Visualization by
Cooperative Distributed Vision (Invited Paper) 252
  Takashi Matsuyama, Toshikazu Wada, and Shogo Tokai
  (Kyoto University)

Videoplex: A New System Framework for Constructing Video-Based
Three-Dimensional Space..289
  Go Nishimura, Tamio Kihara, and Ryoji Kataoka (NTT)

Construction of Virtual Environment from Video Data with Forward
Motion 301
  Xiaohua Zhang, Hiroki Takahashi, and Masayuki Nakajima
  (Tokyo Institute of Technology)

Spatial Browsing for Video Databases 313
  Masatoshi Arikawa and Tetsu Kamiyama
  (Hiroshima City University)

Video Databases

AI-STRATA: A User-Centered Model for Content-Based Description
and Retrieval of Audiovisual Sequences 328
  Yannick Prié (LISI - INSA-Lyon), *Alain Mille* (LISA - CPE-Lyon),
  and Jean-Marie Pinon (LISI - INSA-Lyon)

Use of Action History Views for Indexing Continuous Media Objects 344
  Kaoru Katayama (Kyoto University), *Osami Kagawa*
  (Hiroshima Denki Institute of Technology), *Yasuhiro Kamiya*
  (Toyota Automatic Loom Works, Ltd.), *Hideki Tsushima,*
  Takuya Yoshihiro, and Yahiko Kambayashi (Kyoto University)

Semantic Structures for Video Data Indexing 356
  Koji Zettsu (Kobe Research Center, TAO), *Kuniaki Uehara,*
  and Katsumi Tanaka (Kobe University)

Interactive Content Creation

A Study of Emergent Computation of Life-like Behavior by Indefinite
Observation...370
    Michita Imai and Tsutomu Miyasato (ATR)

An Interactive Digital Fishtank Based on Live Video Images 386
    Toshihiro Masaki, Tetsuya Yamaguchi, and Yoshifumi Kitamura
    (Osaka University)

Contents Creation for Interactive Media (Invited Paper)...397
    Ryohei Nakatsu (ATR)

Visual Modeling for Multimedia Content

Visual Modeling for Multimedia Content (Invited Paper).......................................406
    Demetri Terzopoulos (University of Toronto)

Automatic Generation of Moving Crowds in the Virtual Environment....................422
    Naoki Saiwaki, Toshiaki Komatsu, and Shogo Nishida
    (Osaka University)

Extracting Facial Motion Parameters by Tracking Feature Points 433
    Takahiro Otsuka and Jun Ohya (ATR)

Dynamic Media Contest Session

Immersion Reconsidered: Virtual Reality Projects of Art+Com
(Invited Talk)...445
    Steffen Meschkat (Art+Com)

Synthetic Characters: Behaving in Character (Invited Talk).....................................451
    Bruce M. Blumberg (MIT, Media Lab.)

Author Index 453

Video Summarization Based on
Semantic Representation

Rafael Paulin Carlos[1] and Kuniaki Uehara[2]

[1] Department of Computer & Systems Engineering, Kobe University
rpaulin@jedi.cs.kobe-u.ac.jp
[2] Research Center for Urban Safety & Security, Kobe University
uehara@kobe-u.ac.jp

Abstract. Summarization of video data is of growing practical importance because the more expanding video databases, inundate users with
vast amounts of video data, the more users need reduced versions which
they can assimilate with limited effort in shorter browsing time. In recent
days, many researchers have investigated summarizing techniques, such
as fast-forward play back, and skipping video frames at fixed intervals
of time. However, all these techniques are based on syntactic aspects of
the video.

Another idea is to present summarized videos according its semantic
representation. The critical aspect of compacting a video is context understanding, which is the key to choosing "significant scenes" that should
be included in the summarized video. The goal of this work is to show
the utility of semantic representation method for video summarization.
We propose a method to extract significant scenes and create a summarized video without losing the content of the video's story. The story is
analyzed by its semantic content and is represented in a structured graph
where each scene is represented by affect units.

Keywords
Summarization, video semantic representation, skim video, video content.

1   Introduction

With the increase use of multimedia applications, video-on-demand as well as
video libraries will make thousands of hours of video available to users. Unlike a
full-length video that may not be always useful in retrieval or searching, video
libraries should provide a brief but full-content segments that may allow users to
understand the whole of the video. Since the early days of computing, researchers
have attempted to automate these scenarios. Even with some video skimming
techniques already developed[3][4][5], context understanding of the entire video
is required.

When the system is asked to summarize a story, vast amounts of information
within the selection are selectively ignored in order to produce a distilled version
of the original. During summarization, an information object is reduced to a

S. Nishio, F. Kishino (Eds.): AMCP'98, LNCS 1554, pp. 1–16, 1999.

smaller size and to its most important points. The critical aspect of summarizing a video is context understanding, which is the key to choosing "significant scenes" that should be included in the final summarized video.

We concentrate our work on summarizing traditional Japanese folktales. To create a semantic representation of these stories, we use the method of Lehnert [1], which structurally represents knowledge aquired from video data. For summarization, we have established two rules on which the importance of each scene is evaluated. These rules work as filters to finally obtain a full content summarized video.

The goal of this research focuses in the design of a system that will let the user get a clear and objective idea of a full video story just showing about 20% to 30% of the whole video data. The system allows the user to retrieve significant scenes from a video, in a procedure called "best" summarization.

In addition, the user can perform a "personal" summarization to create a summarized video with user preferences such as favorite characters, or specific relationships between characters in the story, cause-effect relation, etc. Furthermore, the user can control the length of the summarized video by adjusting "summarization rate," while keeping the most important scenes. With all these tools, the user is able to create a variety of summarized videos according to his or her necessities.

2   Affect Units

Through techniques in story analysis, it is possible to represent a video by its semantic representation[2]. In this research, we attained semantic representation by analysis of the video content through affect units. An affect unit represents a mental state or event of a single character at a specific moment in the story. All affect units are constructed from primitive units called "affect state." Three kinds of affect states are designed to represent mental states, positive events, and negative events. Each state has the following representation:

> OO : Representing a positive (or desirable) event.
> XX : Representing a negative (or undesirable) event.
> MM : Representing mental states that a character may incur.

With these affect units, simple events or situations of a story can be represented. A combination between these affect units helps to represent more complex situations and have a clear understanding of the story. To make these combinations possible, affect units will work as building blocks for more complicated affect configurations. These configurations do not provide all the recognition abilities needed, but an expanded set of these primitive units will describe more complex situations.

For instance, let us consider the combination using three affect units. It is possible to create a complex affect unit called "intentional problem resolution." One segment from "The Golden Ax Fable" was taken to represent this example.

"The woodsman's ax felt into the lake, he decided to ask for help to the God of water. Finally, he recovered his ax."

The semantic representation of this segment is shown in Fig. 1.

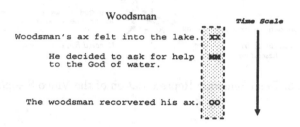

Fig. 1. Woodsman's Semantic Representation.

Affect units with more than one character require cross-character affect link. Diagonal segments between affect states will represent these, where the higher affect state precedes the lower affect state in time. For instance, consider the same story mentioned above. Now including the God of water for semantic representation.

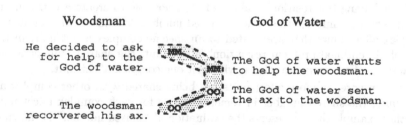

Fig. 2. Semantic Representation of the Woodsman and God of Water.

In Fig. 2, four affect units create the complex affect configuration called "honored request," where the God of water honorably answered Woodsman's request. If we repeat this process with all the characters, it is possible to create a semantic representation of the whole story based on affect units. More detailed information about affect units and their relation with scenes will be described in the following sections.

3 Semantic Representation

To achieve a high level analysis of activities and interrelations within the narrative, we need to recognize the affect unit of each character at each scene. To

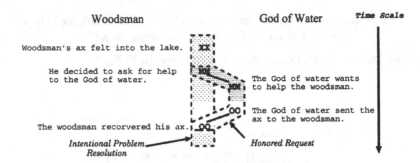

Fig. 3. Final Semantic Representation of the Video Segment.

establish interrelation between the characters, we need to link affect units with others, creating units that share the same affect state within the narrative.

For instance, consider the complex units analyzed in Fig. 1 and Fig. 2. A superposition between both complex units creates the full semantic representation of the story's segment. This can be observed in the video description graph shown in Fig. 3, both complex affect units are sharing two of their affect units, these units are called shared units. The superposition creates a relation between the complex affect unit "honored request" and the complex affect unit "intentional problem resolution".

Once having this relation established, it is possible to create a new concept called "shared number," the value of shared number depends on the number of complex affect units that are related to an specific complex affect unit. In the case of "intentional problem resolution" affect unit, the shared number is one because its affect units are related with just one complex affect unit.

Furthermore, as more affect units are being shared with other complex affect units, the shared number increments in one for each complex affect unit. Another example that represents the evaluation of "shared number" is shown in Fig. 4. Where the complex affect unit "unsolicited help" shares units with two

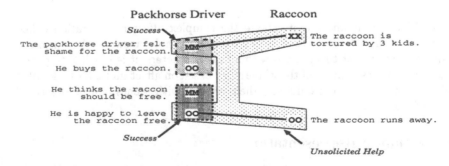

Fig. 4. Another Example of Semantic Representation.

different complex affect units, then shared number for "unsolicited help" is two. Furthermore, shared number for "success" is just one.

Taking these concepts, all the necessary tools to create a video semantic representation can be established. First, it is necessary to set up affect unit's structure, each unit is composed from its owner, affect state, description, and video information as it is represented in Fig. 5. This video information includes file name and start-end frame number of the video segment that the affect unit is representing. Each unit also includes information about the links which the unit is related. With these definitions it is possible to create a video scene, concatenating video segments from each unit. The final step is to describe the relation between scenes; this is made by the shared units they have in common.

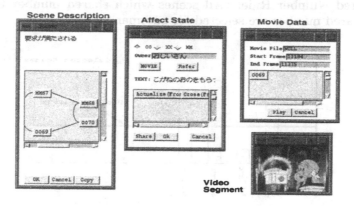

Fig. 5. Scene's Representation.

4   Summarization Rules

Once the connectivity of the affect units has been established, we can drive a summarization process based on affect analysis. In order to evaluate the importance of a scene, some rules have been proposed. These rules identify a set of scenes and units that play an important role in the summarization of videos. Detail information of each rule is given in the following subsections.

4.1   Shared Units Rule

By analyzing the nature of scenes by affect unit occurrences in term of their relationship, we begin to see that some units are central (representative) in driving inferences about other scenes. The identification of central scenes will be very important in the actual process of summarization. "Shared number rule" has been established according to the theory that as more units a scene shares with other scenes its importance becomes higher. It is not feasible to establish a fixed shared number over which the scene should be considered important or not. In a semantic representation of a story, the shared number varies according to the story content.

To solve this problem the system creates a list of shared numbers of each scene. Then, the system finds the maximum and minimum value and finally calculate the mean. This mean value is called "best shared number". Fig. 6 shows how this mean is evaluated. Finally, to concretize this rule, scenes that exceed the best shared number are treated as important scenes of the video. In Fig. 6 the 2nd., 4th., and 5th. scenes are selected by this rule because their shared number values are higher than the mean. Finally, shared number rule is defined as follows:

Shared Number Rule: "All scenes which shared number is over best shared number are selected for summarization."

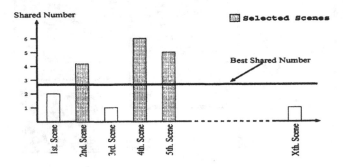

Fig. 6. Selected Scenes by Shared Number Rule.

4.2   Cause-Effect Rule

By shared number rule it is possible to obtain central (important) scenes within the story. However we need to identify scenes over all the story which may help to get chronological aspects. Most of stories tend to begin an intrigue, problem, or situations that are solved just at the end of the story. Generally, these scenes are not tightly related with others. Then the shared number rule is not feasible for summarization. Another way to select important scenes from a video is considering cause-effect relation. As an empirical theory, two scenes far apart in time with each other, but sharing at least one unit are important for its cause-effect relation.

To avoid ambiguity in the definition of cause-effect number, the system creates a list with the separation value of each scene. To evaluate this distance, the precedent and subsequent farthest related scenes are selected and the distance among them gives the cause-effect number. An example of how to obtain a cause-effect number is given in Fig. 7.

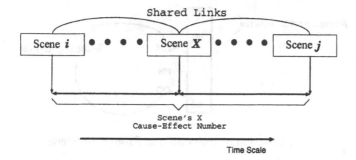

Fig. 7. Cause-Effect Number.

For instance, let us consider the Japanese folktale "The Golden Ax". At the beginning of the story, the humble and honest man appears with his wife (scene i). Then in the middle of the story, he receives the golden ax (scene X); finally, at the end of the story a scene shows that he lived happily forever (scene j). Scenes i and j may not represent important points of the story but may help for a clear understanding of the context of the story. Furthermore scenes i and j are not tightly related with others, this means that these scenes may not be selected for summarization by their shared numbers. However, these scenes are critical to represent scenarios and cause-effects aspects of the story. In Fig. 7, we calculate cause-effect number of scene X by adding the number of scenes between scene i and scene j. This process is repeated for each scene on the story's semantic representation.

Now with the created list it is possible to define the maximum and minimum cause-effect number, and obtain its mean. This mean value is called "best cause-effect number." To concretize this rule, scenes that are over the best shared number represent important parts of the video, having as final rule.

Cause-Effect Rule: "All scenes which cause-effect number is over the best cause-effect number are selected for summarization."

4.3 Prominent Scenes

Now having a set of scenes in which shared number is greater then best shared number and another set of scenes in which cause-effect number is greater than best cause-effect number, the final set of units can be created. The final rule identifies scenes which satisfy both rules. These scenes are called "prominent scenes."

Prominent scenes imply that the units (not necessarily the scenes) around these scenes are also highly important. The system creates a new list, which includes all the units around the prominent scenes. Fig. 8 shows how these units are selected. Only affect units directly related by a link with the prominent scene are selected for summarization. In Fig. 8, MM and OO affect units are related with a cross-character link with the prominent scene, so both of them are selected for summarization.

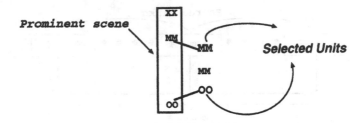

Fig. 8. Selection of Units around Prominent Scenes.

The final summarization set includes all the scenes that come over shared number rule, cause-effect rule, and the units from the prominent scenes.

5   Summarization Procedures

The system provides three different ways to summarize a video; the characteristics and performance of each process will be described in the following subsections. Each process invokes a function called *summarized_list*, which obtains the list of the scenes that satisfies the parameters established for each process. More specifically, the search algorithm for a whole match search of scenes is as follows:

Algorithm 1 Search for scenes.

1. For each scene, the system traces the characters on the scene. If at least one of the requested characters is in the scene, then goes to 2.
2. The shared number of each scene is calculated. If it is greater than or equal to the requested, then goes to 3.
3. The cause-effect number of the scene is calculated. If it is greater or equal to the requested, then the scene is included in the resulting list (*scene_list.*)
4. If the units around the prominent scenes where requested, the system trace these units and its corresponding scene is added to *scene_list.*
5. The *scene_list* is returned from *summarized_list* function.

The algorithm presented above was implemented in Amulet[6], a new user interface development environment for C++ that runs on X11, the source file is as follows:

```
Am_Value_List
summarized_list(Am_Value_List character, int fix_shared_num,
                int fix_cau_eff_num, Am_Boolean prominent_scenes)
{Am_Value_List scene_list;
 for (all_scenes.Start();!all_scenes.Last();all_scenes.Next()){
   Am_Object tmp_scene = all_scenes.Get();
   if (unit_owner(character,tmp_scene)== 1){              /*1
     if (shared_num(tmp_scene) >=  fix_shared_num ){       /*2
       if (cau_eff_num(tmp_scene) >=  fix_cau_eff_num){    /*3
```

```
       if (prominent_scenes) scene_list.Add(unit_around.Get());    /*4
       scene_list.Add(tmp_scene);
       }
    }
  }
}
 return scene_list;                                                /*5
}
```

The Function *summarized_list* is called with its respective parameters, then the system proceeds to trace all the scenes (*all_scenes*) from the video description graph. First, the system traces if at least one of the requested characters is on the scene (*unit_owner*). Then the system evaluates the shared number and cause effect number of the scene (*shared_num* & *cau_eff_num*). If the scene approves the conditions is added to *scene_list*. If the units around prominent scenes were requested, the system traces these units' scenes (*unit_around*) and adds them into *scene_list*. Finally, the function returns the list *scene_list* as result.

5.1 "Best" Summarization Procedure

"Best" summarization procedure extracts significant scenes from a video, attempting to obtain the essential content of the story with a short synopsis of the original. For "best" summarization procedure, all scenes greater than best shared number and best cause-effect number are selected, including the units from the prominent scenes. The algorithm that represents how this procedure works is given below.

Algorithm 2 Best Summarization

1. The system obtains from all the scenes its shared number and calculates the mean, called *best_shared*.
2. In the same way the mean of cause-effect number is calculated, this is called *best_cau_eff*.
3. The system retrieves all scenes over *best_shared* and *best_cau_eff*, and stores them in a list called *best_scenes*.
4. Finally the scenes from the units around the prominent scenes are retrieved and appended to the list.
5. The system returns the list of scenes for later browse.

The algorithm implemented in Amulet is as follows.

```
Am_Value_List
best_summarization
{Am_Value_List best_scenes;
 int shared,cau_eff,x,best_shared,best_cau_eff;
 for (all_scenes.Start();!all_scenes.Last();all_scenes.Next()){
    Am_Object tmp_scene = all_scenes.Get();
    shared=shared+shared_num(tmp_scene);
    cau_eff=cau_eff+cau_eff_num(tmp_scene);
```

```
    x++;
    }
best_shared=shared/x;                                      \*1
best_cau_eff=cau_eff/x;                                    \*2
best_scenes=summarized_list(ALL,best_shared,1,FALSE);      \*3a
best_scenes.Append(ALL,1,best_cau_eff,FALSE);              \*3b
best_scenes.Append(ALL,best_shared,best_cau_eff,TRUE);     \*4
return best_scenes;                                        \*5
}
```

5.2 "Summarization Rate" Procedure

"Summarization rate" procedure is designed to obtain a summarized video with fixed rate. Until now, the system supports rates between 10% to 40%, giving the user flex control of the summarized video length. After the user selected the desired rate, the system determines which combination between shared number, cause-effect number, and units around gives the closest rate. Then the system displays the final combination with its respective rate. The algorithm that represents this procedure is given below.

Algorithm 3 Summarization Rate.

1. The system starts retrieving scenes with the highest shared number and cause-effect number, with and without the scenes from the units around the prominent scenes.
2. The total length of the scenes is calculated and compression rate is evaluated.
3. If the compression rate is achieved,the system returns the actual group of scenes.
4. Else the system decreases cause-effect number and retrieves a new set of scenes. Later on, shared number is also decreased.

This algorithm is represented in Amulet as follows.

```
Am_Value_List rate_summarization (int rate)
{Am_Value_List rate_scenes;
 int tmp=MAX_CAU_EFF;
 while(MAX_SHARED >= 1){                                           \*1
    while(MAX_CAU_EFF >=1){
       rate_scenes=summarized_list (ALL,MAX_SHARED,MAX_CAU_EFF,FALSE);
       if(skim_rate(rate_scenes) >= rate)                          \*2
          goto complete;                                           \*3
       rate_scenes=summarized_list (ALL,MAX_SHARED,MAX_CAU_EFF,TRUE);
       if(skim_rate(rate_scenes) >= rate) goto complete;
       MAX_CAU_EFF--;                                              \*4a
       }
    MAX_CAU_EFF=tmp;
    MAX_SHARED_NUM--;                                              \*4b
    }
 complete: return rate_scenes;
}
```

5.3   "Personal" Summarization Procedure

"Personal" summarization procedure is designed to give the user a full range of possibilities for video summarization creation. In personal summarization the user can select characters, shared number, cause-effect number, etc. to create his own summarized video. It is important to point out that just scenes that fulfill all the rules established by the user are selected for the summarized video.

The "personal" summarization algorithm works as the same way as Algorithm 1. All the parameters are established by the user, then the system just trace the scenes that fulfill these parameters. This procedure represents a very important tool for browsing scenes that users may need. For instance, the users can browse all the scenes where a specific character appears, or obtain central scenes from the movie. Some experiments showing the importance of this tool are presented in the next section.

6   Experimental Results

Until now, two Japanese folktales have been used for experiments. Each story has been analyzed to create its video semantic representation. The total video length of each story is about 10 minutes and is compressed in SGI movie format. We use Movie Player to browse the stories. The system is being developed on a server Silicon Graphics Origin 2000 and implemented in Amulet [6].

Amulet is a user interface development environment for C++ and is portable across X11; Amulet helps to create graphical, interactive user interfaces for software. Amulet includes features specifically designed to make the creation of highly-interactive, graphical, direct manipulation user interfaces significantly easier. Furthermore, it includes a prototype-instance object model, constraints, high-level input handling including automatic undo, built-in support for animation and gesture-recognition, and a set of widgets.

Fig. 9 shows our system's user interface. These windows include all the necessary tools for summarization. Detail description of each button and selector is given below.

Summarization Procedure Selector
This selector includes the three different ways that summarization can be done: "best," "personal," and "summarization rate." For summarization rate, it has been implemented rates between 10% and 40%.

Character Selector
Names for all the characters of the story are listed in appearance order. The user can select which characters have to be included for summarization.

Shared Number Selector
This selector includes a list with all the different values of shared number, obtained at summarization time. These values are displayed in ascendant order from one to the highest shared number found in summarization procedure.

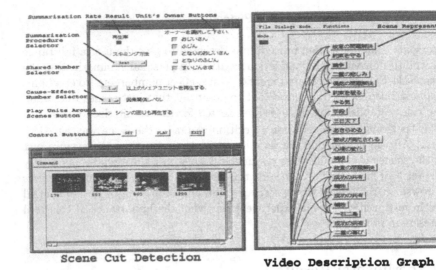

Scene Cut Detection   Video Description Graph

Fig. 9. User Interface.

Cause-Effect Number Selector

This selector includes a list with all the different values of cause-effect number, obtained at summarization time. As same as shared number selector, these values are displayed in ascendant order.

Play Units around Scenes Button

This button activates the system to trace units around scenes which value is over the selected shared number and cause-effect number. These units are included at video composition time.

Control Buttons

SET Shows the summarization rate result according to conditions established by the user. Also stores the file "final_sum_video," with the composed video information.

PLAY Retrieves "final_sum_video" file and sends the information to "serial_player." Serial_player is a modified program from the IRIS Digital Media Library, which displays only video segments established by the user.

EXIT Exit summarization window.

Best Summarization

Both stories had been analyzed using best summarization procedure. We have been able to reduce each story in less than 30% of the original, without apparent loss in the content of the story. Results by using "best" summarization procedure is shown in Table 1.

Story A shows a high value of best shared number due to high inter-relation between scenes. Also best cause-effect number has a high value due to scenes with

Table 1. "Best" Summarization Results.

Story Title	Best Shared Number	Best Cause-Effect Number	Selected Characters	Summarization Rate	Understanding
Story A	5	7	All	22%	GOOD
Story B	3	4	All	27%	GOOD

Story A: The Golden Ax Fable
Story B: The Packhorse Driver and the Raccoon Fable

Table 2. "Summarization Rate" Results.

Story Title	Shared Number	Cause - Effect Number	Units around	Selected Characters	Summarization Rate Required	Result
Story A	4	23	ON	All	10%	13%
Story A	3	6	OFF	All	20%	22%
Story A	1	1	ON	All	30%	24%
Story B	4	5	ON	All	10%	15%
Story B	3	5	OFF	All	20%	22%
Story B	1	2	ON	All	30%	31%

Story A: The Golden Ax Fable
Story B: The Packhorse Driver and the Raccoon Fable

shared units that appear at the beginning and the end of the story. However, story B has lower values in both cases.

A special characteristic of story B is that one of main characters is replaced at the middle of the story (The raccoon gives his son to the packhorse driver). This originates almost no relation between the first and last part of the video's story description. Because of low values in shared and cause-effect numbers, more scenes are selected for summarizing and the summarization rate becomes higher.Experimental results show in both cases, that the summarized video kept the essential content of the story, and irrelevant scenes were dismissed.

Summarization Rate

Some experiments by using "summarization rate" procedure were done. Rates between 10% to 30% were requested and results are presented in Table 2.

Experimental results show that no matter the story, the summarized rate is relatively close to the requested rate. Only in the fourth experiment the resulted rate differs in 5%, this means there is no combination lower than 15% which include all the characters at the same time. By eliminating some characters' participation, the summarization rate would decrease to reach much closer values to the requested.

Table 3. "Personal" Summarization Results.

Story Title	Shared Number	Cause-Effect Number	Units around	Selected Characters	Summarization Rate	Understanding
Story A	5	7	ON	All	18%	FAIR
Story A	1	1	OFF	God of water	9%	NULL
Story A	9	1	ON	All	11%	NULL
Story B	3	4	ON	All	27%	FAIR
Story B	1	1	ON	Mr.Yoshida	18%	NULL
Story B	1	16	OFF	All	7%	NULL

Story A: The Golden Ax Fable
Story B: The Packhorse Driver and the Raccoon Fable

The scenes presented are selected by its importance rank. However, at present time the system can not assure if the content of the summarized video is enough for a clear understanding.

Personal Summarization

Personal summarization represents an important tool for browsing scenes that users may need. Some experiments showing the potential of this procedure are shown in Table 3.

In the first and fourth experiment, we used the same conditions obtained from best summarization procedure, however the summarization rate is lower. As it was explained above, only the scenes that fulfill the established rules, are selected for summarization, this means that only prominent scenes and units around these ones are selected for summarization.

Characters can be selected to create a summarized video, which includes all the scenes where the character appears. This can be observed in the experiment second and fifth, where just one of the characters of the story is selected for summarization. This means that specified character's scenes can be obtained from the video, without matter of the relation with other characters. Due to the low participation of other characters in the summarized video, the understanding of story's content is almost null.

Furthermore, scenes can be selected according to the importance of each rule. The third experiment creates a summarized video with the most important scene according to shared value rule. Finally, the sixth experiment creates a summarized video with the most important scene according to cause-effect rule.

7   Related Work

A recent work about video semantic representation (Marc Davis [7]) has been developed. This approach uses Media Stream, a prototype for video representation and retrieval. With an iconic visual language designed for video representation, users create stream-based representations of video content. These icons are grouped into descriptive categories and are structured to deal with special semantic and syntactic properties of video data.

Media Stream constitutes an easy and simple way to represent semantic and syntactic aspects of the video. However, content of the story can not be included. This means that co-relation between characters within the story is not established. By this approach, summarizing video can just be applied to the syntactic aspect of the video. Our approach makes a deep analysis of the story emphasizing on actions between each character. With this, it is possible to establish which parts of the video are the most important according the content of the story. Nevertheless, this approach represents an extensive analysis of the story which has to be done manually, this means high cost and the exactitude can not be assured.

A second approach to create skims videos is made by Informedia; a recent report[8] presents the usage and evaluation data for abstractions implemented by Informedia Digital Video Library. The approach currently under study by Informedia project members for creating skims utilizes speech recognition, image processing and character recognition to extract most important content from a video and uses that content as components of the skim. However, this approach represents syntactic aspects of the video. Thus, videos with story can not be skimmed by the story content. This approach could have excellent usage in documentaries, news, etc. dislike the results that can be obtained in videos like movies, TV dramas, etc.

8   Conclusion

We have concentrated on semantic aspect of video data, especially in the story content. The use of the presented method to represent story content was shown to be efficient for summarization. Experimental results proved that by this method, videos can be compacted without loss in the content of the narrative. Compaction rate of a video as high as 22% has been achieved without apparent lost in content.

Full capabilities of the system rely on the representation of the story content. As more affect units are used, the system was shown to be more precise. However, to create a full detailed representation and find shared units between affect units requires time and effort. Until now, all this work has to be done manually which is not practical, represents high cost, and the exactitude and veracity can not be assured.Improvements to provide a more consistent summarized video have to be done. Some of the problems and possible solutions to improve the system are listed below:

First, we have implemented two rules on the system; "shared number rule," which helps to evaluate the importance of the scene according to its relation with others, and "cause-effect rule," which helps to evaluate the cause-effect relation of a scene. To improve the content of the summarized video, we need to evaluate scenes in other aspects such as character's relationship, etc.

Second, "summarization rate" procedure is restricted to select scenes just included in the story representation. Transition scenes such as people walking, sunsets, etc., are not included in the story representation. Transition scenes, as well as video segments not included on affect units could not be included in the summarized video, even if the user wants a large version of the video. Some procedures to include frames around affect units will help to create long summarized videos.

At present time, we have been working on two Japanese folktales, because the story's content nature is easy to represent. Furthermore testing is planned with different folktales and other types of videos, such as feature films. We believe that it is possible to provide a wider range of tools for summarizing, but we need experience that will provide feedback to accurate summarization rules.

Acknowledgement

This project is supported in part by the Japanese Ministry of Education Grant-in-Aid for Scientific Research on Priority Area: "Research and Development of Advanced Database Systems for Integration of Media and User Environments." This work is also supported in part by Research for the Future Program of Japan Society for the Promotion of Science under the Project "Researches on Advanced Multimedia Contents Processing."

References

1. Lehnert, W. G.: Affect Units and Narrative Summarization, Research Report #179, Department of Computer Science, Yale University (1980).
2. K. Uehara and N. Horiuchi: Semantic Representation of Video Content based on Prototype-Instance Model, *Proceedings of International Symposium on Digital Media Information Base*, pp. 167–175 (1997).
3. M. A. Smith and T. Kanade: Video Skimming and Characterization through the Combination of Image and Language Understanding, Technical report CMU-CS-95-186, Carnegie Mellon University (1995).
4. A. Hauptmann and M. A. Smith: Text, Speech, and Vision for Video Segmentation: The Informedia Project, *AAAI Fall 1995 Symposium on Computational Models for Integrating Language and Vision* (1995).
5. M. Smith and T. Kanade: Video Skimming for Quick Browsing Based on Audio and Image Characterization, Technical report CMU-CS-97-111, Carnegie Mellon University (1997).
6. Myers, B. A., et al.: The Amulet Environment: New Models for Effective User Interface Software Development, Technical Report CMU-CS-96-189, School of Computer Science, Carnegie Mellon University (1996).
7. Marc Davis: Knowledge Representation for Video, *Proceedings of the 12th National Conference on Artificial Intelligence*, pp. 128–134 (1994).
8. M. Christel, D. Winkler and C. Taylor: Multimedia Abstractions for a Digital Video Library, *Proceedings of the ACM Digital Libraries '97 Conference* (1997).

Valbum:
Album-Oriented Video Storyboard for Editing and Viewing Video

Yasuhiro Niikura[1], Yukinobu Taniguchi[1], Akihito Akutsu[1], and
Yoshinobu Tonomura[1]

Advanced Video Processing Laboratory, NTT Human Interface laboratories,
1-1 Hikarinooka, Yokosuka-Shi, Kanagawa, 239-0847, Japan,
{yas,taniguti,akutsu,tonomura}@aether.hil.ntt.co.jp

Abstract. This paper proposes Valbum, a new environment for editing
and viewing video. The Valbum has friendly interface which is based on
Album and Storyboard. It allows the novice to easily and efficiently edit
and enjoy video.

1 Introduction

With the increasing popularity of handy video cameras and the advance of high
performance personal computers, it seems that the age of consumer desk-top
video publishing has begun. Moreover, digital video cameras that can be directly
connected to personal computers are now available at reasonable cost.

However, it is still difficult for ordinary people to edit videos and the only
way of enjoying them is to simply watch them. Most people haphazardly shoot
video and need some easy way of editing the resulting footage.

One basic problem is that editing and viewing are totally different process
in the traditional video production style because editors are professionals while
the viewers are ordinary people (Fig.1-(a)). In the traditional style, professional
video producers create high quality video for TV programs or cinemas. They
create a new video stream by designing a storyboard at the planning step: record
objects with video cameras, edit films and add sound information or special
effects to them. Using a lot of knowledge and techniques, these video streams are
created to be attractive when viewed sequential. Because current video creation
tools were designed to support the professional user, the operations needed are
too complex for novices. They are, therefore, not suitable for ordinary people.

For example, Adobe Premiere[4] and Avid MCIExpress[5] are famous video
tools for ordinary people. These tools use similar editing interfaces based on the
time-line window because they are developed from professional video tools. The
requirements for using these video tools are;

1. playing back videos (to understand the contents),
2. setting IN/OUT points on the source video (for selecting parts),
3. placing the selected parts in the time-line window (for editing),

S. Nishio, F. Kishino (Eds.): AMCP'98, LNCS 1554, pp. 17–29, 1999.

4. adding splicing effects to them (for arranging video stream in detail).
5. adding sound or music (for arranging overall video stream).

If ordinary people become skilled in using these tools and learn a lot of techniques to present their ideas, they can easily create high quality video stream. But it is difficult to do. We think that an easier video tool for editing and viewing video is required for ordinary people.

The essential point of our proposal is that both editing and viewing videos use the what-you-see-is-what-you-get style. We propose an album-oriented video storyboard called Valbum as the interface for editing and viewing videos; it is easy, efficient, and enjoyable to use (Fig.1-(b)). This tool supports users in creating attractive video albums without expert knowledge. This tool automatically creates a video stream from each video album which can be regarded as a storyboard. When viewing the video, users can select various styles like watching video streams, viewing the video via storyboard, and so on.

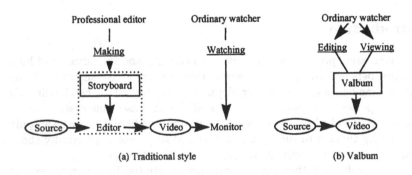

Fig. 1. Video environment.

We will describe related useful tools in Section 2. In Section 3, we introduce the Valbum system and explain its concept. We discuss related works in section 4. Section 5 explains our system in detail. A prototype system is described and examined in Section 6.

2   Useful Tools: Storyboard and Photo Album

The proposed interface is based on storyboards and photo albums.

A storyboard, used by the director at the planning step when making a movie, is a kind of video draft with key images defined for each shot, telops, and words inserted to improve comprehension. It is a good tool to create, enhance,

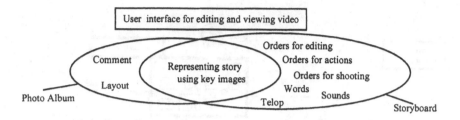

Fig. 2. Concept of combining Storyboard and Album.

and refine the story. Harada et al. proposed "Anecdote" which has an interface that takes advantage of these features in making multimedia presentation materials[2].

A photo album is a book on whose pages we can arrange photographs together with comments. Photos are usually laid out to preserve temporal order but can be arrange to also indicate importance or surprising juxtapositions. Photo albums are easy to arrange and annotate and encourage efficient browsing.

Taniguchi et al. proposed "PanoramaExcerpts" which is a kind of photo album automatically created from an input video stream. Extracted key images and generated panoramic images are laid out automatically to assist in the rapid comprehension of the video.

They have a common feature in that they present an outline of a video stream using key images. We want to make use of this nice feature of both storyboard and photo albums for the purposes of simple editing and enhancing the enjoyment and comprehension of videos (Fig.2).

3   Valbum Interface

We propose the new tool called Valbum (Album-oriented Video Storyboard). The basic concepts of Valbum are, (1) it offers the user a simple and efficient interface based on key image icon operations for the editing and viewing processes, (2) the processes and the related information are described in scripts, (3) the edited video stream is automatically generated by icon manipulation.

The user can easily and effectively edit and view videos with Valbum. Valbum generates image icons from the input source video. To edit the video, the user selects and lays out image icons to design the storyboard for creating the desired video title. Valbum automatically creates the final video stream by analyzing of the image icons of the completed storyboard. Users can watch and view the video streams in the storyboard by accessing the key image icons in various ways.

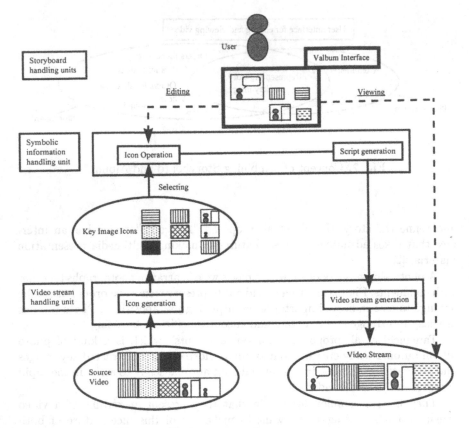

Fig. 3. Valbum system: three components.

4   Related Work

Some video handling approaches that use image icons already exist. Most of the video interfaces proposed to date were intended to help rapid understanding, browsing, and viewing desired parts of input video streams [1], [6], [7], [8].

The image icons consist of segmented frames and representative images. The extraction of representative images can be made by filtering using various conditions and rules based on an analysis of typical videos. For example, the simplest way is to extract the first frame of each shot as extracted by shot-boundary detection. The other approaches use the start or the end of a set of camera motions, object motions, telops, or conversations, for extraction.

These existing approaches can be applied to the system proposed herein for generating image icons. Also, these browsing interfaces can be applied to our system for selecting source image icons.

Some researchers have used image icons to create video streams. Ueda et al. proposed "IMPACT" and the Tree-structured-story-editor which automatically generate image icons by scene separation and whose editing interface uses image icons [8],[9]. The Tree-structured-story-editor provides a method to edit video structure at several different levels and avoids the constraint of limited window size. It can create structured videos like dramas and documentaries.

However, it creates video streams in the traditional style and is difficult for novices to use because the user needs expert knowledge to create effective video streams. The video tool proposed herein provides a novice-friendly interface.

5 Valbum System

Valbum consists of three components: graphical user interface, symbolic information handling unit, and video stream handling unit. All are controlled through scripts.

5.1 Editing

There are three steps in editing a video: automatic generation of draft key image icons for the storyboard, user manipulation of key image icons such as selecting, editing, and laying them out, and generation of a new video stream.

Material Stock: Key Image Icon Generation It is laborious if the user must manually select the key images that will be used to realize the storyboard. In Valbum, the key images are automatically generated and displayed in the material window.

It is important to have a comprehensive set of key images at the start with no essential images missed. The system must extract video segmentation and create a key image for the frames in each segment. We will apply the proposed methods described in Section 4.

Source videos recorded by handy cameras often include frames of bad image quality cased by unstable or too quick camera work. The system avoids selecting such frames as key image candidates. Also, according to standard editing grammar, relatively still frames should proceed and follow shot boundaries excepting particular effects. People usually feel uncomfortable at watching the video which include camera work or object motion before and after shot boundaries.

Therefore, the system excludes frames containing excessive movement and sets shot boundaries that have relatively still frames at the start and end. To find these boundaries, the system first detects shot boundaries in the input source video. The system then calculates spatial motion vectors between two contiguous frames to estimate the camera parameters in each shot. When the average motion vectors in the two contiguous frames are very different, the system ignores the frames. Similarly, the system estimates camera parameters at the start and end of each video segment to assess the existence of still frames. The system sets the start and end of the segment by locating the still frames nearest the original boundaries.

Moreover, if a shot contains the camera operations of panning or tilting, it is difficult to summarize using a single key frame. We will use a panoramic icon to represent all visible contents of the shot [1].

Storyboard Design: Laying Out Key Image Icons and Adding Information In this step the user picks up key images from the material window and lays them out according to his/her design policy in the material window.

The important point is that the user can perform these operations in what-you-see-is-what-you-get manner.

The user can add, delete and alter a key image. The user also can modify several attributes of each icon such as position and size, similar to what is possible with a photo album. For example, the image size may be set larger when the shot is important and the corresponding shot is longer than the default length.

Moreover, the user can insert telops, onomatopoeia (for example, quack, glug, buck, cluck, cuckoo, grrr), and speech via balloons into the key image icon. When inserting speech words with balloons, users first indicate the related person or object. One of the most interesting features of our system is that information is described visually as is common in texts like comic books. We think that this is important for a user to quickly comprehend what is in the shot. In particular, word balloons emphasize that the target person or object speaks these words.

The balloon words and telops are generated and inserted at the appropriate frames. The user indicates the object to which the balloon is to be connected and the system uses color information to trace the object across the frames in the shot. The word balloon tails are made to point to the target object (Fig.4).

Inserting onomatopoeia allows pre-defined sounds to be added to the video segment. For example, if the inserted onomatopoeia is "vroom", real sounds of a moving automobile are added to the corresponding video shot.

The user can easily add orders for special effects like the original storyboard, too. For example, users can select the way of connecting two shots. The default connection is an abrupt change at the shot boundary. By adding a pre-defined symbol between two sequential image icons, special video effects such as curtain peels can be set.

The results of the user operations are stored as scripts defining the Valbum data. The scripts include the definition of key images, and the various attributes related to the video editing process.

Video Stream Generation In Valbum, a new video stream is automatically generated from the scripts corresponding to the storyboard the user designed. This process is performed based on pre-defined generation rules.

It is important that the video stream generation rules be intuitive.

The stream generation rules include (1) association of the key image to the actual shot, (2) words to appear in the related frames, (3) icon position which determines shot sequence.

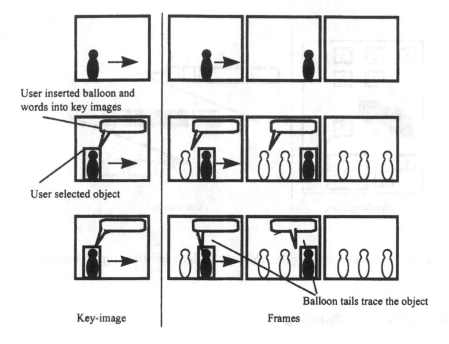

Fig. 4. Word balloon tails must point to the target object. If the object moves, the balloon tail automatically traces the target.

The default shot definition is the entire shot output by the key image generation step.

The position of key images is converted into the sequence of shots (Fig.5). In the default rule, the sequence is from left to right and from top to bottom, which is the obvious temporal order.

There can be various kinds of bridging rules between key image representation and video stream generation. For example, one rule is that the size of key image icons on the storyboard determines the length of the video shot. If the size of a key image icon is reduced, the corresponding video shot is shortened by reducing the number of frames. Other rule is that the size of image icons determines the level of sounds when sound is added to video streams. If the size of key image icons is expanded, the level of sounds is increased in the corresponding term. User will be able to select and modify these rules.

When a transition is used to link key images, the pre-defined operation is applied to the corresponding video shots. Video stream generation and actual playing can be performed at any time during the storyboard design process in an iterative manner.

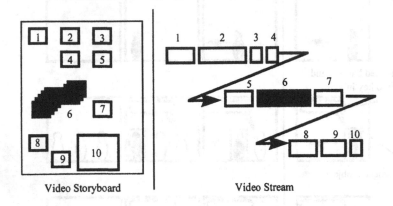

Video Storyboard   Video Stream

Fig. 5. Default conversion rule: from left to right and from top to bottom.

5.2   Viewing and Enjoying

Valbum gives users an interface for flexible viewing.

The user-created storyboard represents the outline of the video desired. The user can play the entire video stream or individual shots as desired by selecting the key image icons in the storyboard. The video stream can also be selectively replayed using user-defined characteristics such as icon size, balloon words, and telops as retrieval keys. Valbum also has an interface for changing the characteristics for selective replay.

The generated video stream is replayed in an individual window and the individual shots are replayed within the window surrounding the key image.

5.3   Example

Fig.6 and Fig.7 show examples of Valbum in use. Source videos were recorded by a home video camera.

Fig.6 shows examples of editing in the Valbum system. User inputs source video in Fig.6-(a). The system analyzes source video and generates image icons and displays source key images in the material window Fig.6-(b). The user picks up key images and lays them out to design the video storyboard. The user then inserts text information as telops and word balloons into key-images Fig.6-(c). The system automatically generates a video stream according to the designed video storyboard.

Fig.7 shows examples of viewing using the Valbum system. Key images are laid out with some word balloons in the storyboard (Fig.7-(a)). The generated video stream can be played to enjoy it.

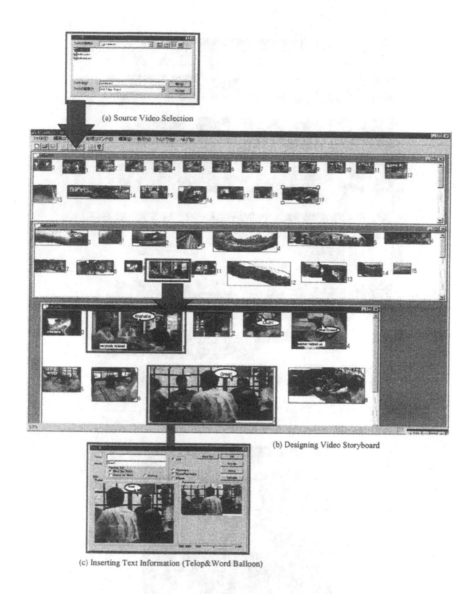

(a) Source Video Selection

(b) Designing Video Storyboard

(c) Inserting Text Information (Telop&Word Balloon)

Fig. 6. Example of Editing : (a) Input Source Video:Valbum automatically analyzes source video and generates image icons, (b) Designing Video Storyboard: Editor picks up key image icons and lays them out for designing video storyboard, (c) Adding text information: Adding telops or word balloon to key-images.

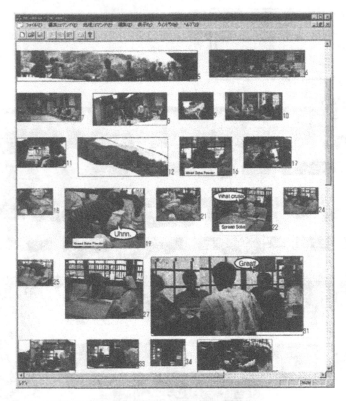

(a) Video Browser based on the designed Video Storyboard

(b) Monitor

Fig. 7. Example of Viewing : (a) Valbum Viewer : it is designed as video storyboard by editor. (b) Monitor : Playing back video as the user desires.

6 Discussion

We developed a prototype based on the proposed concept and examined it in operation. This section discusses the results of the examination.

Fig.6, Fig.7 show example of Valbum in use. Source videos were recorded by a home video camera.

The designed storyboard more easily creates the final video stream compared to the other video browsers described in Section 4 because Vablum allows the creator's ideas to be accurately reflected in the final video. The user can easily create the different video stream by their different idea. Inserting text information as telops or word balloons is useful for clarifying the user's ideas.

Initial tests indicate that Valbum is easy and fun for people to use. But we recognized some problems in this prototype.

In the image icon generation step, the input source video is segmented and some sequences are omitted. However, the user may have some particular interest in these sequences so our idea is to rearrange the material window to show all shots and sequences using some indication of stability. Figure shows an example of this. In the current version of Valbum, sequences S2 and S6 would be omitted. Allowing the user to access and utilize these two sequences may enhance video quality.

When making a key image, the key frame is important for rapidly understanding the video shot for editing and viewing. A simple approach is to use the first frame in each shot. When a shot contains the camera operations of panning or tilting, a panoramic icon can represent the entire video contents of the shot. However, when a shot contains object motion, the proposed approaches cannot represent shot content. Further work is needed to represent the shot contents that include object motion.

When laying out key image icons on the storyboard, the size of the storyboard is limited by the computer display. In the sample of Fig.6, only 9 key image icons were used and the length of the generated video stream was about a minute; storyboard size was 987 x 627 pixels. A storyboard with 30 icons of resonable size can be displayed on a 17 inch monitor, and the length of generating stream is shorter than 5 minutes.

However, it is possible that a user may want to create longer videos using more icons. In the traditional style, the long video stream is divided into short video units that are connected. We can adopt the same concept. The user can create several storyboards that are connected to create the final video stream. The storyboards will be displayed like pages in a photo album. We believe that these interfaces will yield a good browser for video streams. Of course, user may switch the storyboards in the design step.

We think the quality of the video streams that are automatically generated by the current storyboard are lower than the those created by professional producers. This is because the generated video stream is a simple cut and paste edit of home videos. For example, a word balloon is displayed from the start to end in the corresponding shot, and the viewer may be shocked when the text information appears at shot boundaries. Professional producers usually fade telops

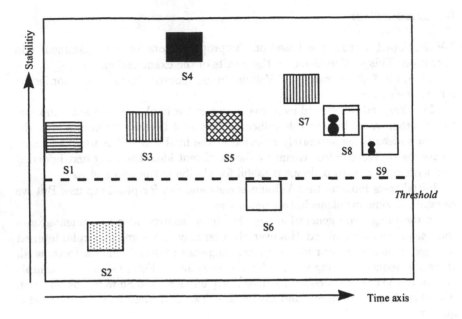

Fig. 8. Sample of material window.

and word balloons in and out at shot boundaries, so we should implement these functions. We will analyze the professional techniques and enhance our system to easily create high quality video streams.

7   Conclusion

We proposed Valbum, a new video environment for editing and watching video that is significantly easier to use than traditional video editing tools.

In the Valbum system, user

1. watches key-images in material windows (to understand the contents),
2. picks up image icons (for selecting parts),
3. drops image icons into storyboard (for editing),
4. adding some information to video storyboard (for arranging video).

Because Valbum automatically generates image icons from source video and generates a new video stream according to the designed video storyboard, the user needs no complex operations.

Moreover, in the Valbum system, users can present their ideas more effectively than traditional tools. This is because, Valbum allows the viewer to access the components of the video in a more natural style.

With Valbum, we can easily create video streams and enjoy viewing and watching video. In fact, we can create interesting video contents from source videos haphazardly recorded by handy video camera.

We will refine the prototype by collecting feedback from test users and will add the functions of panoramic icons and others.

Acknowledgments The authors would like to thank Dr. Yukio Tokunaga and Dr. Harutaka Nakano for their support of our reseach. We would like to thank Dr. Takashi Satoh for reading the draft and making a number of helpful suggestions. We also wish to thank our colleagues for their constant help and discussions.

References

1. Y. Taniguchi, A. Akutsu, Y. Tonomura, "PanoramaExcerpts: Extracting and packing panoramas for Video Browsing", Proc. ACM Multimedia'97
2. Y. Harada,E. Tanaka,R. Ogawa and Y. Hara, "Anecdote : A multimedia Storyboarding System with seamless Authoring Support", Proc. ACM multimedia'96
3. D. Kurlander, T. Skelly, D. Salesin, "Comic Chat", SIGGRAPH'96
4. "Adobe Premiere": Adobe Co. Ltd.
5. "Avid MCIExpres": Avid Co. Ltd.
6. M. A. Smith, T. Kanade, "Video Skimming and characterization through the Combination of image and language Understanding Techniques", IEEE CVPR, 1997
7. P. Aigrain, H. Zhang, D. Petkovic, "Content-Based representation and Retrieval of Visual Media: A State-of-the-Art Review", Multimedia Tools and Applications 3, 179-202(1996)
8. H. Ueda, T. Miyatake, S. Yoshisawa, "IMPACT: An interactive natural-motion-picture dedicated multimedia authoring system," Proc. CHI*91, ACM, 1991, pp.343-350
9. H. Ueda, T. Miyatake, "Automatic Scene Separation and Tree Structure GUI for Video Editing," ACM Multimedia 96, pp. 405-406

Beyond the Desktop Metaphor: Toward More Effective Display, Interaction, and Telecollaboration in the Office of the Future via a Multitude of Sensors and Displays

Henry Fuchs

Department of Computer Science
University of North Carolina at Chapel Hill, USA

Abstract.

We are engaged in a long-term project to improve personal productivity for computer-related activities and tele-collaboration in an office environment of the future. Personal computer-related activities, we believe will be enhanced by capability to project imagery on any surface in the office, that together with precise head and eye-tracking, will enable head-tracked stereo imagery to be added to the user's views of his/her office environment --- creating a 3D immersive generalization of the now ubiquitous 2D desktop metaphor as the principal human-computer interface. We plan to realize this kind of system by mounting many video projectors and video cameras around the room, especially around the ceiling. The projectors may provide the only source of light in the room and will allow detailed imagery to be projected (almost) everywhere in the office. In order to generate the appropriate imagery, however, a detailed 3D map of the changing office environment needs to be acquired. This will be acquired by measuring, with synchronized cameras and projectors, the precise 3D location(s) of the surface(s) light up by each pixel of each projector. Local collaboration will be enhanced by tracking each of several individuals in the office and generating (by time-division multiplexing or by other means), a stereo image pair appropriate for each individual. Objects under design may be displayed, for each individual, from his/her own perspective and to his/her own specifications of interest. TELE-collaboration activities, we believe will be enhanced by having such an enhanced office environment for each of the small group of distant collaborators, and displaying for each participant, in addition to the shared objects under discussion, some combination of the remote scenes that include the changing 3D images of each of the participants and 3D images of physical objects of joint interest. To realize many of these capabilities, each user may need to wear polarized eyeglasses to perceive proper stereo imagery. Although initial results are encouraging, numerous difficult problems remain -- how, for example, can imagery be projected onto a dark-colored surface in the room. The cost of such systems, with many projectors and cameras, image generators and image acquisition devices, may initially be prohibitively expensive, but is expected to decrease as the cost of such off-the-shelf equipment naturally decreases with increased market size. The positive psychological effects of working and interacting in such an immersive

S. Nishio, F. Kishino (Eds.): AMCP'98, LNCS 1554, pp. 30-43, 1999.

environment within a "standard" office will be so compelling, we believe, that users will not readily wish to return to working within the constraints of a 21" monitor. Much of this work is being carried out as part of a collaboration among the five sites of the NSF Science and Technology Center in Computer Graphics and Scientific Visualization (Brown, Caltech, Cornell, UNC, and Utah) and is also being carried on in collaboration with the GRASP Lab at University of Pennsylvania, and is part of the National Teleimmersion Initiative sponsored by Advanced Network and Services.

Introduction

This project originated from two influences in our department: 1) many years of research into virtual reality technologies and applications, and 2) a long tradition of working in collaboration with others near and far (rather than working alone). Long-distance collaboration in particular has been at the core of the Science and Technology Center for Computer Graphics and Visualization ("the Graphics STC"), of which UNC is one of the five sites. The five sites are widely separated within the USA, so the distant participants have no choice but to rely on whatever tools and techniques are available to enable the long-distance collaboration.

Several of the virtual reality applications have in recent years involved "augmented reality" -- visualization and interaction not just with a completely synthetic, computer-generated environment, but visualization and interaction with an environment composed of *a fusion of* the computer-generated one and the user's own physical environment (the user's environment *augmented* with a synthetic one). In several medical applications, for example, the user (a physician) sees the patient on the examining table AND this patient's image data properly registered inside. In this way the physician could, for example, identify the precise location of tumors during needle biopsy. About five years ago (1993), this vision of augmented reality was expanded to include augmentation of the user's local environment by (3D) images of distant environments, as in the linking of teachers with students in separate classrooms, and establishing remote surgical consultations between medical experts based at certain sites and physicians handling cases (such as trauma) in others.

Support for collaboration within the Graphics STC began with standard tele-conferencing technology. Later developments included an experimental shared virtual environment with "live" video images of participants' faces on crude rectangular avatars in the shared virtual environment. From these experiments, some of us concluded that remote participants would benefit not just from "live" video of the faces of the other participants, but also from 3D reconstructions of each participant and his or her nearby environment.

Recently these two interests have come together in a project supported by the National Tele-immersion Initiative funded by Advanced Network and Services. Their support, along with that of NSF and DARPA for related research, have enabled us to begin building a prototype office in the UNC graphics and image lab (built along the dimensions of a standard 12 foot x 16 foot university office) to try out these ideas of remote tele-collaboration. The near term objective is to link two rooms on the same floor for telecollaboration experiments — a small conference room and the prototype office (We are also working on linking the conference room to a distant lab.)

Design Considerations for the Office of the Future

Given that the future of tele-collaboration technology will be of most use to individual researchers working alone or with local colleagues rather than by teams working at great distances, whatever tele-collaborative system that is built is to be optimized for solo and local work. Additionally, in a world where increasingly individuals are basing their work environment at home, *every* interaction will be remote.

For display technology, we would like to use head-mounted displays, but these are not well enough developed at the present time to give a comfortable, stable image with sufficient image quality. Thus we are "forced" to consider other display technologies, ones with displays *not* mounted on the head. Among these remaining possibilities are projection-based displays: front- and rear-projection. Since rear-projection arrangements take up impractically large amounts of space for an office environment, we have been focusing on front-projection.

Next, we have had to decide *where* to project. Our answer was, onto well-specified surfaces such as walls or onto more general surfaces. At. UNC we are beginning with display onto walls, but we are eager to project images anywhere in the environment. In Raskar, *Siggraph98*, we show a method for pre-distorting the projected image in order to cancel out distortion effects due to the geometry of the surface. This approach fits in well with concepts developed here since 1993, such as extracting 3D scene descriptions from a "sea of cameras," so that in theory every available surface location can be extracted. This, in turn, has led us to an interest in real-time depth extraction techniques that take advantage of lighting control provided by the projectors. In turn, this has led us to a consideration of techniques for dynamically structured light that are not too distracting to the participants. Other considerations are support for display in stereo mode, and support for multiple participants.

Implementation and Problems

We are eager to try various parts of such a system as early as possible — in order to benefit from them and to learn the problems and limitations. Two such problems are listed below:

1) shadowing created by the user's body getting in the way of the projector paths. This results in parts of the image not reaching the intended surface. Although this getting-in-the-way is necessary in order to do the 3D scanning of the user's own body

surface, we think it may be very irritating to have NO displayed image on the some surface location of interest, such as the desk surface directly under a user's hand. We hope to lessen this difficulty by the careful placement of projectors that aim onto the surfaces and by allocating multiple projectors to project onto regions of particular importance and particularly susceptible to this shadowing problem.

2) interreflections—significant image degredation may occur in the Office from (projected) light bouncing from one surface to another, reducing contrast and readability. This effect is especially noticeable near inside lit corners of the Office. Since we intend to control all the light in the room through projectors, we can predict some of these scattering effects, but we can't control them

We have begun to use these concepts in simplest form, using projectors mounted on the ceiling and projecting onto walls (actually, wall-mounted white boards). Professor Gary Bishop (UNC) has already converted his private office this way (see his UNC web page photos at http://www.cs.unc.edu/~gb/office.htm) and uses it daily. In the future, for telecollaboration, we envision a system consisting of only a few sites (4-5, each with only 1 or 2 people) sending images to one another. Each collaborator at work would have the impression of two rooms from different sites being adjacent to a picture window in between. With more than two sites involved, this impression would be expanded, in that the offices would appear to be in grid arrangement and meeting at the corners. Each site, therefore, can see the others with the impression of intimacy, with their tables and chairs pushed up to the viewing surfaces.

Extractions of 3D suirface points are done by imperceptible structured light and cameras. We hope to acquire surface information (location and color) of every millimeter of lighted surface in the scene and send any changes (from people moving, etc.) to each of the other sites. In order to render the proper image, the precise location of the participant's eyes must be known.

Current Status of Project

We have just begun operating eight projectors simultaneously in our 12-foot by 16-foot prototype office, and experimenting with (very near) "telecollaboration" between a single participant in that office and a group of participants in a small conference room just a few meters away. The projectors present a panorama of images from the conference room acquired by a single cluster of 10 to 12 cameras. This specialized cluster of cameras was developed in collaboration with the Alpha_1 Project and the Advanced Manufacturing Lab at the University of Utah [Chi, et al, 1998].

We have also developed proof-of-concept demonstrations for imperceptible structured light and for correction of geometric distortion induced by nonplanar geometry of the display surfaces. We have also demonstrated stereo display using micro-mirror based projectors — called Digital Light Projection (DLP) by its developer, Texas Instruments.

Our collaborators at the University of Pennsylvania have developed techniques to extract precise depth and color maps from increasingly complex room environments.

Conclusions

Although various pieces of technologies are just beginning to be demonstrable, we see no fundamental barrier to realizing the long-term plans described above. For example, the idea of outfitting a room with dozens of projectors and digital cameras appears daunting at first, but we are encouraged that both of these items (cameras or projectors) are widely available, with steadily declining costs. We foresee a time when such configurations will be widely affordable. Initial results of displaying many coordinated images simultaneously encourage us in the belief that, once an individual experiences such a rich, interactive environment, he will never again be satisfied with a 19" display screen. Once telecollaboration can be experienced in a convincing, surround fashion, today's conventional teleconferencing with a small window on a single screen, showing a distant participant's "talking head", will appear as appealing as a console teletype for computer interaction after experiencing a multi-windows system.

Acknowledgement

This continuing project is the result of active collaboration among many individuals; Greg Welch coordinated much of the initial work in his capacity as the UNC site coordinator of the Graphics STC.   Herman Towles has led most of the implementation effort since joining us early this year; Graduate research assistants Matt Cutts, Ramesh Raskar, Adam Lake, and Lev Stesin developed most of the initial ideas and built many of the first system components. Graduate research assistants Ruigang Yang, Aditi Majumder, David Marshburn and Wei-Chao Chen have joined the team this fall 1998 and have already contributed significantly to various camera calibration, projector calibration and audio subsystems. Visiting Professor Brent Seales and graduate research assistant Mike Brown from the University of Kentucky are considerably strengthening our computer vision, image processing and calibration capabilities.

Much of this work is being carried out as part of a collaboration among the five sites of the NSF Science and Technology Center in Computer Graphics and Scientific Visualization (Brown, Caltech, Cornell, UNC, and Utah) and is also being carried on in collaboration with the GRASP Lab at University of Pennsylvania. Recent support from Advanced Network and Services, as part of their National Tele-immersion Initiative, has enabled us to transition this year from developing proof of concept components to building a full office-sized prototype.

Bibliography

Aliaga, J., D. Cohen, A. Wilson. H. Zhang, C. Erickson, K. Hoff, T. Hudson, W. Stüerzlinger, E. Baker, R. Bastos, M. Whitton, F. Brooks, D. Manocha.1998. "A Framework for the Real-Time Walkthrough of Massive Models," UNC Computer Science Technical Report TR98-013, University of North Carolina at Chapel Hill, March 1998.

Azuma, R., Bishop, G. Improving Static and Dynamic Registration in an Optical See-through HMD. Proceedings of SIGGRAPH 94 (Orlando, Florida, July 24-29, 1994). In *Computer Graphics* Proceedings, Annual Conference Series, 1994, ACM SIGGRAPH, pp. 197-204.

Bajaj, C.L., F. Bernardini, and G. Xu. "Automatic reconstruction of surfaces and scalar fields from 3D scans," SIGGRAPH 95 Conference Proceedings, Annual Conference Series, ACM SIGGRAPH, Addison-Wesley, pp. 109-118, August 1995.

Bowen, Loftin, R. "Hands Across the Atlantic," IEEE Computer Graphics and Applications, Vol. 17, No. 2, pp. 78-79, March-April 1997.

Bryson, Steve, David Zeltzer, Mark T. Bolas, Bertrand de La Chapelle, and David Bennett. "The Future of Virtual Reality: Head Mounted Displays Versus Spatially Immersive Displays," SIGGRAPH 97 Conference Proceedings, Annual Conference Series, ACM SIGGRAPH, Addison-Wesley, pp. 485-486, August 1997.

Buxton, W., Sellen, A. & Sheasby, M. "Interfaces for multiparty videoconferencing," In K. Finn, A. Sellen & S. Wilber (Eds.). Video Mediated Communication. Hillsdale, N.J.: Erlbaum, pp. 385-400, 1997.

Capin, Tolga K., Hansrudi Noser, Daniel Thalmann, Igor Sunday Pandzic and Nadia Magnenat Thalman. "Virtual Human Representation and Communication in VLNet," IEEE Computer Graphics and Applications, Vol. 17, No. 2, pp. 42-53, March-April 1997.

Chi, Vern, Matt Cutts, Henry Fuchs, Kurtis Keller, Greg Welch, Mark Bloomenthal, Elaine Cohen, Sam Drake, Russ Fish, Rich Riesenfeld. 1998. "A Wide Field-of-View Camera Cluster", University of North Carolina at Chapel Hill, Dept of Computer Science, Technical Report TR98-018.

Chien, C.H., Y.B. Sim, and J.K. Aggarwal. "Generation of volume/surface octree from range data," The Computer Graphics Society Conference on Computer Vision and Pattern Recognition, pp. 254-260, June 1988.

Conner, D.B., Cutts, M., Fish, R., Fuchs, H., Holden, L., Jacobs, M., Loss, B., Markosian, L., Riesenfeld, R., and Turk, G. "An Immersive Tool for Wide-Area

Collaborative Design," TeamCAD, the First Graphics Visualization, and Usability (GVU) Workshop on Collaborative Design. Atlanta, Georgia, May 12-13, 1997.

Connolly, C.I. "Cumulative generation of octree models from range data," Proceedings, Int'l. Conference Robotics, pp. 25-32, March 1984.

Cruz-Neira, Carolina, Daniel J. Sandin, and Thomas A. DeFanti. "Surround-Screen Projection-Based Virtual Reality: The Design and Implementation of the CAVE," Computer Graphics, SIGGRAPH Annual Conference Proceedings, 1993.

Cutts, Matt, Henry Fuchs, Adam Lake, Ramesh Raskar, Lev Stesin, and Greg Welch. 1998. "The Office of the Future: A Unified Approach to Image-Based Modeling and Spatially Immersive Displays," SIGGRAPH 98 Conference Proceedings, Annual Conference Series, ACM SIGGRAPH, Addison Wesley, July 1998, Orlando, FL, USA

Curless, Brian, and Marc Levoy. "A Volumetric Method for Building Complex Models from Range Images," SIGGRAPH 96 Conference Proceedings, Annual Conference Series, ACM SIGGRAPH, Addison-Wesley. pp. 303-312, 1996.

DePiero, F.W., and Trivedi, M.M., "3-D Computer Vision using Structured Light: Design, Calibration, and Implementation Issues," Advances in Computers(43), 1996, Academic Press, pp.243-278

Dias, José Miguel Salles, Ricardo Galli, António Carlos Almeida, Carlos A. C. Belo, and José Manuel Rebordã. "mWorld: A Multiuser 3D Virtual Environment," IEEE Computer Graphics and Applications, Vol. 17, No. 2., pp. 55-64, March-April 1997.

Fuchs, Henry, Gary Bishop, Kevin Arthur, Leonard McMillan, Ruzena Bajcsy, Sang Lee, Hany Farid, and Takeo Kanade. "Virtual Space Teleconferencing Using a Sea of Cameras," Proceedings of the First International Symposium on Medical Robotics and Computer Assisted Surgery, (Pittsburgh, PA.) Sept 22-24, 1994.

Gajewska , Hania , Jay Kistler, Mark S. Manasse, and David D. Redell. "Argo: A System for Distributed Collaboration," (DEC, Multimedia '94)

Gibbs, Simon, Constantin Arapis and Christian J. Breiteneder. "TELEPORT-Towards Immersive Copresence," to appear in ACM Multimedia Systems Journal, 1998.

Hilton, A., A.J. Toddart, J. Illingworth, and T. Windeatt. "Reliable surface reconstruction from multiple range images," In Fouth European Conference on Computer Vision, Volume 1, pp. 117-126. April 1996.

Holloway, R. "Registration Errors in Augmented Reality Systems," PhD Thesis. University of North Carolina at Chapel Hill, 1995.

Holmes, Richard E. "Common Projector and Display Modules for Aircraft Simulator Visual Systems," Presented at the IMAGE V Conference, Phoenix, AZ, June 19-22, pp. 81-88, 1990.

Hornbeck, Larry J., "Deformable-Mirror Spatial Light Modulators,"Proceedings SPIE, Vol. 1150, Aug 1989.

Hornbeck, Larry J., "Digital Light Processing for High-Brightness High-Resolution Applications," Available from http://www.ti.com/dlp/docs/business/resources/white/hornbeck.pdf, 1995.

Ichikawa, Y., Okada, K., Jeong, G., Tanaka, S. and Matsushita, Y.: "MAJIC Videoconferencing System: Experiments, Evaluation and Improvement'," In Proceedings of ECSCW'95, pp. 279-292, Sept. 1995.

Ishii, Hiroshi, Minoru Kobayashi, Kazuho Arita. "Iterative Design of Seamless Collaboration Media," CACM, Volume 37, Number 8, pp. 83-97, August 1994.

Kanade, Takeo and Haruhiko Asada. "Noncontact Visual Three-Dimensional Ranging Devices," Proceedings of SPIE: 3D Machine Perception. Volume 283, Pages 48-54. April 23-24, 1981.

Kanade, Takeo, Hiroshi Kano, Shigeru Kimura, Atsushi Yoshida, Kazuo Oda. "Development of a Video-Rate Stereo Machine," Proceedings of International Robotics and Systems Conference (IROS '95). pp. 95-100, Pittsburgh, PA., August 5-9, 1995.

Lamotte, Wim, Eddy Flerackers, Frank Van Reeth, Rae Earnshaw, Joao Mena De Matos. Visinet: Collaborative 3D Visualization and VR over ATM Networks. IEEE Computer Graphics and Applications, Vol. 17, No. 2, pp. 66-75, March-April 1997.

Lehner, Valerie D., and Thomas A. DeFanti. "Distributed Virtual Reality: Supporting Remote Collaboration in Vehicle Design," IEEE Computer Graphics and Applications, Vol. 17, No. 2, pp. 13-17, March-April 1997.

Lyon, Paul. "Edge-blending Multiple Projection Displays On A Dome Surface To Form Continuous Wide Angle Fields-of-View," pp. 203-209. Proceedings of 7th I/ITEC, 1985.

Macedonia, Michale R. and Stefan Noll. "A Transatlantic Research and Development Environment," IEEE Computer Graphics and Applications, Vol. 17, No. 2, pp. 76-82, March-April 1997.

Mandeville, J., T. Furness, M. Kawahata, D. Campbell, P. Danset, A. Dahl, J. Dauner, J. Davidson, K. Kandie, and P. Schwartz. "GreenSpace: Creating a Distributed

Virtual Environment for Global Applications," Proceedings of IEEE Networked Virtual Reality Workshop, 1995.

Milgram, P and Kishino, F. "A taxonomy of mixed reality visual displays", *IEICE (Institute of Electronics, Information and Communication Engineers) Transactions on Information and Systems*, Special issue on Networked Reality, Dec. 1994.

Milgram, P., Takemura, Utsumi and F. Kishino. "Augmented Reality: A class of displays on the reality-virtuality continuum". *SPIE Vol. 2351-34, Telemanipulator and Telepresence Technologies*, 1994.

Nayar, Shree, Masahiro Watanabe, Minori Noguchi. "Real-time Focus Range Sensor," Columbia University, CUCS-028-94.

Neumann, Ulrich and Henry Fuchs, "A Vision of Telepresence for Medical Consultation and Other Applications," Proceedings of the Sixth International Symposium on Robotics Research, Hidden Valley, PA, Oct. 1-5, 1993, pp. 565-571.

Ohya, Jun, Kitamura, Yasuichi, Takemura, Haruo, et al. "Real-time Reproduction of 3D Human Images in Virtual Space Teleconferencing," IEEE Virtual Reality International Symposium. Sep 1993.

Raskar, Ramesh, Matt Cutts, Greg Welch, Wolfgang Stüerzlinger. "Efficient Image Generation for Multiprojector and Multisurface Displays," University of North Carolina at Chapel Hill, Dept of Computer Science, Technical Report TR98-016, 1998.

Raskar, Ramesh, Henry Fuchs, Greg Welch, Adam Lake, Matt Cutts. "3D Talking Heads : Image Based Modeling at Interactive rate using Structured Light Projection,"University of North Carolina at Chapel Hill, Dept of Computer Science, Technical Report TR98-017, 1998

Segal, Mark, Carl Korobkin, Rolf van Widenfelt, Jim Foran, and Paul E. Haeberli. 1992. "Fast shadows and lighting effects using texture mapping," SIGGRAPH 92 Conference Proceedings, Annual Conference Series, ACM SIGGRAPH, Addison Wesley, volume 26, pages 249-252, July 1992.

Slater, M. and M. Usoh. (1994). "Body-Centered Interaction in Immersive Virtual Environments," in N. Magnenat Thalmann and D. Thalmann (Eds.), *Artificial Life and Virtual Reality* (pp. 125-148). London: Wiley

State, A., Hirota, G., Chen, D.T., Garrett, W.F., Livingston, M.A. "Superior Augmented Reality Registration by Integrating Landmark Tracking and Magnetic Tracking". Proceedings of SIGGRAPH '96 (New Orleans, LA, August 4-9, 1996).

In *Computer Graphics* Proceedings, Annual Conference Series, 1996, ACM SIGGRAPH.

Tsai , Roger Y. "An Efficient and Accurate Camera Calibration Technique for 3D Machine Vision," Proceedings of IEEE Conference on Computer Vision and Pattern Recognition, Miami Beach, FL, pp. 364-374, 1986.

Underkoffler, J. "A View From the Luminous Room," Personal Technologies, Vol. 1, No. 2, pp. 49-59, June 1997.

Underkoffler, J., and Hiroshi Ishii. "Illuminating Light: An Optical Design Tool with a Luminous-Tangible Interface," Proceedings of CHI '98, ACM, April 1998.

Fig. 1. Artist conception of the Office of the Future, illustrating a single local participant collaborating with two participants at one remote site and a single participant at one remote site collaborating on the design and manufacture of a Head Mounted Display. The geometric model of therelevant parts of a face are shown next to the boxlike Head Mounted Display so the esigners remain aware of the physical constraints on the design imposed by the contours of the human face.

Fig. 2. The Office scene during projector calibration procedure, showing the results of automatic blending of overlapped projected image regions.

Fig. 3. The Office scene during display of CAD data but without images of distant collaborators.

Fig. 4. The Office during 3D scene scan showing projection of structured light onto the user. We expect to use imperceptible structured light techniques so that the checkerboard pattern will not be perceptible to the user.

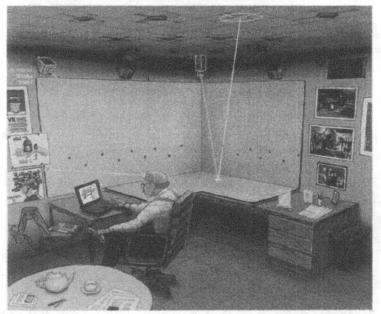

Fig. 5. The Office during extraction of points of particular interest as part of 3D scene scan, also using imperceptible light techniques so the points will not be perceptible to the user.

Fig. 6. The Office after a scene scan outlines regions of particular interest -- in this case, points which have changed since the previous night.

Fig. 7. Another application in which there are multiple users. If done with stereo, then 2n images for n users, a distinct stereo pair for each user's point of view.

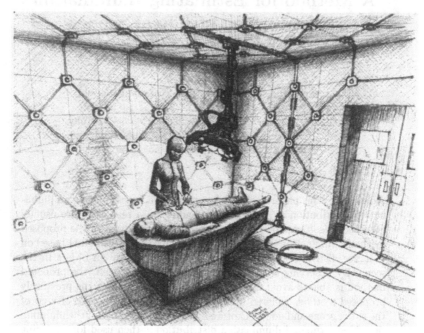

Fig. 8. A medical application with a local surgeon showing the sea of cameras around the operating room.

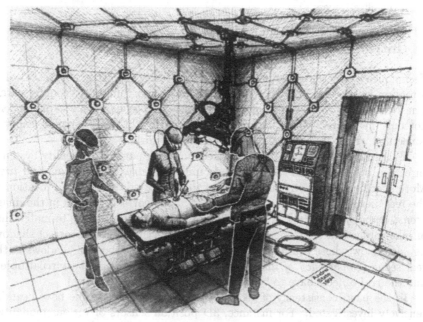

Fig. 9. The same application in which the remote consultants are viewed by Head Mounted Displays, so that each participant sees the others.

A Method for Estimating Illumination Distribution of a Real Scene Based on Soft Shadows

Imari Sato, Yoichi Sato, and Katsushi Ikeuchi

Institute of Industrial Science, The University of Tokyo
7-22-1 Roppongi, Minato-ku, Tokyo 106-8558, Japan
{imarik, ysato, ki}@iis.u-tokyo.ac.jp

Abstract. This paper describes a new method for estimating an illumination distribution of a real scene. Shadows in a real scene are usually observed as soft shadows that do not have sharp edges. In the proposed method, illumination distribution of the real scene is estimated based on radiance distribution inside the soft shadows cast by an object in the scene. By observing shadows and not illumination itself, the proposed method is able to avoid several technical problems which the previously proposed methods suffered from: how to capture a wide field of view of the entire scene and how to capture a high dynamic range of the illumination. The estimated illumination distribution is then used for rendering virtual objects superimposed onto images of the real scene. We successfully tested the proposed method by using real images to demonstrate its effectiveness.

1 Introduction

Techniques for merging virtual objects with a real scene attract a great deal of attention in the field of both computer graphics and computer vision research. The synthesized world called augmented reality enables us to handle phenomena not only in the real world but also in a virtual world, while virtual reality is concerned with only a fully computer generated world.

To achieve a high quality synthesized image in augmented reality systems, three aspects have to be taken into account: *geometry*, *illumination*, and *time*. More specifically, the virtual object has to be located at a desired location in the real scene, and the object must appear at the correct location in the image (*consistency of geometry*). Also, shading of the virtual object has to match that of other objects in the scene, and the virtual object must cast a correct shadow, i.e., a shadow whose characteristics are consistent with those of shadows in the real scene (*consistency of illumination*). Lastly, motions of the virtual object and the real objects have to be coordinated (*consistency of time*).

In the past, consistency of geometry and consistency of time have been intensively investigated.[1] For instance, 3D position sensors of various modalities

---

[1] For a good survey of augmented reality technologies, see [1].

S. Nishio, F. Kishino (Eds.): AMCP'98, LNCS 1554, pp. 44–58, 1999.

are used for measuring the position and orientation of a user's viewpoint in real time, so that a virtual object can be superimposed onto the image that the user is looking at (for instance, [2,12,3]).

On the other hand, few methods have been proposed for using correct illumination to superimpose virtual objects onto an image of a real scene. This is because real scenes usually include both direct and indirect illumination distributed in a complex way and it is not easy to obtain correct illumination models to be used for augmented reality systems.

Pioneering work in this field was proposed by Fournier et al. [7]. Fournier et al.'s method takes into account not only direct illumination but also indirect illumination by using the radiosity algorithm [4], which is commonly used for rendering diffuse interreflection. This method is effective for modeling subtle indirect illumination from nearby objects. However, this method requires a user to specify the 3D shapes of all objects in the scene. This object selection process could be tedious and difficult if a scene is full of objects. Also, since this method computes global illumination using pixel values of an input image, it is required that the image have a reasonably wide field of view. Even so, this method cannot model a direct illumination from outside of the input image unless a user specifies the positions of all lights.

Later, Drettakis et al. [6] extended Fourier et al.'s work. Drettakis et al.'s method made the creation of the 3D model much easier using computer vision techniques. They also introduced the use of a panoramic image built by image mosaicing to enlarge the field-of-view of the input image, and the use of hierarchical radiosity for efficient computation of global illumination. However, this method still requires a user to define vertices and topology of all objects in the scene, and it is often the case that the achieved field-of-view is not wide enough to cover all surfaces in the scene. This causes the same limitation on direct illumination outside the input image as in Fournier et al.'s method.

Recently, Debevec [5] introduced a framework of superimposing virtual objects onto an image of a real scene with correct illuminations. The method first constructs a light-based model: a representation of a scene which consists of radiance information of the surfaces of the scene. The model is constructed by mapping the reflections on a spherical mirror onto a geometric model of the scene specified by a user. Then global illumination is computed among three scene components: distant scene, local scene, and synthetic objects. It was reported that the method was able to superimpose virtual objects onto the image with convincing shadings. However, the method required several series of input images for constructing the light base model. To accurately record the brightness of the reflections, a high dynamic range image of the scene was generated from a series of images taken with different sensor integration time. To cover the entire geometric model by the reflections on the mirror, multiple images viewed from different angles were also necessary. These additional images cost more processing time, and therefore would not be welcomed in some augmented reality applications. Also, defining a geodesic model of the scene often requires a great deal of manual processes.

A similar approach was also proposed by Sato et al. [11]. Their method measures an illumination distribution in a real scene from a pair of omni-directional images taken by a camera with a fisheye lens. Unlike the previously proposed methods which require a user to specify a geometric model of the entire scene, Sato et al.'s method uses an omni-directional stereo algorithm for automatically generating the geodesic model of the scene. Then the radiance of the scene is computed from a sequence of omni-directional images taken with different shutter speeds and mapped onto the generated geometric model. For synthesizing images, the illumination distribution of the scene is approximated by sampling the distribution at an equal solid angle specified by the node directions of a geodesic dome. Using the sampled illumination radiance, the method is able to render virtual objects and shadows cast by the virtual objects onto a real object. However, in Sato et al.'s method, the indirect illumination between synthetic objects and nearby real objects was not modeled. The method was tested using real images of both in indoor and outdoor scenes.

To summarize, the previously proposed methods attempted to measure illumination distribution of the scene directly from images. As a result, they suffered from two technical problems: how to record high dynamic range of the scene, and how to capture a wide field of view of the scene. To overcome the first problem, a high dynamic range image generated by a series of images taken with different shutter speeds was used. To overcome the second problem, special equipment such as a spherical mirror and a camera with fisheye lens or a panoramic image built by mosaicing was used. However, these solutions tend to increase the number of input images, and therefore would not be desirable in real applications.

The purpose of this study is to present a new method for estimating an illumination distribution of a real scene from radiance distribution inside soft shadows. Shadows in a real scene are caused due to the occlusion of incoming lights as illustrated in Figure 1. It is thus reasonable to assume that shadows contain various pieces of information about the illumination of the scene. From this point of view, this proposed method estimates illumination distribution of a real scene by observing shadows cast by a real object, instead of directly measuring illumination of the scene as in the previously proposed methods. Since we observe shadows and not the illumination itself, any effort to capture a wide field of view of the scene or to capture high dynamic range of the scene is not necessary. In most cases, an image taken by an ordinary CCD color camera is enough to capture the shadows of an object. Furthermore, the previously proposed method required a great deal of manual processes such as specifying object shapes and line segments in order to construct a geometric model of the scene, while our method requires only a shape and a location of a real object which casts shadows onto the scene. The proposed method is thus simple and seems to be applicable in most augmented reality applications.

1.1   Overview

We estimates an illumination distribution of a real scene from the following three images of the scene taken by a color CCD camera without changing the camera setting. A typical example of input images is shown in Figure 4.

- *Original image* : An image of the scene onto which virtual objects are superimposed.
- *Shadow image* : We place an object of known shape at an arbitrary location so that shadows of the object appear in the image. We call this object occluding object.
- *Calibration image* : We place a calibration board with regularly spaced dots in the scene so that all dots on the calibration board appear in the image. We establish the correspondence between the 3D world coordinate system in the scene and the 2D image coordinate system in the image by using the camera calibration algorithm proposed by Tsai [14]. From the calibration process, a plane of $z = 0$ is also defined on the calibration board.

The rest of the paper is organized as follows. In Section 2 and Section 3, we explain how to estimate an illumination distribution of a real scene from image irradiance of a *shadow image*. We first obtain a formula which relate an illumination distribution of a real scene with the image irradiance of the *shadow image*(Section 2). Second, by assigning the image irradiance of the *shadow image* to the formula, we obtain a set of linear equations with unknown illumination radiance values sampled at an equal solid angle. Finally, we solve the set of linear equations for a unique illumination radiance solution set which represents the illumination distribution of the scene(Section 3).

Section 4 explains how to superimpose virtual objects onto the real scene by using the estimated illumination distribution. Section 5 shows experimental results of the proposed method applied to real images. Section 6 presents concluding remarks.

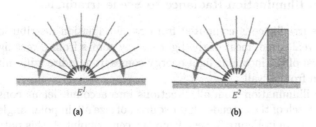

Fig. 1. Total irradiance: (a) without occluding object (b) with occluding object

2   Formula for Relating Illumination Radiance with Image Irradiance

In this section, we present a formula which relates an illumination distribution of a real scene with the image irradiance of a *shadow image*. Based on the image formation as illustrated in Figure 2, the formula is obtained as follows:

1. Find a relationship between an illumination distribution of a real scene and irradiance at a surface point in the scene.
   (**illumination radiance** to **scene irradiance**)
2. Compute how much of the incoming lights are reflected from the surface toward an image plane.
   (**scene irradiance** to **scene radiance**)
3. Find a relationship between the reflected light from the surface and image irradiance at a corresponding point on the image plane.
   (**scene radiance** to **image irradiance**)

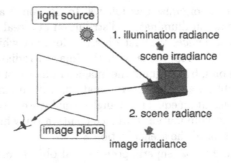

Fig. 2. Image formation

2.1   From Illumination Radiance to Scene Irradiance

First, scene irradiance is computed from an illumination distribution of a real scene. In a real scene, most of the light energy comes from direct light sources, while the rest of the incoming light energy comes from indirect illumination such as reflection from a wall.

To take illumination from all directions into account, let us consider an infinitesimal patch of the extended light source, of size $\delta\theta_i$ in polar angle and $\delta\phi_i$ in azimuth as shown in Figure 3. Seen from the center point A, this patch subtends a solid angle $\delta\omega = \sin\theta_i\delta\theta_i\delta\phi_i$. If we let $L_0(\theta_i, \phi_i)$ be the illumination radiance per unit solid angle coming from the direction (θ_i, ϕ_i), then the radiance from the patch is $L_0(\theta_i, \phi_i)\sin\theta_i\delta\theta_i\delta\phi_i$, and the total irradiance of the surface point A is [8]

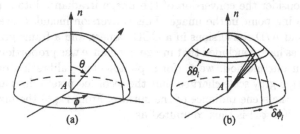

Fig. 3. (a)the direction of incident and emitted light rays (b)infinitesimal patch of an extended light source)

$$E = \int_{-\pi}^{\pi} \int_{0}^{\frac{\pi}{2}} L_0(\theta_i, \phi_i) \cos\theta_i \sin\theta_i d\theta_i d\phi_i \tag{1}$$

Then occlusion of the incoming light by the occluding object is considered as

$$E = \int_{-\pi}^{\pi} \int_{0}^{\frac{\pi}{2}} L_0(\theta_i, \phi_i) S(\theta_i, \phi_i) \cos\theta_i \sin\theta_i d\theta_i d\phi_i \tag{2}$$

where $S(\theta_i, \phi_i)$ are occlusion coefficients; $S(\theta_i, \phi_i) = 0$ if $L_0(\theta_i, \phi_i)$ is occluded by the occluding object; Otherwise $S(\theta_i, \phi_i) = 1$.

2.2 From Scene Irradiance to Scene Radiance

Some of the incoming lights at point A are reflected toward the image plane. As a result, point A becomes a secondary light source with scene radiance, which can be computed from scene irradiance at point A.

The bidirectional reflectance distribution function (BRDF) $f(\theta_i, \phi_i; \theta_e, \phi_e)$ is defined as a ratio of the radiance of a surface as viewed from the direction (θ_e, ϕ_e) to the irradiance resulting from illumination from the direction (θ_i, ϕ_i).

Thus, by integrating the product of the BRDF and the illumination radiance over the entire hemisphere, the scene radiance $Rs(\theta_e, \phi_e)$ viewed from the direction (θ_e, ϕ_e) is computed as

$$Rs(\theta_e, \phi_e) = \int_{-\pi}^{\pi} \int_{0}^{\frac{\pi}{2}} f(\theta_i, \phi_i; \theta_e, \phi_e) L_0(\theta_i, \phi_i) S(\theta_i, \phi_i) \cos\theta_i \sin\theta_i d\theta_i d\phi_i \tag{3}$$

2.3 From Scene Radiance to Image Irradiance

Finally, the illumination radiance of the scene is related with image irradiance on the image plane. Since what we actually observe is not image irradiance on the image plane, but rather a recorded pixel value in *shadow image*, it is also necessary to consider the conversion of the image irradiance into a pixel value of a corresponding point in the image. This conversion includes several factors such as D/A and A/D conversions in a CCD camera and a frame grabber.

Other studies have concluded that image irradiance was proportional to scene radiance [8]. In our method, we perform photometric calibration of the CCD camera by using a gray scale chart so that the recorded pixel values also become proportional to the scene radiance of the surface. From Eq. 3 the pixel value of *shadow image* $P(\theta_e, \phi_e)$ is thus computed as

$$P(\theta_e, \phi_e) = k \int_{-\pi}^{\pi} \int_{0}^{\frac{\pi}{2}} f(\theta_i, \phi_i; \theta_e, \phi_e) L_0(\theta_i, \phi_i) S(\theta_i, \phi_i) \cos\theta_i \sin\theta_i d\theta_i d\phi_i \quad (4)$$

where k is a scaling factor between scene radiance and a pixel value.

Due to the scaling factor k, we are able to estimate unknown $L_0(\theta_i, \phi_i) (i = 1, 2, .., n)$ up to scale.

However, since we are interested only in superimposing virtual objects onto an image of the scene, what we are required to do is to respect the same proportionality when we synthesize images.

3   Estimation of Illumination Distribution Using Image Irradiance

After obtaining the formula which relates the illumination radiance of the scene with the pixel values of the *shadow image*, illumination radiance is estimated based on the recorded pixel values of the *shadow image*.

First, the double integral in Eq. 4 is approximated by discrete sampling over the entire surface of the extended light source. Second, by assigning the pixel values of the *shadow image* to the new formula, we obtain a set of linear equations where unknowns are the illumination radiance sampled at an equal solid angle. By solving the set of linear equations, we obtain the distribution of illumination radiance sampled at equal solid angles; this distribution approximates the illumination distribution of the scene.

3.1   Approximation of Illuminate Distribution by a Geodesic Dome

Node directions of a geodesic dome are used for approximating the illumination distribution of the scene as a summation of illumination radiance sampled at equal solid angles.

Nodes of a geodesic dome are known to be uniformly distributed over the surface of a sphere. Therefore, by using n nodes of a geodesic dome in a northern hemisphere as a sampling direction, the double integral in Eq. 4 is approximated as a sampling at an equal solid angle $\delta\omega = 2\pi/n$.

$$P(\theta_e, \phi_e) = \sum_{i=0}^{n} f(\theta_i, \phi_i; \theta_e, \phi_e) L(\theta_i, \phi_i) S(\theta_i, \phi_i) \cos\theta_i \quad (5)$$

where $L(\theta_i, \phi_i)$ is the illumination radiance per solid angle $\delta\omega = 2\pi/n$ coming from the direction (θ_i, ϕ_i), which also includes the scaling factor k between scene radiance and pixel values. The number of the nodes n can be adjusted by changing the sampling frequency of a geodesic dome.

The important point to note is that the recorded pixel value $P(\theta_e, \phi_e)$ is computed as a function of the illumination radiance $L(\theta_i, \phi_i)$ and the BRDF $f(\theta_i, \phi_i; \theta_e, \phi_e)$ in Eq. 5. We thus take two different approaches depending on whether BRDF of the surface is given. We explain the case where the BRDF is given in Section 3.2 and Section 3.3, and the other case in Section 3.4.

3.2 Known Reflectance Properties: Lambertian Model

Suppose the surface is a Lambertian surface, BRDF $f(\theta_i, \phi_i; \theta_e, \phi_e)$ for a Lambertian surface is known to be a constant. From Eq. 5, an equation for a Lambertian surface is obtained as

$$P(\theta_e, \phi_e) = \sum_{i=0}^{n} K_d L(\theta_i, \phi_i) cos\theta_i S(\theta_i, \phi_i) \qquad (6)$$

where K_d is a diffuse reflection parameter of the surface.
From Eq. 6, a linear equation is obtained for each image pixel of *shadow image* as

$$a_1 L_1 + a_2 L_2 + a_3 L_3 + \cdots + a_{1n} L_n = P \qquad (7)$$

where L_i $(i = 1, 2, .., n)$ are n unknown illumination radiance specified by n node directions of a geodesic dome. The coefficients $a_i (i = 1, 2, .., n)$ represent $K_d \cos\theta_i S_i$ in Eq. 6; we can compute these coefficients from the 3D geometry of a surface point, the occluding object and the illumination direction. P is the values of the image pixel $P(\theta_e, \phi_e)$.

If we select a number of pixels, say m pixels, a set of linear equations is obtained as

$$a_{11} L_1 + a_{12} L_2 + a_{13} L_3 + \cdots + a_{1n} L_n = P_1$$
$$a_{21} L_1 + a_{22} L_2 + a_{23} L_3 + \cdots + a_{2n} L_n = P_2$$
$$a_{31} L_1 + a_{32} L_2 + a_{33} L_3 + \cdots + a_{3n} L_n = P_3$$
$$\cdots \qquad \cdots$$
$$a_{m1} L_1 + a_{m2} L_2 + a_{m3} L_3 + \cdots + a_{mn} L_n = P_m \qquad (8)$$

Therefore, by selecting a sufficiently large number of image pixels, we are able to solve for a unique solution set of unknown L_i's. Note that, since each pixel consists of 3 color bands (R, G, and B), each band of radiance L_i is also estimated from the corresponding color band of the image.

3.3 Known Reflectance Properties: Non-Lambertian Model

Our method is limited not only to the Lambertian reflection model, but it can also be extended to other reflection models. As shown in the previous case, our method requires a set of linear equations with unknown illumination radiance. Hence, any reflection model is applicable to our method as long as such a set of linear equations is obtained.

Take a simplified Torrance-Sparrow reflection model [10,13] for example; the pixel value of shadow image $P(\theta_e, \phi_e)$ is computed as

$$
\begin{aligned}
P(\theta_e, \phi_e) &= K_d \sum_{i=0}^{n} L(\theta_i, \phi_i) S(\theta_i, \phi_i) cos\theta_i \\
&+ K_s \sum_{i=0}^{n} L(\theta_i, \phi_i) S(\theta_i, \phi_i) \frac{1}{cos\theta_r} e^{\frac{-\gamma(\theta_i, \phi_i)^2}{2\sigma^2}} \\
&= \sum_{i=0}^{n} (K_d cos\theta_i + K_s \frac{1}{cos\theta_r} e^{\frac{-\gamma(\theta_i, \phi_i)^2}{2\sigma^2}}) S(\theta_i, \phi_i) L(\theta_i, \phi_i)
\end{aligned}
\tag{9}
$$

where θ_i is the angle between the surface normal and the direction to each point light source, θ_r is the angle between the surface normal and the viewing direction, $\gamma(\theta_i, \phi_i)$ is the angle between the surface normal and the bisector of the light source direction and the viewing direction, K_d and K_s are constants for the diffuse and specular reflection components, and σ is the standard deviation of a facet slope of the Torrance-Sparrow reflection model.

From Eq. 9, we obtain a linear equation for each image pixel where $L(\theta_i, \phi_i)$ $(i = 1, 2, .., n)$ are unknown illumination radiance, and $(K_d cos\theta_i + K_s \frac{1}{cos\theta_r} e^{\frac{-\gamma(\theta_i, \phi_i)^2}{2\sigma^2}}) S(\theta_i, \phi_i)$ $(i = 1, 2, .., n)$ are known coefficients. Again, if we use a sufficiently large number of pixels for the estimation, we are able to solve for a unique solution set of unknown illumination radiance $L(\theta_i, \phi_i)(i = 1, 2, .., n)$.

3.4   Unknown Reflectance Properties: Lambertian Model

Even in the case where the BRDF is not given, we are still able to estimate an illumination distribution of a real scene if the surface is a Lambertian surface.

The question we have to consider here is how to cancel the additional unknown number K_d in Eq. 6. Image irradiance of the *original image* is used for this purpose. In the *original image*, since no occluding object exists in the scene, the shadow coefficients $S(\theta_i, \phi_i)$ always become $S(\theta_i, \phi_i) = 1$. Using Eq. 6, the image irradiance $P'(\theta_e, \phi_e)$ of the *original image* is computed as

$$
P'(\theta_e, \phi_e) = K_d \sum_{j=0}^{n} L(\theta_j, \phi_j) cos\theta_j
\tag{10}
$$

From Eq. 6 and Eq. 10, the unknown K_d is canceled as

$$
\begin{aligned}
\frac{P(\theta_e, \phi_e)}{P'(\theta_e, \phi_e)} &= \frac{K_d \sum_{i=0}^{n} L(\theta_i, \phi_i) cos\theta_i S(\theta_i, \phi_i))}{K_d \sum_{j=0}^{n} L(\theta_j, \phi_j) cos\theta_j} \\
&= \sum_{i=0}^{n} \frac{L(\theta_i, \phi_i)}{\sum_{j=0}^{n} L(\theta_j, \phi_j) cos\theta_j} cos\theta_i S(\theta_i, \phi_i)
\end{aligned}
\tag{11}
$$

Finally, we obtain a linear equation for each image pixel where $\frac{L(\theta_i,\phi_i)}{\sum_{j=0}^{n} L(\theta_j,\phi_j)cos\theta_j}$ $(i = 1, 2, .., n)$ are unknowns, $cos\theta_i S(\theta_i, \phi_i)$ $(i = 1, 2, .., n)$ are computable coefficients, and $\frac{P(\theta_e,\phi_e)}{P'(\theta_e,\phi_e)}$ is a right-hand side quantity. Again, if we use a sufficiently large number of pixels for the estimation, we are able to solve the set of linear equations for a unique solution set of unknown $\frac{L(\theta_i,\phi_i)}{\sum_{j=0}^{n} L(\theta_j,\phi_j)cos\theta_j}$ $(i = 1, 2, .., n)$.

We should point out that the estimated radiance from these equations is a ratio of $L(\theta_i, \phi_i)$ to $\sum_{j=0}^{n} L(\theta_j, \phi_j)cos\theta_j$. In other words, it is a ratio of the illumination radiance in one direction $L(\theta_i, \phi_i)$ to scene irradiance at the surface point $\sum_{j=0}^{n} L(\theta_j, \phi_j)cos\theta_j$. Hence, without knowing the ratio of the scene irradiance among color bands, there is no way to relate the estimated radiance over the color bands. Our method avoids this problem based on initial camera calibration. Since we use a white board with regularly spaced dots as a calibration board, the recorded color of the board directly shows the ratio of the scene irradiance among color bands.

4 Superimposing Virtual Objects onto a Real Scene

In the previous sections, we have described how to estimate an illumination distribution of a real scene. In this section, we explain how to superimpose virtual objects onto the real scene by using the estimated illumination distribution.

For superimposing virtual objects onto an image of a real scene, the ray casting algorithm is used as follows.

For each pixel in the input image of the real scene, a ray extending from the camera projection center through the pixel is generated. Then it is determined whether a ray intersects a virtual object or a real object in the scene.

If the ray intersects a virtual object, we consider that the pixel corresponds to a point on the virtual object surface. Then we compute a color to be observed at the surface point, and store it in the pixel as the surface color of the virtual object. In our method, a simplified Torrance-Sparrow reflection model from Section 3.3 is used for this purpose. From the model, a color to be observed at the surface point $Rs(\theta_e, \phi_e)$ is computed using the estimated illumination distribution of the real scene as

$$Rs_c(\theta_e, \phi_e) = K_{d,c} \sum_{i=0}^{n} L_c(\theta_i, \phi_i)cos\theta_i + K_{s,c} \sum_{i=0}^{n} L_c(\theta_i, \phi_i)\frac{1}{cos\theta_r}e^{\frac{-\gamma(\theta_i,\phi_i)^2}{2\sigma^2}} \quad (12)$$

$$c = R, G, B$$

where $L_c(\theta_i, \phi_i)$ $(i = 1, 2, .., n)$ are the estimated illumination radiance values.

We are not concerned here with the problem of occlusion because it is beyond the scope of this paper to measure accurate 3D shapes of real objects.[2] we assume

---

[2] Techniques for determining correct occlusion between virtual and real objects using the shapes of the real objects are called Z-key. For instance, see [9].

that a virtual object always exists between the camera projection center and real objects in the scene.

If a ray through an image pixel does not intersect with a virtual object, the color of a real object surface corresponding to the image pixel needs to be modified so that a shadow cast by the virtual object is created on the real object surface. A shadow cast by a virtual object is created as follows:

1. Obtain a 3D coordinate of a surface point where a ray through an image pixel intersects; to do this, use a transformation between a 3D world coordinate system and a 2D image coordinate system.
2. Compute total irradiance E_1 at the surface point using the estimated illumination distribution. In this case, a virtual object does not occlude any incoming light (Figure 1.a).

$$E_{1,c} = \sum_{i=0}^{n} L_c(\theta_i, \phi_i) cos\theta_i \qquad c = R, G, B \qquad (13)$$

where $L(\theta_i, \phi_i)$ $(i = 1, 2, .., n)$ are the estimated illumination radiance values.
3. Compute total irradiance E_2 at the surface point in the case where the virtual object occludes some of the incoming light (Figure 1.b). As a result, the total irradiance E_2 becomes smaller than the total irradiance E_1.

$$E_{2,c} = \sum_{i=0}^{n} L_c(\theta_i, \phi_i) cos\theta_i S(\theta_i, \phi_i) \qquad c = R, G, B \qquad (14)$$

where $S(\theta_i, \phi_i) = 0$ if the virtual object occludes illumination radiance $L(\theta_i, \phi_i)$; otherwise, $S(\theta_i, \phi_i) = 1$.
4. Compute the ratio of E_2 to E_1. The ratio represents how much of the irradiance at the intersection would be lost if the virtual object were placed in the scene. Then, by multiplying the ratio E_2/E_1 to the observed color of the image pixel I, we obtain the color I' that would be the color of the image pixel if there were a virtual object.

$$I'_c = I_c \frac{E_{2,c}}{E_{1,c}} \qquad c = R, G, B \qquad (15)$$

5   Experimental Results

We have tested the proposed method by using real images of indoor environments. First, in Section 5.1, we describe experimental results in the case where reflectance properties of a reflected surface are known. Then, in Section 5.2, we describe experimental results in the case where reflectance properties of the surface are unknown.

5.1 Experimental Results for Known Reflectance Property

We took an image of a surface onto which a virtual object was to be superimposed (*original image*). Then, without changing the camera setting, we took two more images. One was the image of the surface with an occluding object (*shadow image*), and the other was the image of the surface with a calibration board (*calibration image*). The input images are shown in Figure 4.

First, the illumination distribution of the scene was estimated using image irradiance inside shadows in *shadow image* as explained in Section 3.2. Then a virtual object was superimposed onto the surface of *original Image* by using the estimated illumination distribution as explained in Section 4. The *calibration image* was used to obtain a transformation between the 3D world coordinate system and the 2D image coordinate system.

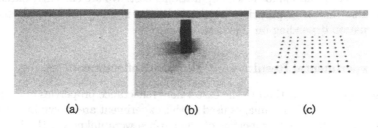

(a)   (b)   (c)

Fig. 4. Input images : (a) *original image* (b) *shadow image* (c) *calibration image*

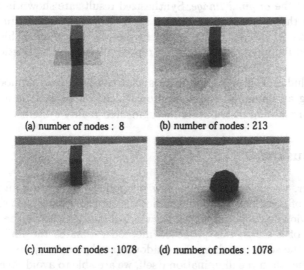

(a) number of nodes : 8   (b) number of nodes : 213

(c) number of nodes : 1078   (d) number of nodes : 1078

Fig. 5. Synthesized images: known reflectance property

To evaluate the accuracy of the illumination distribution estimated by our method, we superimposed a virtual object with the same shape as that of the occluding object. Synthesized results are shown in Figure 5 (a), (b), and (c). Also, we superimposed another virtual object of a different shape onto the scene in Figure 5(d). The number of nodes of a geodesic dome used for the estimation is shown under the resulting image.

It was found through our experiments that, the larger number of nodes we used, the more the shadows of the virtual object resembled those of the occluding object in *shadow image*. Especially in the case of 1078 nodes, the shadows of the virtual object are indistinguishable from those of the occluding object: this shows that the estimated illumination distribution represents the characteristics of that of the real scene well.

Also, the resulting images indicate that it is required to adjust the number of nodes of a geodesic dome depending on the complexity of a scene to obtain a reasonably good estimation for less computational cost. We are currently extending our work so that an appropriate number of nodes is automatically selected for the estimation depending on the scene complexity.

5.2   Experimental Results for Unknown Reflectance Property

We also applied our method to the case where reflectance properties of a surface were unknown. The input images used in this experiment are shown in Figure 6.

Since the reflectance properties of the surface were unknown, the image irradiance of both *original image* and *shadow image* were used for estimating the illumination distribution of the scene as explained in Section 3.4. In the same way as in the previous case, a virtual object with the same shape as that of the occluding object and another object of a different shape were superimposed onto the surface of the *original Image*. Synthesized results are shown in Figure 7.

Again, in the case of 1078 nodes, the shadows in the resulting image strongly resemble those of the occluding object in the *original image*. This shows that the estimated illumination distribution represents the characteristics of the real scene well.

We concluded from our experiments that the proposed method is effective for providing an illumination distribution which can be used as a substitution for a real illumination distribution in augmented reality systems.

6   Conclusions

In this paper, we have proposed a new method for estimating an illumination distribution of a real scene from radiance distribution inside soft shadows. Unlike the previously proposed methods, which directly measure the illumination distribution of a scene, this proposed method estimates the illumination distribution of the scene by observing shadows cast by real objects. By observing shadows rather than the illumination itself, we are able to avoid several technical problems which the previously proposed methods suffered from: how to capture a wide field of view of the entire scene and how to capture a high dynamic range of the illumination. Also, the proposed method is simple and requires only a shape and a location of a single real object in the scene, while the previously proposed method required a great deal of manual processes such as specifying shapes of all objects in order to construct a geometric model of the scene. As a result, from only three images taken by a ordinary CCD color camera, *original image*, *shadow image*, and *calibration image*, our method is able to estimate the illumination distribution of the entire scene automatically and to superimpose virtual objects onto the image of the scene with correct shadings. To demonstrate the effectiveness of the proposed method, we have successfully tested our method by using sets of real images with different surface materials.

(a)   (b)   (c)

Fig. 6. Input images : (a) *original image* (b) *shadow image* (c) *calibration image*

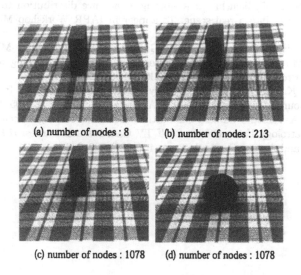

(a) number of nodes : 8   (b) number of nodes : 213

(c) number of nodes : 1078   (d) number of nodes : 1078

Fig. 7. Synthesized images: unknown reflectance property

References

1. Azuma, R.T.: A survey of augmented reality. Presence: Teleoperators and Virtual Environments, vol. 6, no. 4. (1997) 355–385
2. Azuma, R.T., Bishop, G.: Improving static and dynamic registration in an optical see-through HMD. Proceedings of SIGGRAPH 94. (1994) 197–204
3. Bajura, M., Fuchs, H., and Ohbuchi, R.: Merging virtual objects with the real world: seeing ultrasound imagery within the patient. Proceedings of SIGGRAPH 92. (1992) 203–210
4. Cohen, M.F., Chen, S.E., Wallace, J.R., Greenberg, D.P.: A progressive Refinement Approach to Fast Radiosity Image Generation Proceedings of SIGGRAPH 88. (1998) 75–84
5. Debevec, P.E.: Rendering Synthetic Objects into Real Scenes: Bridging Traditional and Image-based Graphics with Global Illumination and High Dynamic Range Photography. Proceedings of SIGGRAPH 98. (1998) 189–198
6. Drettakis, G., Robert, L., Bougnoux, S.: Interactive Common Illumination for Computer Augmented Reality. Proceedings of 8th Eurographics Workshop on Rendering.(1997) 45–57
7. Fournier, A., Gunawan, A., Romanzin, C.: Common Illumination between Real and Computer Generated Scenes. Proceedings of Graphics Interface 93. (1993) 254–262
8. Horn, B.K.P.: Robot Vision. The MIT Press, Cambridge, MA, (1986)
9. Kanade, T., Yoshida, A., Oda, K., Kano, H., Tanaka, M.: A Video-Rate Stereo Machine and Its New Applications. Proceedings of IEEE Conference on Computer Vision and Pattern Recognition 96. (1996) 196–202
10. Nayar, S.K., Ikeuchi, K., Kanade, T.: Surface reflection: physical and geometrical perspectives. IEEE Transactions on Pattern Analysis and Machine Intelligence, vol. 13, no.7. (1991) 611–634
11. Sato, I., Sato, Y., Ikeuchi K.: Acquiring a radiance distribution to superimpose virtual objects into a real scene. To appear in IAPR Workshop Machine Vision and Application. (1998)
12. State, A., Hirota, G., Chen, D.T., Garrett, W.F., Livingston, M.A., Superior augmented-reality registration by integrating landmark tracking and magnetic tracking. Proceedings of SIGGRAPH 96. (1996) 429–438
13. Torrance, K.E., Sparrow, E.M.: Theory for off-specular reflection from roughened surface. Journal of Optical Society of America, vol.57. (1967) 1105–1114
14. Tsai, R.: A Versatile Camera Calibration Technique for High Accuracy Machine Vision Metrology Using Off-the-Shelf TV Cameras and Lenses. IEEE Journal of Robotics and Automation, vol. 3, no. 4. (1987) 323–344

Integrating Real Space and Virtual Space in the 'Invisible Person' Communication Support System

Masahiko Tsukamoto

Department of Information Systems Engineering, Graduate School of Engineering
Osaka University, 2-1 Yamadaoka, Suita, Osaka 565-0871, Japan
tuka@ise.eng.osaka-u.ac.jp

Abstract. The real space computing technologies, such as the mobile computing technology, enable users to make use of computers anywhere in the world. On the other hand, the virtual space computing technologies enable users to use remote computer resources from their desktop environments through intuitive operations. By combining these two kinds of computing technologies, we can construct a more flexible and general platform for computing in either space. Based on this viewpoint, we have realized a communication environment, called the 'invisible person' environment, where virtual space and real space are strongly associated. In this paper, we discuss the system architecture of this environment. The policies that we took in its design are 1) reduction to a feasible design at present, 2) wide-spread popularity to become an invisible person, and 3) emphasis on the realization of communication rather than the concrete analysis and accurate presentation of the real space. These policies are reflected on our system design where we provide users with several kinds of browsers for the flexibility of their operations.

1 Introduction

The recent advancement of the computer hardware technologies has facilitated the trend of computer downsizing, which enables us to use computers anytime and anywhere. Such computer usage is called mobile computing and there have been a lot of researches conducted in this field. Mobile computing has several characteristics that are different from the conventional desktop computing [2]. One of the most remarkable points that mobile computing has is the fact that it is closely related with our daily life. As a result, researches on mobile computing focus not only on the use of the conventional desktop applications but also on the support of human activities in the real space. Researches on supporting human activities have been exhaustively done especially in the areas, so called the nomadic computing [1], the ubiquitous computing [7], and the wearable computing [5]. We may say that they are trying to establish new workflow models of human activities.

When we build applications for supporting human activities in the real space, it is generally difficult to realize interactions between the real space and the computer

S. Nishio, F. Kishino (Eds.): AMCP'98, LNCS 1554, pp. 59-74, 1999.

systems. These difficulties include the difficulty in performing actions on the computer system from the real space. This is partly because there is no common and efficient way to obtain concise status of a number of objects in the real space. This is also because most approaches for supporting activities in the real space associate some operations in the real space with some commands to the computer. Here, these associations are defined in an application-specific manner, and therefore, a user does not know any appropriate sequence of operations to achieve his/her intention in an application that he/she is not familiar with. Moreover, it is difficult to perform actions in the real space from the computer system environment. Although a user may operate electronic devices in the real space by using a computer, most other objects in the real space will not be affected without being equipped with special devices. As a result, it is hard for most computer applications to cover a wide diversity of human activity in the real space.

In order to solve this problem, we consider that the introduction of the virtual space computing technologies can be a feasible solution. These technologies provide us with two important functions. One is a 3D model with the associated language, hardware, and software, such as VRML and Direct 3D. The other is a 3D user interface, such as a VRML browser, with its navigation model and special equipment such as a head-mount display and a data glove. Here, it should be noted that the virtual space computing extends the applicability of each application via the Internet technology. As a result, by combining mobile computing and the virtual reality (VR) technologies, it is expected that we can construct a basis for flexible computing styles in the real space.

Based on this viewpoint, we have realized a communication support system, called the 'invisible person' system [6], where virtual space and real space are strongly associated. That is, the events occurred in the real space are immediately reflected in the virtual space and vice versa. A user who exists in either space can use this system and can communicate with other users in either space. In this paper, we discuss the system architecture of this environment. The policies that we took in its design are 1) reduction to a feasible design at present, 2) wide-spread popularity to become an invisible person, and 3) emphasis on the realization of communication rather than the concrete analysis and accurate presentation of the real space. These policies are reflected on our system design where we provide users with several kinds of browsers for the flexibility of their operation.

The rest of this paper is organized as follows: firstly, we discuss the integration of real space and virtual space and we explain the notion of 'invisible person' in section 2. Then, we show the design of our system in section 3. In section 4, we discuss two kinds of browsers which are used by a person who virtually visits remote place as an 'invisible person'. In section 5, we show the design of a browser used in real space working on a handy computer. Here, we also discuss an enhancement of the browser as a platform for augmented reality. In section 6, we discuss several methods and perspectives on the integration of some components shown in sections 4, 5, and 6. Finally, in section 7, we summarize the paper.

2   The Invisible Person Environment

2.1   Integration of Real Space and Virtual Space

By using recent network technologies, we can easily build various kinds of virtual space that are accessible via the Internet. In such virtual space, a user who uses a computer on his/her desk can communicate with other users in a similar situation even if they are separated by a long distance. They can make use of such an environment for disseminating, gathering, and exchanging several kinds of information. Moreover, the development of computer interface technologies makes it possible for a computer user to achieve a wide diversity of social activities with more reality.

In general, activities in virtual space do not directly affect the real space and it has been pointed out that users hardly know the direct effect of their activities in virtual space onto the real space. For example, when a user wants to buy something on a Web page, it is generally difficult for him/her to know any appropriate sequence of operations for fulfilling his/her intention. Such intention includes the purchase of multiple goods at a time, the abortion of the current transaction, and the cancellation of the recent purchase. If an application system completely supports all users' intention, users' operations may inevitably become complicated. Furthermore, when activities in virtual space become more realistic in the future, the gap between activities in the real space and those in the virtual space will be harmful in the social level: someone may injure other person by his/her misunderstanding on the computer operations. Another person may fall into mental illness due to his/her excessive immersion in virtual space.

By integrating mobile computing and virtual space computing, it is expected that we can overcome many limitations of the conventional style of computing. There are several possible approaches for the integration as follows:

- Enhancing the real space by computer support. This approach is sometimes called augmented reality [3]. A typical approach is to put marks everywhere in the real space that can be detected by a computer and then associate them with computing elements. This approach combines both kinds of space with more emphasis on the real space.
- Enhancing the virtual space by using several kinds of information obtained from the real space, such as location information and real-time video images taken by a camera. This approach is sometimes called augmented virtuality. This approach combines both kinds of space with more emphasis on the virtual space.
- Mixing the real space and the virtual space. We can construct a virtual space that fully reflects the associated real space and also influences it. This approach is the combination of both kinds of space with an even emphasis.

Among these approaches, since the infrastructure must provide a lot of functions for both the real space and the virtual space in order to develop a diversity of application on the top of a single infrastructure, the last approach is the most appropriate one for building the future computing infrastructure.

2.2 The Concept

Based on the observation described above, we are constructing a communication support system, called the 'invisible person' system. In the rest of this section, we describe our ideal goal of the system.

In the 'invisible person' system, virtual space is constructed based on a part of real space, where the changes in the real space are instantly reflected in the virtual space. In general, an invisible person is a person who cannot be seen by other people around him/her [8]. In this paper, we call a person, who visits a virtual space modeling a part of real space via a computer network using his/her high-performance computer, an *invisible person* because he/she cannot be seen by anyone in that part of the real space. A person, who is going to be an invisible person, first runs a program on his/her computer. This computer is called an *invisible person host* (*IPH*). The program is similar to common 3D browsers that are used for visiting virtual malls and cyber cities, but different from them in mainly one point: the 3D space where the user is visiting is almost congruent with a part of the real space. For example, when a user visits a place X with this program, he/she will see a similar situation to the corresponding part of the real space.

In order to grasp the changing situation in the real space, which should be reflected in the virtual space in real-time, the system uses several kinds of sensors placed in various locations in the real space. Such sensors may include video cameras. The data taken by these sensors are analyzed to generate the associated virtual space. Note that the scene each invisible person sees cannot be congruent with a complete real space, but should have reduced information. In this paper, when an invisible person visits a virtual place X, we call the situation "he/she is (virtually) at the (real) place X." On the other hand, of course, any person who really exists in the real space, called a *real person*, cannot directly see the people who virtually visit there, and therefore we call them 'invisible' persons.

A real person can use a handy computer called an *existence-sensitive pad* (*ESP*) to see the invisible persons. An ESP is equipped with a small video camera and shows an image which is composed by the image taken by the camera and the virtual images of the invisible persons visiting the place: it shows as if the invisible persons were there. Moreover, an ESP transmits the voice of each real person to the invisible persons and vice versa to realize a conversation. We call a real person holding an ESP an *esper*. An ESP may be equipped with a transparent head-mount display. In this case, the esper may see as if invisible persons were really in the real space. It should be noted that an invisible person can be an imaginary person generated by a computer, not an avatar of a remote person.

2.3 Applications

The invisible person system is useful because we can realize communication between a user in real space and a user in virtual space (i.e., a remote place). In particular, the reality of an invisible person and a real person provide each of them with a

Fig. 1. An Application Example: Secretary

sense of existence and affinity of the other and this is an important point in communication. In this subsection, we show several useful situations in our daily life.

A person who cannot go out from an office due to his/her tight schedule in the office can be an invisible person to visit other places. He/she can meet other people to negotiate some business issues in his/her short spare time. He/she can also go home to do household matters. If he/she uses conventional tools such as the electronic mail, a remote conference system, or a remote operation tools, it is generally difficult to negotiate with other people or to achieve complicated operations in his/her home. Therefore, in a practical sense, he/she has to go to the particular place to directly see the people or to directly operate something. On the contrary, we may achieve most of such matters by using the invisible person system. If the real persons can feel as if each invisible person exists around in the place, the meeting or the negotiation will smoothly proceed. Moreover, if the virtual space reflects the real space in considerably detail, we will be able to achieve various kinds of complicated activities as an invisible person.

An employee who lives apart from other members of his/her family can have dinner every night with them as an invisible person. A person who cannot take a long vacation can go abroad for a moment as an invisible person. Joining a telecommution or remote learning system as an invisible person will have a better effect than using other systems. An invisible person has a chance to have occasional communication with those who are in the real place that he/she is currently visiting. It is important for a guard of a building, a foreperson of construction, a tour guide for sightseeing, and a participant of an auction to be at the specific places, which can be effectively achieved through becoming invisible persons.

Fig. 1 shows an example usage of the invisible person system where a secretary of an office does her job in her home. Anyone in the room is aware of the existence of the secretary in the office if he/she brings an ESP, and therefore he/she can easily communicate with the secretary.

Fig. 2. An Application Example: Tour Guide

Fig. 3. An Application Example: Home Automation

Fig. 2 shows another usage of the invisible person system. Here, the tour guide is an invisible person who may use her computer at home. She may say "you can see over there the Kansai tower" by indicating a tower by his/her finger in an ESP screen, and she will navigate tourists by taking the lead in walking around a tourist resort. Moreover, she can do her job in London, Paris, New York, San Francisco, and Osaka in a very short period as an invisible person.

Fig. 3 also shows an example usage of the invisible person system. This is a kind of home automation where a user is virtually coming home from her office to turn on the switch of a rice-cooker. In this situation, she can intuitively know the location and the action she should take to achieve her objective and recognize the effect of her behavior in this virtual space. Moreover, if she finds her husband in the room, she has an opportunity to communicate with him.

As described before, an invisible person may be a virtual person generated by a computer, not an avatar of a remote person. Such virtual people will play various kinds of human activities in the real space on behalf of the real persons. Although a tour guide and a secretary of an office have been realized as software systems in computers, more flexible systems can be constructed in the invisible person environment. They may be more easily-understood by everyone.

3   Overview of the System Design

As for the realization of the invisible person system, it should be firstly noted that it is impossible for the current computer technologies to realize the environment with sufficient reality. There are many technical bottlenecks to achieve it, such as the concise recognition of real space, the user interfaces in virtual space, the immediate reflection of the virtual space in the real space, and the real-time processing of huge amount of data. Therefore, we focus especially on the following points [4]:

- Reduction to a feasible design at present. We assume a common environment and we do not use special or expensive hardware as long as possible. By using special hardware, we may enrich the reality of the environment, but the generality for easy deployment is lost.
- Wide-spread popularity to become an invisible person. A user can become an invisible person without using special hardware and software. Moreover, through the Internet, which has already been constructed in the world-wide scale, anyone can visit the environment from anywhere in the world as an invisible person, to communicate with other people in the world. To make the system usable on the Internet, the standard protocols and tools should be used.
- Emphasis on realization of communication rather than concrete analysis and accurate presentation of the real space. Since facial expression and gesture are important in human communication, we rather prefer the use of real-time video images.

Based on this policy, we have designed the system in the following way: In the real space, we fix a graphic workstation and equip it with CCD cameras, microphones, speakers, and a wireless LAN interface. This computer is called the *management server*. The video image of real persons is taken by one or some of the CCD cameras, and a minimal bounding box of each real person is cut from them. This is then used for composing the virtual space seen by an invisible person. Here, we assume that objects, except people, do not move, and should be modeled by a system administrator beforehand. The management server communicates with all IPHs and all ESPs. Several microphones are fixed at some locations in the real space to catch the sound. The caught sound is transmitted to each IPH.

A microphone, a speaker, and a CCD camera are attached to an IPH. The CCD camera is used to capture the users' facial image. The captured video images are transferred to the management server. The voice of each invisible person is captured by the microphone, and then transmitted to the management server. It is played by one of the speakers that are nearest to his/her imaginary location in the real space.

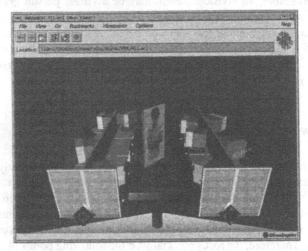

Fig. 4. An IPH Screen Using a VRML Browser

An ESP is a notebook-type personal computer equipped with a CCD camera, a wireless LAN card, and a sensor to obtain the location information of the ESP. The location information and the image data of each invisible person are sent by the management server via the wireless communication. Based on the obtained information, the ESP calculates the x/y coordinate values for each invisible person to paste its images in the ESPs' display. Then, it displays the images taken by its CCD camera composed with the facial images of the invisible persons.

4 IPH: Two Alternatives

In the design of the IPH, we consider that the use of standard tools is important for an easy access to the system. Thus, we designed and implemented two alternatives for realizing IPH. One is based on VRML and the other is based on the dynamic HTML. In a broader sense, the former is geometry-based and the latter is image-based.

4.1 A VRML-Based Approach

VRML provides users with a facility to describe 3D space. There are several browsers to show and navigate the 3D space described in the language. Therefore, in order to reconstruct real space as virtual space in a VRML browser, the real space should be described in VRML. In this case, dynamic aspects such as the display of real persons can be implemented by using Java. The management server transmits programs described in VRML and Java to each IPH at the beginning of the session.

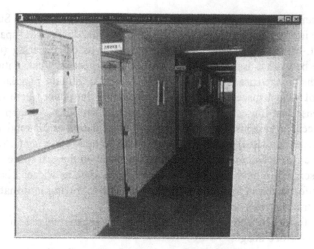

Fig.5. An IPH Screen Using a WWW Browser

Fig. 4 shows an example of the VRML-based IPH screen. Here, a CCD camera attached to the management server takes video images of the real space and the obtained video images are consecutively analyzed by the management server to detect the minimal bounding rectangular region that encloses each real person. The extraction is done by comparing the taken image with the referential image, the image taken beforehand by the same camera. Each rectangular region of the image is sent to each IPH with its location. The remaining area of the image is used in updating the referential image. An invisible person operates a VRML browser, as shown in this figure. In this browser, a virtual space is composed by a VRML description and the video images which are sent by the management server. In this picture, there is a plate in which a real-time video image of a real person is pasted.

4.2 An Image-Based Approach

An image-based approach provides a more convenient way for the system constructors and users. In this approach, pictures or video images taken in the real space are used directly. Each scene is composed of a single resource, i.e., a picture image or a video image. Multiple scenes are used to change the users' view to represent a wide range of space. Here, we use an avatar in the browser to represent the user.

By using an avatar, richer communication with more complex behavior of the user can be achieved. Note that an avatar of a user is used in most TV games. Among these games, there are several games employing the image-based approaches, e.g., 'Biohazard 2' (or 'Residential Evil') by CAPCOM; 'Parasite Eve' and 'Final Fantasy VII' by SQUARE. In these game programs, complex behaviors of a user should be represented in the screen to make them exciting.

This approach is also valid to express a pseudo 3D space based on the real space. A user can distinguish several actions such as the action 'to sit on a chair' and the action 'to lean on the chair' by seeing his/her avatar. In the browser, the image of the avatar is expanded or reduced, and drawn in an appropriate location. Such simple control on the image provides a user with the feeling of the depth of 3D space.

The VRML-based approach is called the geometry-based rendering (GBR) in a broader sense because it uses a geometric model and a geometry calculation. On the other hand, the image-based rendering (IBR) has been gaining popularity for the construction of virtual space with reality. The approach introduced here is a kind of IBR in the sense that it uses an image, but different from other IBR approaches because it enforces the systems with little amount of calculation or 3D rendering. Thus, we call this (rather classical) approach the *image-based non-rendering* (*IBNR*).

In order to construct a large-scale virtual space based on the real space, each scene should be described in an independent manner as possible as it can. Therefore, we manage the information in a scene-based manner. The following information is necessary to construct each scene:

- A resource for the background image: it may be a camera image, a drawn picture, or a real-time video image.
- The geometry information of the floor shape: for example, the shape can be expressed by a rectangle (i.e., a width and a height), the scale of the avatar, or the association with the above resource (e.g., X/Y coordinates in the picture).
- Links to neighboring scenes that include the direction and the scene ID. We employ URL for scene ID, which makes it possible to integrate WWW resources with the scene information.

Fig. 5 is an example screen of the implemented IBNR system. The browser is a standard WWW browser (Microsoft Internet Explorer 4.0, in this case). The dynamic HTML is used to get users' keyboard input and to put the avatars' picture standing in the middle of the corridor. Users' keyboard input is interpreted as the command to the avatar: 't' to forward, 'f' and 'h' to turn, and 'b' to backward. We use eight pictures of a person facing to every 45 degree directions to show the 'turn' operation. If a user inputs a command key, the avatar's picture is appropriately enlarged or shrunk, and moved. If the avatar goes out of a predefined rectangular region of the floor, another scene is displayed based on the linkage information defined for the original scene.

4.3 Discussions

The VRML-based approach provides users with flexible operations in browsing the virtual space. Users can freely move around the scene and they can see the scene from any viewpoint they like. This approach is also valid in the future advanced VR environment: it is possible to generate 3D vision scenes using the parallax effect in this approach. However, it is generally difficult to construct virtual space that faithfully recreates the original real space because the real space is too complicated to model. Moreover, even if someone can concretely model the real space, the modeled data will be too large to be promptly processed by a typical computer of a user.

On the other hand, IBNR is good in the following points:

- It is easy to construct the virtual space based on the real space.
- It can be shown in considerably high speed.

Fig. 6. An ESP Screen

- An avatar makes it possible for a user to perform a complicated action and communicate with other users.

In this sense, we may say that the IBNR approach is more practical in the current status of the network and the hardware technologies.

5   ESP and Its Enhancement to Handle Invisible Objects

An ESP is a handy computer that a real person uses in real space to see invisible persons. In designing an ESP, it is important to consider the lack of system resources in such a handy computer. It usually has a low power CPU and little memory. Therefore it will not be able to handle geometry-based model in sufficiently high speed.

5.1   Design of ESP Using Magnetic Sensor

Fig. 6 shows a situation where a real person uses an ESP. The ESP screen is composed of the video image obtained by the attached CCD camera and the images of invisible persons within the viewpoint of the camera. The location of the latter image in the screen changes when a user moves the ESP and when the invisible person moves in the virtual space.

An ESP uses a sensor to obtain its direction and location. In our prototype system, we employ a magnetic sensor. Although it gives concise information about the direction and the location of the sensor in real-time, the use of a magnetic sensor is against our policy in the point of 'reduction to a feasible design' because it is not a popular tool and is usually expensive. Therefore, we are currently considering the possibility to exclude it from the design of ESP. One possible solution is to analyze the image taken by the camera to extract the direction and the location of the ESP.

5.2 Invisible Objects: An Extension of ESP

There is one important extension of ESP. It is closely related to the area so called the augmented reality. There are a lot of researches done in this area. Most of them take similar approaches to our ESP approach. That is, they use camera images to show imaginary objects on the screen which do not really exist in the real space. Although our ESP only shows imaginary people on its screen, we consider that the approach to show imaginary 'objects' can be incorporated into the ESP. As an extension of the invisible person environment, it is possible to create an 'invisible object'. Here, invisible objects are virtual objects directly handled by computers and, of course, we cannot directly see them. We can virtually put them in the real space and see them by using an ESP. This notion will be good in several situations.

Digital cash, which has been gaining in popularity, can be associated with invisible bills and coins in our environment. Note that, usually, monetary trade based electronic commerce is complicated because some processes should be achieved as real actions in the real space while the others as virtual actions in computer. Contrary to the conventional style of such trade, a user will intuitively know the concise situation in every stage of the trade in this environment.

As for the E-mail, a user should learn in advance how to write, send, and receive mails in his mail system. Such a system can be replaced by a system using invisible letters, invisible posts, invisible post offices, and so on, which can be seen by a user using an ESP. The user can intuitively recognize the operations and the effects of his/her behavior.

Various kinds of desktop metaphors used in the conventional window systems can be extended to objects in the integrated space. For example, an invisible garbage can is a place to dispose invisible objects such as invisible letters and invisible books. Tools in the file system can be associated with invisible books and invisible bookshelves. If there is a little room for invisible books in the real room, we may construct an invisible room where we can visit only as an invisible person. Conventional operations, such as opening a text file by an editor are replaced by intuitive actions of an invisible person, such as holding an invisible book in a bookshelf and opening it. As a side effect, a user may encounter a person who has a similar interest in front of the bookshelf, which does not occur in the usual file system.

Sounds and video files can be represented by invisible CDs and invisible video cassettes. Programs to play these files are naturally associated with a CD player and a video deck.

The access authorization to special data can be realized by using an invisible safety box in which an authorized user can put in the invisible object representing the data. The notion of a key or a password can be associated with an invisible key or buttons attached to the safety box.

We may consider that our approach is a kind of visualization in the sense that we are visualizing everything in 3D space and associating them with several locations in the real space.

Fig. 7. An Example of the Extended Desktop Screen

5.3 The Extended Desktop System

In order to realize the invisible object environment, it is important to consider the interaction with the conventional computing environment such as the X-Window and Microsoft Windows. From this viewpoint, we have designed and implemented an extended desktop environment using video camera images. In this approach, real space taken by a camera can be used as a part of the desktop on a computer where an invisible object is shown as an icon. We also use the conventional icon-based operations, e.g., the drag-and-drop operation in this environment, to put an invisible object to the real space. Moreover, an extended drag-and-drop operation is employed to handle the icons virtually placed on the real space. We call it the real space drag-and-drop. In this method, we can move each icon placed in the real space by moving the computer.

The implementation is based on the technology called the image association memory. The image information about a small fragment of the obtained image is stored in the system associated with the icon information. When showing the real space, the associated memory is retrieved to find the fragment that is best-match to every part of the image taken by the camera in each period of time. By sharing the memory via the network, multiple people can share each invisible object.

In Fig. 7, there is a window which shows a real-time image taken by the camera attached to the notebook computer. In this window, a user can drop an icon onto the desktop to place it virtually on the real space. The user can directly open the resource placed on this window. Printers, computers, speakers, and other devices just in front of a user can be easily used by showing them in this window and by the drag-and-drop operation.

6 Discussions

In section 4 and section 5, we described several mechanisms of browsers. The VRML browser and the IBNR browser can be used selectively depending on the system environment and the network conditions. Two or more users can see simultaneously and consistently a single scene by using both kinds of browsers. Moreover, a user can change his browser depending on the target place. Further, by adding the extended desktop functionality, the ESP can be a platform of the future augmented reality environment.

By integrating these systems and their information, more variety of advanced uses become possible. A possible integration is to automatically generate the image information of IBNR by using the camera images and the location information of an ESP. Conversely, an ESP may sense its location by partially matching retrieval of its camera image with the stored image of IBNR.

Another possibility is to generate IBNR images from VRML resources. If there is only a VRML description for a certain place and a user cannot use the IBNR browser to visit the place, the system can generate a picture and floor information that can be used by the IBNR browser. In this case, a user can have alternatives of the browser to be used in visiting the place.

Some real person (esper) may desire to be an invisible person to visit another place. A person may want to attend two meetings simultaneously. In such cases, an ESP and an IPH are used in a single handy computer. A more effective system design is necessary to integrate these two kinds of systems because some information may be shared by these two components.

It may be useful for a real person (esper) to use an IPH to communicate with invisible persons around him/her. In such a case, there are some little mismatches in the views of the real person and invisible persons. Note that the original environment produces a large mismatch since different resources are used to show users with the space such as the VRML program in the VRML-based IPH browser, the image and floor information in the IBNR-based IPH browser, and the real-time video in ESP.

The environment where virtual space and real space are tightly bound makes it possible for a computer to handle real space objects as computing targets. In such an environment, we can find everything we lost in the real space by a search command of the computer. For example, if a user says "where is my house key?" then a drawer of a piece of furniture in which the house key is put will be blinking on his/her ESP.

As described in section 3, there is a gap between the scenario described in section 2 and our current system design. Rich reality, which is difficult to realize in the current stage of computer technology, involves the following two sides:

- Reality for an invisible person, i.e., "Can he/she have a feeling of being invisible?" To answer this question affirmatively, we will make use of the recent advanced GUIs to traverse 3D space which are rapidly evolving in 3D games and VR technologies because they will provide users virtual space experience with more reality. We will also make use of real space sensing technologies such as magnetic sensor, GPS, and infrared sensor to provide the systems with better presentation of the real space in a virtual space.

- Reality for a real person, i.e., "Can he/she believe the existence of invisible persons?" To achieve this, rapid real-time image processing is necessary in ESP. Such processing includes detection of the location and the direction of each ESP, real-time image analysis, and real-time image composition with the hidden side elimination. Such processing will give espers stronger feeling of existence of each invisible person. We should also point out the importance of the interactions between an invisible person and a real person. Such interaction will be done by motion commands by a real person and remote controls by an invisible person, and will raise the value for a user to be an invisible person.

As a result, we may say that more and more advanced realization will become possible using rapidly growing recent technologies. In this sense, the current design is only a starting point to future realization of more realistic 'invisible person' system.

In addition, the recent development of the WWW technologies such as SMIL and Servlet will make it possible to construct such integrated system in a more general manner. P3P will be useful for the security matter. LivingWorlds will enable efficient realization of the management server for the VRML-based IPH browser.

7 Conclusion

In this paper, we have shown our approach, i.e., the 'invisible person' communication support system. By showing the design of our implemented system, we have shown that a feasible design for current technologies is possible.

Among several similar approaches to support computing and communication in virtual space or real space, our approach realizes the interaction between the real space and the computing space. We treat things in the real space as the objects of computing. Moreover the system realizes the interface that is easy to understand. In general, it is essential for the computing that is closely related to our social life to be understood easily by anyone. Our approach uses a metaphor with a well-known keyword 'invisible person'. The system provides flexible support for human communication. Our approach opens up human communication opportunity to both remote virtual users and users in the real space. Furthermore, it achieves the application independency, i.e., our approach provides a computing metaphor which can be used by a wide range of applications.

Some parts of our implemented system are working all the time and any users in the world can access them (http://aries.ise.eng.osaka-u.ac.jp/invisible-person.html, http://aries.ise.eng.osaka-u.ac.jp/~tuka/ibnr) where he/she can be an invisible person.

74   Masahiko Tsukamoto

Acknowledgement

The author would like to express his gratitude to Mr. T. Ogawa, Mr. Y. Sakane, and other members of the invisible person project for their contribution to the design and implementation of the system. He also thanks Prof. S. Nishio, Prof. F. Kishino, Prof. Y. Kitamura, and the members of Nishio laboratory for their invaluable comments on this project. This research was supported by "Research for the Future" Program of Japan Society for the Promotion of Science under the Project "Advanced Multimedia Content Processing" (Project No. JSPS-RFTF97P00501).

References

1. Bagroodia, R., Chu, W.W., Kleinrock, L., and Popek, G.: Vision, Issues, and Architecture for Nomadic Computing, IEEE Personal Communications, Vol.2, No.6, pp.14-27 (1995).
2. Imielinski, T. and Badrinath, B.R. : Mobile Wireless Computing, Communications of the ACM, Vol.37, No.10, pp.18-28 (1994).
3. Nagao, K. and Rekimoto, J.: Ubiquitous Talker: Spoken Language Interaction with Real space Objects, Proc. IJCAI'95, pp.1284-1290 (1995).
4. Ogawa, T., Sakane, Y., Yanagisawa, Y., Tsukamoto, M., and Nishio, S.: Design and Implementation of a Communication Support System based on Projection of Real Space on Virtual Space, Proc. of 1997 IEEE Pacific Rim Conference on Communications, Computers and Signal Processing (PACRIM'97), pp.247-250 (1997).
5. Smailagic, A. and Siewiorek, D.P.: Modalities of Interaction with CMU Wearable Computers, IEEE Personal Communications, Vol.3, No.1, pp.14-25 (1996).
6. Tsukamoto, M.: Mobile Computing for Integrating Real Space and Virtual Space, Proc. of the 3rd Workshop on Personal Wireless Communications(PWC'98), pp.191-198 (1998).
7. Weiser, M.: Some Computer Science Issues in Ubiquitous Computing, Communications of the ACM, Vol.36, No.7, pp.74-85 (1993).
8. Wells, H.G.: The Invisible Man (1897).

News Dictation and Article Classification Using Automatically Extracted Announcer Utterance

Yasuo Ariki, Jun Ogata, and Masafumi Nishida

Faculty of Science and Technology, Ryukoku University,
1-5 Yokotani, Oe-cho, Seta, Otsu-shi, 520-2194 Japan,
ariki@rins.ryukoku.ac.jp

Abstract. In order to construct a news database with a function of video on demand (VOD), it is required to classify news articles into topics. In this study, we describe a system which can dictate news speech, extract keywords and classify news articles based on the extracted keywords. We propose that it is sufficient to dictate only the announcer utterance in classifying the news articles and it contributes to reduce the processing time. As an experiment, we compared the classification performance of news articles in two cases; dictating only the announcer utterances which are automatically extracted and dictating a whole speech which includes reporter or interviewer utterances.

1 Introduction

Recently, TV news programs are broadcasted all over the world owing to the broadcast digitization. In this situation, TV viewers require to select and watch the most interesting news. For that purpose, word indexing and article classification are key techniques.

The word indexing is a technique to put the discriminative keywords on the news speech articles in order to retrieve the interesting articles. On the other hand, the article classification is a technique to classify the news speech articles into groups (topics) based on their contents such as politics, economy, science and sports in order to retrieve the same kind of articles[1]. These two techniques are strongly required because manually indexing and classification are almost impossible.

From this viewpoint, we propose in this paper a method to automatically index and classify TV news articles into 10 topics based on a speech dictation technique using speaker independent phoneme HMMs and word bigram. After the dictation of the spoken news articles, pre-defined keywords are searched and the new articles are classified based on the keywords.

In general, news speech includes reporter speech as well as announce speech. The announcer speech is clear but the reporter speech sometimes noisy due to wind or environmental noises so that the dictation accuracy for the reporter speech is lower than for the announcer speech. Therefore if the speech dictation process is applied only to the announcer speech, we can reduce the processing time without decreasing the news classification accuracy.

S. Nishio, F. Kishino (Eds.): AMCP'98, LNCS 1554, pp. 75–86, 1999.
© Springer-Verlag Berlin Heidelberg 1999

From this viewpoint, we propose in this paper a method to automatically divide the TV news speech into speaker sections and then index in real time who is speaking. This can be realized by using a technique of speaker verification[2]. However, the speaker verification technique is sensitive to the time lapse. Namely, the speech characteristics of each speaker is subject to change day by day. To solve this problem, speaker models are not prepared in advance but are constructed through indexing in self-organization mode.

We verified the effectiveness of our proposed methods by carrying out the experiment in extracting and dictating only the announcer speech and then classifying the news articles into 10 topics.

2 Announcer Speech Extraction

2.1 Speaker Verification

Speaker verification is a technique to judge if the input speech belongs to the specified person or not[2]. Fig.1 shows the speaker verification process. When the speaker ID of speaker A and his speech are fed to the verification system, the distance is computed between the model of the speaker A and the input speech. If the distance is smaller than some threshold, the input speaker is accepted as the true speaker A. Otherwise the input speaker is rejected. In our experiment, speaker subspace is constructed as the speaker model and the distance between the input speech and the speaker subspace is computed.

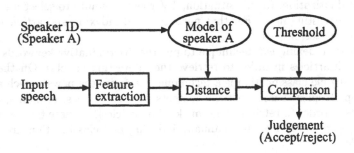

Fig. 1. Speaker verification

2.2 Speaker Subspace

As shown in Fig.2, we observe speech data $X^{(i)}$ of the speaker i and speech data $X^{(j)}$ of the speaker j in an observation space. The speech data are a sequence of spectral feature vectors $x_t^{(i)}$ and $x_t^{(j)}$ obtained at time t by short time spectral analysis. We denote the speech data $X^{(i)}$ as a matrix whose row is a spectral feature vector $x_t^{(i)^T} - \mu^{(i)^T}$ $(1 \le t \le M)$. Here $x_t^{(i)}$ denotes an observed feature vector and $\mu^{(i)}$ is their mean vector. The column of the matrix corresponds to frequency f $(1 \le f \le N)$.

By singular value decomposition, the speech data matrix $X^{(i)}$ is decomposed as

$$X^{(i)} = U^{(i)} \Sigma^{(i)} V^{(i)T} \tag{1}$$

Here $U^{(i)}$ and $V^{(i)}$ are the matrices whose columns are eigenvectors of $X^{(i)} X^{(i)T}$ and $X^{(i)T} X^{(i)}$ respectively. $\Sigma^{(i)}$ is the singular value matrix of $X^{(i)}$.

The eigenvectors of the correlation matrix $X^{(i)T} X^{(i)}$ are the orthonormal bases of the speech data $X^{(i)}$, computed based on a criterion that the total distance is minimized between feature vectors $x_t^{(i)} - \mu^{(i)}$ and the orthonormal bases[3]. Then $V^{(i)}$ is considered as orthonormal bases of the speaker space. This is completely same as the principal component analysis of the speech data $x_t^{(i)}$ [4].

If the large singular values up to r numbers are selected from the matrix $\Sigma^{(i)}$, the matrix $V^{(i)}$ becomes N x r dimension and is considered as the speaker subspace.

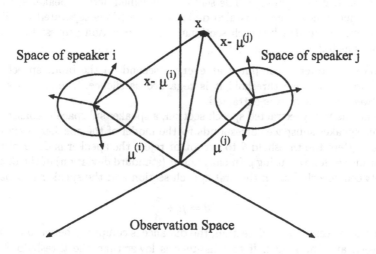

Fig. 2. Speaker subspace

2.3 Verification by Speaker Subspace

The speaker subspace $V^{(i)}$ is composed of orthonormal bases $\{v_1^{(i)}, \cdots, v_r^{(i)}\}$ of the speech data $X^{(i)}$. Speaker verification can be carried out by computing a distance from an input speech vector x_t in the observation space to the speaker subspace $V^{(i)}$.

The distance is presented as follows using a projection matrix $P^{(i)}$ from the observation space to the speaker subspace.

$$Dist(V^{(i)}, x_t) = \|x_t - \{P^{(i)}(x_t - \mu^{(i)}) + \mu^{(i)}\}\|^2$$
$$= \|(I - P^{(i)})(x_t - \mu^{(i)})\|^2 \tag{2}$$

where the projection matrix $P^{(i)}$ is defined as:

$$P^{(i)} = \sum_{k=1}^{r} v_k^{(i)} v_k^{(i)T} = V^{(i)} V^{(i)T} \tag{3}$$

Equation(3) means that the projection matrix $P^{(i)}$ is obtained using the orthonormal bases of the speech data $X^{(i)}$.

The distances computed by Eq.(2) between speech vectors x_t and the speaker subspace $V^{(i)}$ are averaged over time t. The speaker is identified as one with the minimum averaged distance between the speech vectors and the subspace.

2.4   Extraction of Speaker Section

Continuous news speech is divided into sections of respective speaker by using the speaker verification technique. The sections are called here " speaker sections". The continuous news speech is also divided into sections separated by silence. The sections are called " speech sections". The extraction process for speaker sections is as follows;

(1) Averaged power is computed at every 1 second on the input speech. If it is lower than some threshold it is regarded as silence. The speech section between two silences is extracted.

(2) Using the firstly extracted speech section, a speaker subspace is constructed. This speaker subspace corresponds to the model of the speaker A shown in Fig.1. Here the threshold θ to accept or reject the speaker is determined in advance as follows, using μ (mean) and σ (standard deviation) of the distance between speech data in the first speech section and the speaker subspace;

$$\theta = \mu + \frac{\sigma}{3} \tag{4}$$

(3) On the successive speech section, the distance is computed between the input speech and the model. If the distance is lower than the threshold θ, it is judged that the speaker A is still speaking. In this case, the speaker subspace model is updated as well as the threshold θ using all the speech data verified as speaker A.

(4) Otherwise, it is regarded that speaker A has finished his speech and new speaker or previous speaker begins speaking. To judge it, the distance between the input speech section and the previously constructed speaker subspace models are computed. If some speakers have lower distance than threshold θ, then the input speaker is judged to be the speaker with the lowest distance. Otherwise, the input speaker is regarded as a new speaker and step (2) begins starting.

2.5   Experimental Result

We selected 48 news articles which included reporter speech as well as announcer speech from 45 days NHK news program. Everyday news program usually contains 4 or 5 articles and continues for 5 minutes. For these 48 news articles we carried out the experiment to extract the announcer sections. The dimension of speaker subspace was set to 7 after preliminary experiment. The experimental condition is shown in Table.1.

Table 1. Experimental condition

Speech data	48 NHK news articles
Sampling frequency	12kHz
Frame length	20ms
Frame period	5ms
Window type	Hamming window
Features	LPC Cepstrum(16 orders)
Subspace dimension	7
Threshold θ	$\theta = \mu + \frac{\sigma}{3}$

The extraction of announcer sections was evaluated by the extraction rate and the precision rate defined as follows;

$$Extraction\ rate = \frac{\left\{\begin{array}{l}\text{Number of correctly verified} \\ \text{speech sections as announcer}\end{array}\right\}}{\left\{\begin{array}{l}\text{Number of total speech sec-} \\ \text{tions of the announcer}\end{array}\right\}} \tag{5}$$

$$Precision\ rate = \frac{\left\{\begin{array}{l}\text{Number of correctly verified} \\ \text{speech sections as announcer}\end{array}\right\}}{\left\{\begin{array}{l}\text{Number of verified speech} \\ \text{sections as announcer}\end{array}\right\}} \tag{6}$$

Here announcer is judged to be the speaker who spoke the longest time in 1 day 5 minutes NHK news.

The announcer extraction result is shown in Table2. The extraction rate was 92.6% and the precision rate was 82.9%.

3   Speech Dictation

3.1   Experimental Condition

We carried out speech diction for the 48 news articles after the extraction of announcer section. The language model is the word bigram produced from

Table 2. Experimental result(%)

Extraction rate	92.6
Precision rate	82.9

MAINICHI Japanese newspaper of 45 months after morphological analysis. The number of the words in the dictionary is 5,000. The word bigram was back-off smoothed after cutting off at 2 words.

Speaker independent 41 monophone HMMs were constructed. Their structure is 5 states with 3 loops and 8 mixtures for each state. They were trained using 21,782 sentences spoken by 137 Japanese males. These speech data was taken from the database of acoustic society of Japan. The acoustic parameters are 39 MFCCs with 12 Mel cepstrum, log energy and their first and second order derivatives. Cepstrum mean normalization is applied to each sentence to remove the difference of input circumstances. Table3 shows the experimental conditions for acoustic analysis (AA) and HMM.

In the dictation experiment, we used HTK (HMM Toolkit)[5] as the decoder which can perform Viterbi decoding with beam search using the above mentioned language model and the acoustic model.

Table 3. Acoustic Analysis(AA) and HMM

	Sampling frequency	16kHz
	High-pass filter	$1 - 0.97z^{-1}$
A	Feature parameter	MFCC (39th)
A	Frame length	20ms
	Frame shift	5ms
	Window type	Hamming window
H	Number of states	5 states 3 loops
M	Learning method	Concatenated training
M	Type	Left to right continuous HMM
	Number of mixtures	8

3.2 Dictation Result

The dictation was carried out for 48 NHK news articles. They were already divided into two speaker sections by the previously described method; announcer and reporter (interviewer). Table4 shows the property of the 48 news articles in terms of announcer section (Anchor), reporter section (Other) and mixed total section (All). In the reporter section, the ratio of unknown words to the 5,000 dictionary words is high compared with other two sections. This is also reflected in the test-set perplexity which is the measure of task complexity.

Table 4. News articles used for dictation

	Anchor	Other	All
Number of sentences	247	116	363
5K unknown word ratio	13.9%	29.3%	20.6%
Test-set perplexity	153.7	285.2	177.6

The dictation result is shown in table5. In the table, word correct rate and word accuracy are defined as follows;

$$\text{Word correct rate} = \frac{N - S - D}{N} \cdot 100 \tag{7}$$

$$\text{Word accuracy} = \frac{N - S - D - I}{N} \cdot 100 \tag{8}$$

S : The number of substituted words
D : The number of deleted words
I : The number of inserted words
N : Total number of words

The word error rate is defined as $(100 - word\ accuracy)$. From the table, it can be seen that the announcer speaks clearly and grammatically in the clear circumstance. On the other hand, the reporter speaks colloquially in the noisy circumstance. The dictation result is used for topic classification in the successive process.

4   Classification of News Articles

4.1   Classification Flow

Fig.3 shows the classification flow of news speech articles by speech dictation technique. Before the article classification, news articles in the news program are automatically separated each other using the algorithm mentioned in [6]. In the flow, there are following two phases;

Table 5. Dictation result(%)

	Anchor	Other	All
Word error rate	39.7	79.3	54.6
Word correct rate	66.5	23.5	48.7
Word accuracy	60.3	20.7	45.4

(1) Speech dictation phase: word sequence and their probabilities $Ps(w)$ are computed by applying the speech dictation technique.
(2) Article classification phase: articles are classified by integrating the keyword probability $Ps(w)$ and the topic contribution probability of each word $P(n|w)$.

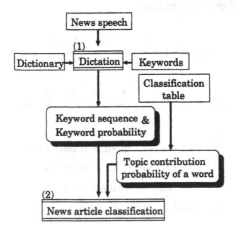

Fig. 3. Flow of news article classification

4.2   Word Probability

Speech dictation by HTK described in 3 can produce the word probability as well as the word sequence for each spoken news sentence. After speech dictation to the 48 news articles, keywords w_i are searched together with their probability $P_s(w_i)$. The keywords were determined in advance as the words included in both "classification indices for ASAHI newspaper article database" and 48 news articles. The number of keywords was 399 words in this experiment.

4.3 Topic Contribution Probability

In article classification using a keyword sequence, topic contribution of a keyword (TCKW) $P(n|w)$ is computed in advance as shown in Eq.(9). The TCKW indicates how the keyword contributes to identify the topic and is defined as the ratio of the occurrence of the keyword w included in the topic n to the occurrence of the keyword w included in all the topics. This definition is *a posteriori* probability of the topic n conditioned by the word w.

$$
P(n|w) = \frac{\left\{\begin{array}{l}\text{The number of occurrence of}\\ \text{the keyword } w \text{ included in the}\\ \text{topic } n\end{array}\right\}}{\left\{\begin{array}{l}\text{The number of occurrence of}\\ \text{the keyword } w \text{ included in all}\\ \text{the topics}\end{array}\right\}} \tag{9}
$$

In this study, we used "classification indices for ASAHI newspaper article database" in computing the TCKW. Fig.4 shows a part of the classification indices.

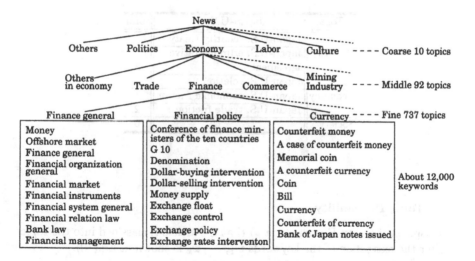

Fig. 4. A part of classification indices

It includes 12,000 keywords and they have links to the related topics. There are three levels in grouping of the topics; coarse, middle and fine level. They have about 10, 92 and 737 kinds of topics respectively. In the fine topics, about 16 indices are prepared at average in each topic. We selected coarse level of 10 topics classification in this study. The 10 topics for classification are Politics, Economy, Labor, Culture, Science, Society, Accidents, Sports, Internationality and Others.

Table6 shows an example of how to compute the TCKW. In the table, there are three complex words which include "Japan-U.S." word in the topic of politics. In the same way, there are two and zero in the topic of economy and society respectively. In total the number of complex words including "Japan-U.S." is five. In this example, the TCKW of the "Japan-U.S." word is computed as follows;

$$P(\text{Politics}|\text{Japan-U.S.}) = \frac{3}{5} = 0.6$$
$$P(\text{Economy}|\text{Japan-U.S.}) = \frac{2}{5} = 0.4$$
$$P(\text{Society}|\text{Japan-U.S.}) = \frac{0}{5} = 0.0$$

The TCKW is computed for all the keywords in advance.

Table 6. Example of topics and keywords

Topic	Japan-U.S.	total
Politics	Japan-U.S. security treaty Japan-U.S. administrative agreement Japan-U.S. relation	3
Economy	Japan-U.S. economic friction Japan-U.S. trade friction	2
Society		0

4.4 Topic Probability

The topic probability $P(n|w_1, \cdots, w_k)$ that the article is classified into the topic n after the extraction of the keywords $w_1, \cdots, w_k$ is shown in Eq.(10).

$$P(n|w_1, \cdots, w_k) = \sum_{i=1,\cdots,k} P(w_i) \times P(n|w_i) \qquad (10)$$

where $P(n|w_i)$ is the topic contribution probability of the keyword w_i. The probability $P(w_i)$ is replaced by the normalized word probability as follows;

$$P(w_i) = \frac{Ps(w_i)}{\displaystyle\sum_{j=1,\cdots,k} Ps(w_j)} \qquad (11)$$

This topic probability is the integration of acoustic word probability $Ps(w)$ and *a priori* knowledge probability TCKW. The news article can be classified into the topics with the highest topic probability $P(n|w_1, \cdots, w_k)$.

5   Classification Result

Table7 shows the classification result of the 48 news articles. In the table, the classification rate is the ratio of the number of correctly classified articles to the total number of articles. The article is judged to be correctly classified if it is classified into the correct topic. The correct topic is determined by setting the word probability $Ps(w_i) = 1$ for the true keywords which are obtained from the text data.

From the table, it can be seen that the classification rate is 63.6% with 60.3% word accuracy for the announcer speech. On the other hand, the classification rate for all utterances is 63.0% with 45.4% word accuracy. This result indicates that the dictation accuracy for reporter speech is lower than that for the announcer speech due to the noisiness and colloquialism. Even though it is true, the classification rate is almost same. This indicates that the keywords are mainly included in the announcer speech and they are well extracted compared with the reporter speech.

Table 7. Classfication result(%)

	Word accuracy	Classification rate
Anchor	60.3	63.6
All	45.4	63.0

6   Conclusion

We have described the automatic classification system of TV news articles. Keywords were extracted from news speech articles after their dictation using word bigram and speaker independent HMM. The acoustic probabilities of the keywords were multiplied with the topic contribution probabilities which were computed from "classification indices for ASAHI newspaper article database" and the topic probability of the article was produced. The news speech articles were classified based on this topic probability.

In order to speed up the processing time, we have omitted the reporter speech section and still kept the same classification accuracy. The highest classification rate was 63.6% and seems to be low in accuracy for real application. We need to improve speech dictation technique in future.

Acknowledgment

This research was partly supported by "Research for the Future" Program of Japan Society for the Promotion of Science under the Project "Advanced Multimedia Content Processing" (Project No. JSPS-RFTF97P00501) and partly supported by the Japanese Ministry of Education Grant-in-Aid for Scientific Research on Priority Area: "Advanced databases," area no. 275(08244103).

References

1. Y.Ariki, M.Sakurai and Y.Sugiyama : " Article Extraction and Classification of TV News Using Image and Speech Processing", CODAS96 (International Symposium on Cooperative Database Systems for Advanced Applications), pp.247-254, 1996.
2. T.Matsui and S.Furui: " Comparison of text independent speaker recognition methods using VQ distortion and discrete/continuous HMMs", Proc.ICASSP, Vol.II, pp157-160, 1992.
3. Y.Ariki and K.Doi, " Speaker Recognition based on Subspace Method", ICSLP'94, pp.1859-1862, 1994.
4. E.Oja,"Subspace Methods of Pattern Recognition, Research Studies Press, England, 1983.
5. Cambridge University Engineering Department Speech Group and Entropic Research Laboratory Inc.:"HTK Hidden Markov Model Toolkit V2.0"
6. Y.Ariki and Y.Saito : " Extraction of TV News Articles based on Scene Cut Detection", ICIP96, pp.III847-III850, 1996.

Automatic Video Indexing Based on Shot Classification

Ichiro Ide, Koji Yamamoto*, and Hidehiko Tanaka

Graduate School of Electrical Engineering, The University of Tokyo
7-3-1 Hongo, Bunkyo-ku, Tokyo, 113-8656 JAPAN
TEL: +81-3-3812-2111 ext.7413,   FAX: +81-3-5800-6922
{ide | kyama | tanaka}@mtl.t.u-tokyo.ac.jp

Abstract. Automatic indexing to video data is in strong demand to cope with the increasing amount. We propose an automatic indexing method for television news video, which indexes to shots considering the correspondence of image contents and semantic attributes of keywords. This is realized by first, (1) classifying shots by graphical feature, and (2) analyzing semantic attributes of accompanying captions. Next, keywords are selectively indexed to shots according to appropriate correspondence of typical shot classes and semantic attributes of keywords. The method was applied to 75 minutes of actual news video, and resulted in indexing successfully to approximately 50% of the typical shots (60% of the shots were classified as typical), and 80% of the typical shots where captions existed.

1 Introduction

As the amount of broadcast video data increases, it is becoming more and more important to store them in a well organized manner considering recycling and searching. Above all, television news programs are worthwhile indexing considering the importance and usefulness. Currently this process is mostly done manually, but automatic indexing is in big demand to cope with the increasing amount and to achieve sufficient precision for detailed searching.

We are trying to accomplish this task by referring to both video data and accompanying natural language data in Japanese television news video. There are several notable attempts to automatically index television news video from this approach. Most of their indexing strategies are based on frequency or just simple occurrence of words or phrases. On the other hand, others search for words in a full text searching manner. These methods are relatively simple, and in that sense quite practical approaches, but the critical point is that they do not necessarily ensure the correspondence of the image contents and the index.

Reflecting these background issues, in this paper we will propose and evaluate an indexing method, which indexes keywords with appropriate semantic attributes to classes of shots with graphically typical feature. The base of this

---

* Currently at Research & Development Center, Toshiba Corporation.

S. Nishio, F. Kishino (Eds.): AMCP'98, LNCS 1554, pp. 87–102, 1999.
© Springer-Verlag Berlin Heidelberg 1999

method lies in the characteristics of television news programs; graphically similar (in a certain perspective) shots contain semantically similar contents. Since keywords are tagged selectively according to the contents of each typical shot class, the correspondence of image contents and keywords is guaranteed to a certain extent.

We will first take an overview of the characteristics specific to television news video and related works in the next Sect., and then introduce the proposed method in Sect. 3. The succeeding Sects. 4, 5 and 6 discuss in-depth matters of the method, and Sect. 7 concludes the paper.

2   Indexing to Television News

First, we will overview the characteristics specific to television news video, and next introduce several related works by other people.

2.1   Structure of Television News Video

News videos have both graphical and semantic structures as described here.

Graphical Structure. The graphical structure that television news videos have are not specific to the genre. As shown in Fig. 1, they are generally hierarchically structured.

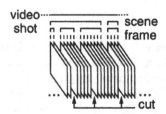

Fig. 1. Graphical structure of video.   **Frame:** still images that constitute a movie, **Shot:** group of graphically continuous frames, **Scene:** group of shots with graphically and/or semantically similar contents, **Cut:** discontinuous point between adjoining shots.

Semantic Structure. As shown in Fig. 2, the semantic structure of television news video is quite unique.

It is very important to detect the boundaries of topics and to grasp the semantic structure of the video before indexing, so this structure could be used as an opportune key. As mentioned in Sect. 1, since news video tend to be taken in similar situations, most shots could be classified to several typical shot classes referring to graphical feature.

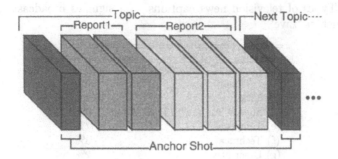

Fig. 2. Semantic structure of news video; Each block represents a shot. Basically, each topic begins with an 'anchor shot', and between them are miscellaneous shots including 'report shots'.

2.2 Natural Language Data in Video

Variety of Natural Language Data Source. There are various natural language data sources accompanying the video; main audio, sub audio, closed caption (mostly same contents as the main audio) and caption. Particularly, captions are usually used to describe important matters in a digestive form, so they could be considered as adequate keyword candidates for indexing. According to our statistics, they appear approximately once every 15 seconds in news programs, which is a moderate frequency for finding keyword candidates. Since main audio (or closed caption) require complicated process to be used as a keyword extraction source, captions that have these characteristics are employed in the proposed system.

Characteristics of Captions. Captions have specific characteristics that differ from normal texts, which makes the analysis employing conventional natural language processing methods difficult. This problem is solved to a certain extent in this application by the method described in Sect. 5.

On the other hand, semantic characteristics of captions could be classified as shown in Tab. 1. Among these types, (a) and (b), which consist about half of the captions, represent the contents in the image directly. This allows (a) and (b) to almost directly become keyword candidates. Although (c) does not necessarily reflect the graphical contents of the video, it is an important information which explains the topic. Thus, nearly 60% of the captions; (a), (b) and (c), could be directly used as keywords.

2.3 Related Works

As a general video database creating and browsing system, Informedia project [8,10,13,21] at CMU is the most significant work in this field. They have created

Table 1. Types of television news captions. (f) signifies broadcast technical captions such as 'Live'.

Types	Ratio
(a) Locational/Organizational	30%
(b) Personal	15%
(c) Title	14%
(d) Speech Summary/Translation	10%
(e) Temporal	7%
(f) Technical	3%
(g) Descriptive	2%
(h) Etc.	19%

an automatic archiving and presenting system for CNN (Cable News Network) news video. It automatically recognizes main audio speech and extract keywords from the text deduced from it by evaluating the rarity of words by the TF-IDF (Term Frequency Inverse Document Frequency) method. Although such statistic approaches are relatively simple, and in that sense quite practical, they do not necessarily ensure the correspondence of image contents and keywords, which is essential for video database.

On the other hand, Nakamura and Kanade [5] have proposed an automatic indexing method that classifies shots into several typical classes, and tag key sentences derived from syntactic and semantic analyses of the closed caption. Although the basic idea of classifying shots into typical classes is similar to our approach, the point that they utilize closed captions and that they execute key sentence extraction, differentiates the two methods.

Similar to this approach, Satoh *et al.* [6] proposed an automatic facial image and personal name associating method that associates facial regions extracted from the image and personal names derived from closed captions analysis. Although this is completely automated and performs fairly well, it concentrates on associating personal faces and names, which is an acceptable limitation, but not sufficient for news video database.

3   Indexing Based on Shot Classification

Considering the issues discussed in Sect. 2, we will propose an automatic indexing method based on shot classification. The basic idea of the method is based on the characteristic specific to television news video; graphically similar images contain similar contents. Based on this assumption, graphically typical shots will be indexed keywords with certain attributes.

The overall indexing scheme is shown in Fig. 3. A simple overview of each phase is introduced in this Sect. In-depth description and evaluation on shot classification, caption analysis, and indexing are discussed in the succeeding Sects.

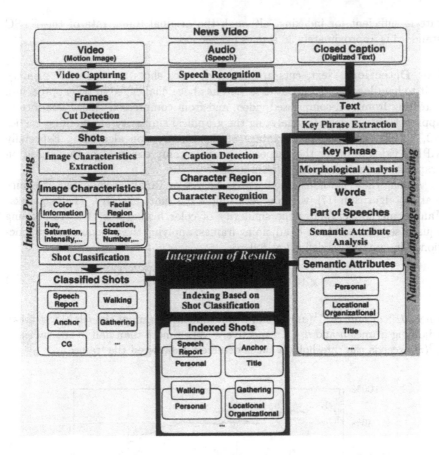

Fig. 3. Overall indexing scheme. The left half shows the flow of the image processing phase, the right half shows the natural language processing phase, and the bottom shows the integration phase. Audio and Closed Caption are not currently used.

Although we do not currently employ main/sub audio and closed caption data, the scheme could be easily extended to handle them as shown in the figure. This may also seem as a mere combination of conventional algorithms, but the overall ability of indexing by integrating video and natural language data is far superior than a simple combination.

3.1 Image Processing Phase

Video Capturing. First, analog video is digitized by an image capture board connected to a PC at a sampling rate of 5 frames per second. We consider this rate is sufficient for indexing, although the original frame rate of the NTSC standard is approximately 30 frames per second.

Cut Detection. Next, cuts are detected, since shots are the most primary unit to handle video data. Cut detection has been challenged in two approaches; detecting from non-compressed video, and from compressed video. The former approach detects cuts by analyzing the graphical similarity of adjoining frames [12,17]. The latter takes advantage of the compression algorithm. Especially MPEG video is fit for the purpose due to its compression algorithm based on graphical correlation between frames [3].

Among these methods, we chose the Nagasaka-Tanaka (segmented χ^2 examination) algorithm [17], which is very simple but quite effective. The Nagasaka-Tanaka method evaluates the similarity of color histograms of corresponding equally segmented blocks of adjoining frames, applying the χ^2 examination function. The function is defined as follows:

$$\chi^2(i) = \sum_{c=0}^{c_{max}} \frac{(H_i^n(c) - H_i^{n+1}(c))^2}{H_i^n(c)} \tag{1}$$

where $H_i^n(c)$ and $H_i^{n+1}(c)$ represent the color histogram of the ith block of two adjoining frames n and $n+1$ respectively. When more than half of the values of $\chi^2(i)$ exceeds the threshold, a cut is detected in between the frames.

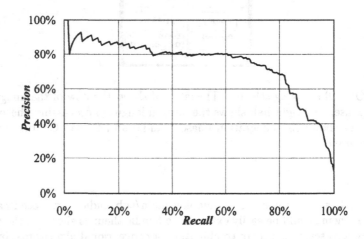

Fig. 4. Recall and precision of cut detection by Nagasaka-Tanaka (segmented χ^2 examination) method applied to 30 minutes of video with 210 cuts. If 60% of recall could be considered sufficient, more than 80% of precision is ensured.

Figure 4 shows the relation between recall and precision of cut detection, when the method was applied to 30 minutes of actual news video with 210 cuts.

After these pre-processes, extraction of graphical feature is performed before the shot classification.

Caption Recognition. As a tributary to the main-stream image processing, caption detection and character recognition should be performed. This is not currently implemented, and captions are written down manually. Although OCR (Optical Character Recognition) softwares with high recognition rates do exist, the resolution of TV captions are limited due to their size and the number of scanning lines (525 lines per frame in the case of NTSC broadcasting standard). Fragments of background image filtering through the characters should be eliminated before character recognition, which is also a difficult process. These restrict the application of conventional OCR techniques to television caption character recognition, especially to complicated Japanese characters. Nonetheless, several attempts are made to accomplish the task [9,14], although their recognition rates have still room for improvement. We may hope for digitized caption texts to be broadcast along with the video, following the future digitalization of television broadcasting.

3.2 Natural Language Processing Phase

First, morphological analysis to digitized texts derived from captions are performed using the Japanese morphological analysis system JUMAN [20]. This is a pre-process for analyzing the semantic attribute of the entire caption, which refers to suffixable nouns.

3.3 Integration Phase

After image classification and caption analysis of the shot are done, the integration phase indexes shots with captions with appropriate semantic attributes for the typical shot class; *i.e.* when the shot is a 'speech shot', the speaker's name is an appropriate keyword. When it is a 'gathering shot', the name of the gathering, say a conference, is considered appropriate.

Such indexing scheme ensures the correspondence of image content and keyword, which is essential for video database.

4   Shot Classification

After the pre-process, each shot is classified based on its graphical feature. Note that the classification rules are based on combinations of relatively simple graphical feature extraction process, which makes the method applicable to large amount of incoming video data.

We have defined five shot classes:

– Speech/Report
– Anchor

- Walking
- Gathering
- Computer Graphics (CG)

These classes covered 57% of the entire news video that we used for experiment. Details on each shot class are described in this Sect.

4.1 Speech/Report Shot

When a person is addressing a speech, or a reporter is reporting from a relay spot, there is usually one person speaking in the middle of a frame. In order to detect such a shot, (1) human face, and (2) lip movement should be detected. Condition (2) is employed to avoid detecting a portrait picture, or a video with a person just standing in the middle of a frame without speaking.

An anchor shot is also detected from these conditions, but they will be separated later.

Face Detection. Face detection is a very popular research field that has developed various algorithms, but the following method is considered sufficient and simple enough to serve our purpose.

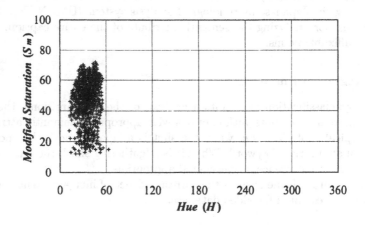

Fig. 5. Skin colored regions on the H - S_m plane.

1. **Skin colored region extraction**
   The modified HSI color system is used to detect skin colored regions [16]. I (Intensity) is used only for excluding dark regions. A certain rectangular region in the H (Hue) - S_m (Modified Saturation) plane was defined as skin color. The distribution of sample skin colored regions on the H - S_m plane is

shown in Fig. 5. Pixels whose H and S_m exist in this region are determined as skin colored. Small regions that consist of less than a certain number of pixels are deleted, and adjoining isolated regions are merged by spreading out their boundaries for a few pixels.

2. **Template matching**

   Template matching is performed to exclude hands, walls, desks and so on that were extracted as skin colored regions. Average faces in several different resolutions are prepared from the I of various facial regions, and are selectively used for matching, according to the size of the extracted skin colored region. In order to decrease the influence of optical states, the I of the extracted region is regularized by the overall I of the frame.

Lip Movement Detection. Once a facial region is detected, it is easy to estimate the mouth location. This is because faces in speech/report shots and anchor shots are usually full faces. If the temporal change of the area around the estimated mouth location is relatively bigger than the change of the entire facial region during a shot, lip movement is detected.

4.2 Anchor Shot

Separation of Anchor Shots Among Speech/Report Shots. Anchor shots initially satisfy the classification conditions for the speech/report shot. One distinctive feature of anchor shots is that they are graphically extremely similar among themselves, and also frequent compared to other speech/report shots. Therefore, after speech/report shot classification, anchor shots are separated by clustering the speech/report shots by evaluating the mutual similarity between all the classified shots. The largest and most dense cluster would be the anchor shots. The similarity is evaluated by the comparison of color histograms applying the segmented χ^2 examination previously used for cut detection. Each shot is regularized so that the facial region should be located in the same position.

Detection of Topic Boundary. The main purpose for detecting anchor shots is to detect boundaries of news topics. However, since anchor shots may appear in the middle of a long topic, just separating them from speech/report shots is not sufficient to fulfill this purpose. A distinctive feature of anchor shots in the beginning of a new topic is the presence of a title caption. As shown in Fig. 6, this could be detected by observing the transition of the overall edge intensity of frames, caused by the superimposion of the title caption CG. Therefore, anchor shots with prominent edge intensity transitions are used to detect topic boundaries.

4.3 Walking Shot

When a person is walking, the upper half of the body oscillates up and down following the steps. A television camera is usually stabilized on a tripod and does not oscillate along the vertical axis. Facial region detection is performed likewise the speech/report shot to the yet unclassified shots. As shown in Fig. 7, a walking shot is classified by detecting the up and down oscillation of the bottom of a facial region.

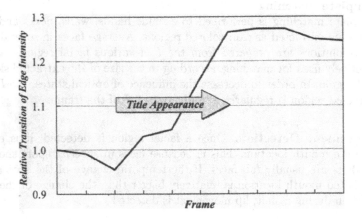

Fig. 6. Relative transition of edge intensity following the appearance of title caption.

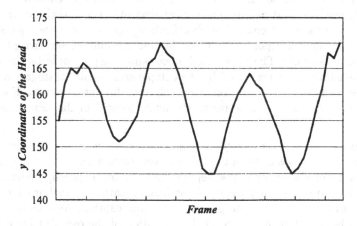

Fig. 7. Oscillation of facial region following the person's step.

4.4   Gathering Shot

When many people are gathering, there are usually more than two similar sized people in a frame. Facial region detection is performed likewise the speech/report shot to the yet unclassified shots. A gathering shot is classified by detecting more than two similar sized facial regions in a frame.

4.5 Computer Graphics (CG) Shot

CG shots are quite tricky, since they may contain a referential portrait picture, or have explanatory figures and texts, which may cause mal-effects to shot classification and caption analysis. Therefore, CG shots need to be detected and excluded from the indexing scheme. A distinctive feature of CG shots is that they are usually motionless. Nevertheless, there are occasional movements, for example in an explanatory flowchart. So, CG shots are classified by the total duration of motionless frames, not by the overall motionlessness. In the following experiment, the duration was set to one second.

4.6 Classification Experiment

Table 2 shows the result of the shot classification applied to 75 minutes of news video. The numbers of true answers ($= N_{Classified} + N_{Unclassified}$) were determined and counted manually for evaluation. Recall and precision are defined as follows:

$$Recall = \frac{N_{Classified}}{N_{Classified} + N_{Unclassified}} \quad (2)$$

$$Precision = \frac{N_{Classified}}{N_{Classified} + N_{Misclassified}} \quad (3)$$

Table 2. Result of shot classification. Most shots were classified well in terms of precision, but recall for some classes were relatively low. This is prominent in 'gathering' shots, since facial regions were not always exposed in the video.

Shot Class	$N_{Classified}$	$N_{Misclassified}$	$N_{Unclassified}$	Recall	Precision
Speech/Report	55	3	12	83%	95%
Anchor (Overall)	53	0	0	100%	100%
Anchor (Topic Head)	37	0	6	86%	100%
Walking	16	0	15	52%	100%
Gathering	70	13	62	53%	84%
Computer Graphics	14	1	0	100%	93%

Major reasons for misclassification and unclassification were:

- Lip movement could not be detected in a speech/report shot, since the face was not a full face.

- Topic boundaries were not detected for some very short topics, which did not have title captions in the beginning.
- Facial regions were not correctly detected in a gathering shot, since the faces were too small and/or hidden by hair. This is often the case with indoor meetings, when a camera shoots from the rear.
- Up and down oscillation of a head could not be detected in a walking shot, since the person was walking too far away from the camera.
- A completely still image of an object was misclassified as a CG shot.

5 Caption Analysis

Caption analysis is necessary to index typical shots with appropriate keywords reflecting their typical contents. As shown in Tab. 1, captions that have (1) personal and (2) locational/organizational attributes are adequate keyword candidates. They are mostly noun phrases (often simple arrays of nouns). In Japanese, the utmost tail nouns, *i.e.* suffixable nouns, define their attributes[1].

As related research on semantic disambiguation of nouns, several methods do exist. Nasukawa [7] proposed a method that determines semantic attributes of proper nouns (*i.e.* whether a proper noun indicates a place or a person) referring to the context of neighboring sentences. On the other hand, Watanabe *et al.* [11] proposed a method that analyzes television news captions by referring to both locations and grammatical characteristics as keys.

Although these methods perform fairly well, the former method is difficult to serve our purpose since captions do not have enough neighboring information to analyze contexts, and also since it is purposed to handle only proper nouns. The latter method is originally designated to serve similar purpose to ours, but is not generally applicable to various news programs, which have different designing policies where layouts of captions vary.

Similarly to our task, the Named Entity task defined for the Message Understanding Conference (MUC) [15] assigns the participants to classify personal, organizational, locational, temporal and numerical phrases. The difference is that our aim is not limited to proper nouns, where the Named Entity task limits the tagging to personal, organizational and locational phrases to proper names.

Considering these issues, we decided to analyze captions on their own by referring to suffixable nouns.

5.1 Collecting Suffixable Nouns

To enable caption analysis based on suffixable nouns, first such suffixable nouns, *i.e.* (1) personal nouns, and (2) locational/organizational nouns, should be collected. These were collected according to certain conditions from two text cor-

---

[1] This is presumably common in some other East Asian languages, but different in major European languages. For example, personal noun 'President' as in 'President Clinton' comes to the head, and locational/organizational noun 'City' as in 'New York City' comes to the tail.

pora that consist of newspaper articles [18,19]. These were manually morphological analyzed beforehand, which ensures the basic reliability of the collection process.

Details are discussed elsewhere [1,4] since they meddle with language specific issues, but as a result, 3,793 nouns were collected as personal nouns, and 11,166 as locational/organizational nouns. Note that the collected suffixable nouns include those that represent people or locations/organizations alone, such as 'volunteer' and 'kitchen', but does not include proper names.

5.2 Semantic Analysis Experiment

Table 3 shows the result of the caption analysis applied to the same 75 minutes of news video used for the shot classification experiment. The numbers of true answers ($= N_{Classified} + N_{Unclassified}$) were determined and counted manually by a third person, and recall and precision are defined as noted in formulae (2) and (3), respectively.

Table 3. Result of caption analysis. The result shows fairly well performance for analyzing 'personal' captions, but analysis for 'locational/organizational' captions needs further more improvement for practical use.

Caption Attribute	$N_{Classified}$	$N_{Misclassified}$	$N_{Unclassified}$	Recall	Precision
Personal	83	23	4	95%	78%
Locational/ Organizational	139	121	23	86%	54%

Major reasons for misclassification and unclassification were:

- Some nouns were essentially applicable to both categories (Semantic diversity).
- Some nouns in the collected dictionary were inappropriate (Noise).
- The collected nouns were insufficient (Lack of vocabulary).

The latter two could be solved by further improvement of the collection scheme, but the semantic diversity is an essential issue when dealing with semantics of words.

The reason for the low rates of locational/organizational captions is due to the loose conditions of the collection rule. It is difficult to tighten the rule without more precise grammatical information tagged beforehand in the corpora.

6 Indexing to Classified Shots

Following shot classification and caption analysis, appropriate keywords are tagged to each classified shots according to their classes.

6.1   Indexing Scheme

Appropriate semantic attributes of keywords are defined to each typical shot class as shown in Tab. 4. Anchor shots with captions are referred to detect boundaries of topics for this process. On the other hand, CG shots are excluded from indexing, considering their uniqueness.

Table 4. Shot class and semantic attributes of indexes. Each shot class (except 'CG shots' and intermediate 'anchor shots') are indexed with captions with corresponding semantic attributes. Titles are extracted from 'anchor shots' referring to the special designs around them.

Shot Class	Semantic Attribute of Index
Speech/Report	personal
Anchor (Topic Head)	title
Walking	personal
Gathering	locational/organizational

Following these principles, indexing is performed to all the classified shots except intermediate anchor shots and CG shots according to the following procedure:

1. Search for a caption with an appropriate attribute inside the shot, and index it if found.
2. If not found, search for it in graphically similar shots inside the topic, and index it.

6.2   Indexing Experiment

Table 5 shows the result of indexing applied to the same 75 minutes of news video used in previous experiments. In order to evaluate "indexing based on shot classification" independently, true answers of semantic attributes of captions given by a third person were used.

The result does not necessarily show practical performance as a whole (/All), but as an evaluation of the proposed method itself (/Indexable), all classes showed more than 75% of recognition rate. The overall performance should improve by employing other natural language source as shown in Fig. 3.

7   Conclusion

In this paper, we have proposed and evaluated an indexing method that indexes television news video, considering the correspondence of image contents and keywords. The overall result is not necessarily practical, but the performance of the method itself is quite promising. Although the techniques applied for shot

Table 5. Result of indexing. The evaluation was done from two points of view: '/All' indicates the ratio to the number of all classified shots, and '/Indexable' the ratio to the number of shots that are manually indexable. This was done since some topics lacked of caption information necessary for indexing. They can not possibly be indexed in any way by the current method without external knowledge, so they were excluded to evaluate the method independently. 'Walking shot' had too small number of samples to discuss the result.

Shot Class	All	Indexable	Indexed	/All	/Indexable
Speech/Report	38	30	25	66%	83%
Anchor (Topic Head)	15	15	14	93%	93%
Walking	4	1	1	25%	100%
Gathering	59	44	33	56%	75%

classification and pre-process for caption analysis were conventional, the effectiveness of integrating image and language information was also shown through the indexing.

The problem is that the numbers of typical shot classes and caption attributes are relatively small, since the classification rules were given in a top-down manner. We are currently examining an automatic classification rule acquisition method based on statistic relations of graphical feature and semantic attributes of captions [2]. The result of a preliminary experiment showed promising performance, although the amount of data was not large enough to discuss statistics. We will apply the acquired classification rules to enable better automatic indexing based on shot classification in the near future.

Acknowledgments

The Japanese Morphological Analysis System JUMAN is a free software developed at and distributed by Nagao Lab., Kyoto University and Matsumoto Lab., Nara Institute of Science and Technology. The Kyoto University Text Corpus is a product of Nagao Lab., Kyoto University. RWC text database is a product of Real World Computing Partnership (RWCP), and is used by the authors under a licensed agreement.

References

1. Ide, I. and Tanaka, H.; "Automatic Semantic Analysis of Television News Captions"; *Proc. 3rd Intl. Workshop on Information Retrieval with Asian Languages*, Oct 1998 (*to appear*).
2. Ide, I., Hamada, R., Tanaka, H. and Sakai, S.; "News Video Classification based on Semantic Attributes of Captions"; *Proc. 6th ACM Intl. Multimedia Conf. -Art Demos-Techinical Demos-Poster Papers-*, pp.60-61, Sep 1998.
3. Kaneko, T., and Hori, O.; "Cut Detection Technique from MPEG Compressed Video Using Likelihood Ratio Test"; *Proc. 14th Intl. Conf. on Pattern Recognition*, Aug 1998.

4. Ide, I. and Tanaka, H.; "Semantic Analysis of Television News Captions by Suffixes"; *Trans. IPS Japan*, Vol.39, No.8, pp.2543-2546, Aug 1998 (*in Japanese*).
5. Nakamura, Y. and Kanade, T.; "Semantic Analysis for Video Contents Extraction –Spotting by Association in News Video–"; *Proc. 5th ACM Intl. Multimedia Conf.*, pp.393-402, Nov 1997.
6. Satoh, S., Nakamura, Y. and Kanade, T.; "Name-It: Naming and Detecting Faces in Video by the Integration of Image and Natural Language Processing"; *Proc. IJCAI'97*, pp.1488-1493, Aug 1997.
7. Nasukawa, T.; "Keyword Categorization based on Discourse Information"; *Proc. 11th Annual Conf. JSAI*, pp.348-349, Jun 1997 (*in Japanese*).
8. Hauptmann, A. G. and Witbrock, M. J.; "Informedia News-on-Demand: Using Speech Recognition to Create a Digital Video Library"; *Proc. AAAI'97 Spring Symp. on Intelligent Integration and Use of Text, Image, Video and Audio Corpora*, pp.120-126, Mar 1997.
9. Kurakake, S., Kuwano, H. and Odaka, K.; "Recognition and Visual Feature Matching of Text Region in Video for Conceptual Indexing"; *SPIE Proc. of Storage and Retrieval for Image and Video Database V*, Vol.3022, Feb 1997.
10. Smith, M. A. and Kanade, T.; "Video Skimming and Characterization through the Combination of Image and Language Understanding Techniques"; *CMU Tech. Rep.* CMU-CS-97-111, Feb 1997.
11. Watanabe, Y., Okada, Y. and Nagao, M.; "Semantic Analysis of Telops in TV Newscasts"; *Tech. Rep. IPS Japan* 96-NL-116, Vol.96, No.89, pp.107-114, Nov 1996 (*in Japanese*).
12. Ariki, Y. and Saito, Y.; "Extraction of TV News Articles Based on Scene Cut Detection Using DCT Clustering"; *Proc. 1996 Intl. Conf. on Image Processing*, pp.847-850, Sep 1996.
13. Wactlar, H. D., Kanade, T., Smith, M. A. and Stevens, S. M.; "Intelligent Access to Digital Video: Informedia Project"; *IEEE Computer*, Vol.29, No.3, pp.46-52, May 1996.
14. Motegi, Y. and Ariki, Y.; "Indexing to TV News Articles Based on Character Recognition"; *Tech. Report IEICE*, PRU-95-240, Vol.95, No.584, pp.33-40, Mar 1996 (*in Japanese*).
15. United States Defense Advanced Research Projects Agency (DARPA), Information Technology Office; "Named Entity Task Definition, Version 2.1"; *Proc. 6th Message Understanding Conference*, pp.317-332, Nov 1995.
16. Matsuhashi, S., Nakamura, O. and Minami, T.; "Human-Face Extraction Using Modified HSV Color System and Personal Identification Through Facial Image Based on Isodensity Maps"; *Canadian Conf. on Elec. and Comp. Eng. '95*, Vol.2, pp.909-912, Sep 1995.
17. Nagasaka, A. and Tanaka, Y.; "Automatic Video Indexing and Full-Video Search for Object Appearances"; *IFIP Trans.*, Vol.A, No.7, 1992.
18. Kurohsashi, S., Saito, Y. and Nagao, M.; "Kyoto University Corpus version 2.0"; Jun 1998.
    Available from http://www-lab25.kuee.kyoto-u.ac.jp/nl-resource/corpus.html
19. Real World Computing Partnership (RWCP); "RWC Text Database"; Mar 1996.
20. Kurohsashi, S. and Nagao, M.; "Japanese Morphological Analysis System JUMAN version 3.5"; Mar 1998.
    Available from http://www-lab25.kuee.kyoto-u.ac.jp/nl-resource/juman-e.html
21. "The Informedia Project";
    http://www.informedia.cs.cmu.edu/.

Mutual Spotting Retrieval between Speech and Video Image Using Self-Organized Network Databases

Takashi Endo[1], JianXin Zhang[2], Masakyuki Nakazawa[1], and Ryuichi Oka[1]

[1] Real World Computing Partnership,
Mitsui Building 13F, 1-6-1 Takezono, Tsukuba, Ibaraki 305, Japan,
enchan@trc.rwcp.or.jp
http://www.rwcp.or.jp/

[2] Mediadrive Co, Ltd., Kumagaya-Ekimae Building 7F, 3-195 Tsukuba, Kumagaya, Saitama 360, Japan,
chou@mediadrive.co.jp

Abstract. Video codec technology like MPEG and improved performance of microprocessors enable environments to be setup in which large volumes of video images can be stored. The ability to perform search and retrieve operations on stored video is therefore becoming more important. This paper proposes a technique for performing mutual spotting retrieval between speech and video images in which either speech or video is used as a query to retrieve the other. This technique makes use of a network that self organizes itself incrementally and represents redundant structures in degenerate form, which makes for efficient searches. As a result, the capacity of a database can be decreased by about one half for speech and by about three fourths for video when expressed in network form. Applying this technique to a database consisting of six-hours worth of speech and video, it was found that a search from video to speech could be performed in 0.5 seconds per frame.

1 Introduction

The amount of text, speech, and graphic information that people can get hold of in digital form is increasing steadily as computer technology advances and the World Wide Web (WWW) expands. In this regard, keyword-based full-text search systems like Altavista[1] and goo[2] have proven to be quite useful as sources of information with respect to the huge volume of data on the Web. At the same time, video codec technology like MPEG and improved performance of microprocessors have led to the creation of environments in which multi-modal data including images can be stored as digital data. It is still quite difficult, however, to search video data even when stored in digital form.

Many techniques[3][4][5] have been proposed over the years for segmenting video into scenes and some of these have been commercialized. Much research has also been performed on video searching, but how best to express search queries still presents a problem.

S. Nishio, F. Kishino (Eds.): AMCP'98, LNCS 1554, pp. 103–118, 1999.
© Springer-Verlag Berlin Heidelberg 1999

On the other hand, speech can be denoted in text form, and if speaker adaptation can be achieved, one's own utterances can be used to present a speech-based query for searching. Considering, therefore, that there are many cases in which video data is accompanied by speech, we can consider the possibility of mutual searching between speech and video in which both video and speech can be used to form queries targeting the other. In other words, a video query can be used to detect a similar video interval in a video database, and the speech for that interval can then be output, and vice versa. In this way, we can say that speech and video are interconnected when it comes to searching. This paper describes a self-organizing technique for efficiently representing speech and video in the form of a network, and proposes a technique for performing spotting retrievals on this network using queries made up of either speech or video. "Spotting" here refers to searching in which neither queries nor speech and video databases need be delimited beforehand. This kind of searching therefore reduces the burden on the user when performing searches.

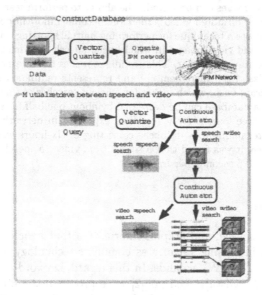

Fig. 1. Concept of mutual spotting retrieval between speech and video

2   IPM Network

It is impossible in practice to prepare all possible input patterns beforehand in the case of general video. A model is therefore needed for constructing structure in an incremental manner with respect to input. The authors have previously proposed the Incremental Path Method (IPM) as a technique for incrementally constructing a network structure with respect to an input stream. In IPM, a network is constructed so that identical input label streams that exist in the label stream have identical traces on the network. This feature makes it possible to express a database in compact form and to perform high-speed searches.

2.1 Converting Speech and Video to Label Streams

To convert video and speech to label streams, we use vector quantization. While it possible to prepare a code book beforehand in the case of speech, it is practically impossible to prepare all possible input patterns beforehand in the case of video, as mentioned above. Consequently, centroid vectors are created during vector quantization whenever quantization error exceeds a fixed value; the input vector is therefore added to this new centroid vector.

2.2 Construction of an IPM Network

A network constructed using IPM consists of nodes and directed arcs. One symbol label is assigned to each arc and one arc is assigned to one pair of nodes. In addition, the number of arcs extending outward from one node is limited, and the limiting value will determine the structure of the IPM network. Specifically, the limiting value can range from one to the number of symbol labels. For a limiting value of one, the network structure will be such that the input stream takes on a single linear formation. This means that sequential information appearing in the database will be preserved. On the other hand, when setting the limiting value to the number of symbol labels, sequential information will be lost and a bigram structure will result. The value halfway between these two values will result in a mixed structure from bigram to n-gram.

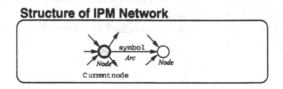

Fig. 2. Basic structure of IPM network

IPM Network is defined as $(S,N,A,L,\lambda,\varepsilon,\omega)$

$$IPM = (S, N, A, L, \lambda, \varepsilon, \omega)$$
$$N = \{n_1, n_2, \cdots, n_k, n_P\}$$
$$A = \{a_1, a_2, \cdots, a_k, a_Q\}$$

$$S = \{s_1, s_2, \cdots, s_k, s_R\}$$

$$\varepsilon(n_i, n_j) = \begin{cases} a_k \text{ if found arc } a_k \text{ in } n_i \text{ to } n_j \\ 0 \quad \text{otherwise} \end{cases}$$

(1)

where set of symbols is S, set of nodes is N, set of arcs is A, maximum number of arcs which one node can have is L, arc transitting from node n_i to node n_j, maximum number of referred symbols in constructing network is λ, each element of N is n_k ($1 \leq k \leq P$), each element of A is a_k ($1 \leq k \leq Q$), each element of S is s_k ($1 \leq k \leq R$), arc from node n_i to node n_j is $\varepsilon(n_i, n_j) = a_k$, symbol binded to arc a_k is $\omega(k)$. When arc $a(k)$ do not have symbol, $\omega(k) = \phi$

Where number of arcs which comes from node n_k is $\rho(n_k)$ and number of arcs which go into node n_k is $\tau(n_k)$, $\rho(n_k)$ and $\tau(n_k)$ are calculated with $\delta(k)$ as

$$\delta(k) = \begin{cases} 1 \text{ if } k \in A \text{ or } \phi \\ 0 \text{ otherwise} \end{cases}$$

$$\rho(n_k) = \sum_{i=1}^{P} \delta\big(\varepsilon(n_k, n_i)\big) \leq L, \quad (1 \leq k \leq P)$$

$$\tau(n_k) = \sum_{i=1}^{P} \delta\big(\varepsilon(n_i, n_k)\big) \leq P, \quad (1 \leq k \leq P)$$

(2)

where number of symbol categories is P.

Now we determine 3 functions that are used for constructing IPM network and then explain algorithm for constructing IPM network.

1. Function for finding a node which do not have an arc
2. Function for finding the node which can be transitted with a symbol s_k from node n_i and can have a new arc
3. Find the furthest node which can reached by transitting with symbols s_k ($1 \leq k \leq \lambda$) from node i

```
(* Function for finding a node which do not have an arc *)
function free_node() : integer;
  var j : integer;
begin
  free_node := 0;
  for j := P downto 1 do begin
    if τ(nj) = 0 and ρ(nj) = 0 then begin
      free_node := j;
    end;
  end;
end;
```

```
(* Function for finding the node that can be transitted with a symbol s_k from node
n_i and can have a new arc *)
function find_node( i, k, L : integer ) : integer;
  var j : integer;
begin
  find_node := 0;
  bf for j := 1 to P do begin
    if ω(ε(n_i, n_j)) = s_k then begin
      if ρ(n_j) ≤ L - 1 then begin
        find_node := j;
      end;
    end;
  end;
end;
```

```
(* Find the furthest node which can reached by transitting with symbols s_k (1 ≤ k ≤
λ) from node i *)
function find_path( i, k, L, p, λ : integer; var l : integer ) : integer;
    var n, j : integer;
  begin
    find_path := 0;
    if p ≤ λ then begin
      j := find_node( i, k, L );
      if j <> 0 then begin
        find_path := find_path( j, k + 1, L, p + 1, λ, l );
        if find_path = 0 then begin
          l := p; find_path := j;
        end;
      end;
    end;
  end;
```

Algorithm for constructing IPM Network is defined as follows with using function
free_node, find_node, find_path

```
(* Algorithm for constructing IPM network *)
procedure make_ipm_network();
  var i, j, k, l, p, q, r, found : integer;
begin
  i := 1; k := 1; r := 1;
  repeat
    found := 0;

    (* strategy 1 *)
    if found = 0 then begin
      p := 1; j := find_path( i, k, L, p, λ, l );
```

```
      if j <> 0 then begin
         found := 1;  i := j;  k := k + l;
      end;
   end;

   (* strategy 3 *)
   if found = 0 then begin
      if s_k = s_{k+1} and ρ(n_i) ≤ L - 1 then begin
         ε(n_i, n_i) := a_r;  ω(a_r) := s_k;
         found := 1;  k := k + 2;  r := r + 1;
      end;
   end;

   (* strategy 4 *)
   if found = 0 then begin
      for q := 1 to P then begin
         if found = 0 then begin
            p := 1;  j := find_path( q, k, L, p, λ, l );
            if j <> 0 then begin
               found := 1;
               ε(n_i, n_j) := a_r;  ω(a_r) := φ;
               i := j;  k := k + l;  r := r + 1;
            end;
         end;
      end;
   end;

   (* strategy 1' *)
   if found = 0 then begin
      for q := 1 to P then begin
         if found = 0 then begin
            p := 1;  j := find_path( q, k + 1, L, p, λ, l );
            if j <> 0 then begin
               ε(n_i, n_j) := a_r;  ω(a_r) := s_k;
               i := j;  k := k + l;  r := r + 1;
               found := 1;  i := j;  k := k + l;  r := r + 1;
            end;
         end;
      end;
   end;

   (* strategy 2 *)
   if found = 0 then begin
      j := free_node();
      if j <> 0 then begin
         ε(n_i, n_j) := a_r;  ω(a_r) := s_k;
         found := 1;  i := j;  k := k + 1;  r := r + 1;
      end;
   end;
```

until $k \leq R$ and *found* $= 1$;
end;

Where i, j, k and r are index of n_i, n_j, symbol s_k and arc a_r respectively.

Function *free_node* find a node that has no arcs and return it. When the function can not find a node, return 0. Function *find_node* find the node that is connected from node i with arc binded with symbol s_k and can have new arc. If function can not find such a node, return 0. Function *find_path* find a node that is reachable from node i with symbol series of up to λ symbols. Function *find_node* calls recursively itself to find the furthest node that can be reached by transitting with symbol series s_k of maximum length up to l ($1 \leq l \leq \lambda$). When return value of function *find_path* is 0, it means that function can not find reachable node.

In *strategy 1* of procedure *make_ipm_network*, find the reachable furthest node from node i by function *find_path*. When the function *find_path* find a reachable node, update current node i by $i := j$ and current input symbol series index by $k := k + l$.

strategy 3: create self loop arc for consecutive same symbol. Create self loop arc by $\varepsilon(n_i, n_i) := a_r$ and assign symbol to the arc by $\omega(a_r) := s_k$

In *strategy 4*, to create a null transition, find a node from that there is a reachable node by transition with symbol series s_k of length up to λ, call function *find_path* using all nodes as a parameter. If the node is found, make new null arc from node i to node j by making arc $\varepsilon(n_i, n_j) := a_r$ and binding ϕ to the created arc.

In *strategy 1'*, find a node from that there is a reachable node by transition with symbol series s_{k+1} of the length up to λ. When such node is found, make an arc to the node j with biding symbol s_k.

At the *strategy 2*, as there are no nodes that can be transit from node i with symbol s_k, create a new node j and make new arc a_r with binding symbol s_k.

Algorithm stops when all bran new node has gone or all symbol series s_k is recorded into network.

3   Spotting

Spotting from IPM network uses the Continuous Automaton method [11] proposed by Oka et al. This method can estimate node transition loci that produce similar symbol series to input symbol series. From the estimated transition loci, categories of input symbol series are determined.

Where input symbol at time t is $u(t)$ and symbol binded to the arc from node n_i to node n_j is $\omega(\varepsilon(n_i, n_j))$, local distance between input symbol at time t and the arc from node n_i to node n_j is defined as $d_{ij}(t) = \omega(\varepsilon(n_i, n_j)) \cdot u(t)$.

Accumulated distance of node j at time t, $P_j(t)$ is calculated as

$$P_j(t) = \begin{cases} \min_i \{(1-\alpha) \cdot P_i(t-1) + \alpha \cdot d_{ij}(t)\} & : \text{if } \omega(\varepsilon(n_i, n_j)) \neq \phi \\ (1-\alpha) \cdot P_i(t-1) & : \text{otherwise} \end{cases}$$

$$(3)$$

where α $(0 < \alpha < 1)$ is oblivion parameter. In the case of $\alpha \simeq 0$, past accumulated distance $P_i(t-1)$ influences on $P_i(t)$ heavily. In the other case of $\alpha \simeq 1$, present local distance $d_{ij}(t)$ influences on $P_i(t)$ heavily.

Matching paths in Continuous Automaton method are corresponding to DP matching path shown in fig. 3. In the Continuous Automaton method, null arc and self-loop arc enables non-linear expansion and contraction in time domain.

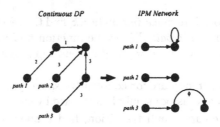

Fig. 3. Spotting of IPM network

Where the threshold for reducing searching space is $h3$ and the node that have larger accumulated distance than h_3 is removed from searching, algorithm for searching by Continuous automaton is as follows.

```
(* Find the node that have lowest cumulated distance *)
function find_min_node( t : integer; var p : integer ) : integer;
  var j : integer;
begin
  find_min_node := 1;
  p := P_1(t);
  for j = 1 to P then begin
    if p > P_j(t) then begin
      p := P_j(t); find_min_node := j;
    end;
  end;
end;
```

```
(* Calculate local distance with symbol c and update a cumulated distance *)
procedure update_score( t, i, c : integer );
  var j : integer;
  var p : real;
begin
  for j = 1 to P then begin
    if ε(n_i, n_j) <> 0 then begin
      if ω(ε(n_i, n_j)) = φ then begin
        d_ij(t) := 0;
      end else begin
        d_ij(t) := ω(ε(n_i, n_j)) · c;
      end;
        p := (1 − α) · P_i(t − 1) + α · d_ij(t);
      if P_j(t) > p then begin
        P_j(t) := p;
      end;
    end;
  end;
end;
```

Function *find_min_node* searches the node that has lowest cumulated distance and store cumulated distance into variable p and return the node.

The procedure *update_score* calculate local distance $d_{ij}(t)$ for each time and update accumulated distance of every node $P_j(t)$ of every node.

According to equation (3), local distance is 0 and global distance is the same as distance at $t - 1$ when the symbol assigned to the arc from node i to node j. Otherwise calculate local distance and update accumulated distance.

On constructing network, category that the input symbol belongs is recorded on the arc corresponding to input symbol as a category information. Categories are defined flexibil. Category can be word category, sentence number or scene number and so on. One arc can be passed by input that belongs to several different category. In recoginition phase, node transition locus is recorded. So the cateory that input query belongs to is estimated by calculating sum total of category information along the arcs in transition locus.

In the spotting, first step is finding the node whose accumulated distance is the lowest, calculating sum total of category information along the locus, and determine the categorie of the input.

W is the set of word categories, U is the number of words categories, $\psi(a_r, \varpi_k) \geq 0$ $(1 \leq r \leq Q, 1 \leq k \leq U)$ is the frequency of word category ϖ_k at the arc a_r. And $\vartheta_m(n_i)$ is the m-th past history of the passed arcs to reach the node n_i.

$$W = \{\varpi_1, \varpi_2, \cdots, \varpi_k, \cdots, \varpi_U\}$$

$$\psi(a_r, \varpi_k) = \begin{cases} \geq 1 & \text{number of symbols} \\ & \text{that belong to the category } \varpi_k \\ & \text{passing through the the arc } a_r \\ 0 & \text{otherwise} \end{cases}$$

$$\vartheta_k(n_i) \in A$$

$$(4)$$

Frequency of arc usage for each categories are recorded in the procedure *make_ipm_network* while constructing IPM network. In retrieving phase, reffering these frequency of arc usage and node transition history, category of input symbol can be determined.

```
(* Algorithm for retrieving by Continuous Automaton *)
function retrieval_ipm( t, c : integer ) : integer;
  var i, j : integer;
  var d, p : real;
  var w : array [1,U] of integer;
begin
  for i = 1 to P then begin
    update_score( t, i, c );
  end;
  i := find_min_node( t, p );
  if p < h₃ then begin
    for m = 1 to Q then begin
      for k = 1 to U then begin
        w[k] := w[k] +ψ(ϑₘ(nᵢ), ϖₖ);
      end;
    end;
    retrieval_ipm := arg{maxⱼ₌₁,...,U  w[j]};
  end;
end;
```

This algorithm gives spotting result for each input. By this algorithm, spotting result can be obtained frame synchronously.

4   Evaluation Experiments

4.1   Speech Spotting by Speech Query

In this experiment, we used ten-minutes worth of a TV news program as a test sample. In this sample, speech was analyzed and subjected to vector quantization to construct an IPM network. In particular, speech was extracted at a frame length of 16 ms and a frame shift of 8 ms, a graduated spectrum vector field was calculated and subjected to vector quantization, and an IPM network for searching was constructed. The speech consisted of 58 utterances over a total of

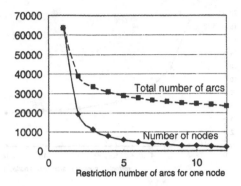

Fig. 4. Number of nodes and arcs in IPM network for TV news program

507 seconds. The speech query that we used for searching here consisted of the word "consumption tax" that we extracted from a database lasting 424 ms. The words "consumption tax" and "consumption tax rate" make up seven of the 58 utterances mentioned above. In this experiment, we made use of an extracted speech interval as a speech query for the sake of simplicity in the evaluation. We point out, however, that the proposed technique does not require the speech query to be extracted.

Experimental Results Figure 4 shows the relationship between the restricted number of arcs for one node, the number of IPM-network nodes, and the total number of arcs. Figure 5 shows the calculation time required for a spotting search versus the restricted number of arcs for one node. Programs for experiment run on SGI indigo2(R10000 195MHz). In the continuous automaton algorithm, the local cost between input and arcs is calculated for each frame, and as a result, calculation time is proportional to the total number of arcs. Figure 6 shows the spotting rate (correct retrieval) versus the restricted number of arcs for one node. It can be seen that as the number of restricted arcs for one node increases, the retrieval rate drops dramatically. This can be attributed to the fact that the network structure approaches a bigram as the number of restricted arcs increases. When speech that corresponds to one scene is treated as one word as in this experiment, the loci on the network will overlap and it will become difficult to distinguish them. Figure 8 shows an example of spotting results.

4.2 Video Spotting by Video Query

Figure 7 shows structure of sample IPM network construnted from video image of 10 minutes news program. It can be seen that cut structure forms some structure in IPM network. In this experiment, we used six-hours worth of a TV program consisting of 648,002 frames as the test sample. The video was subjected to vector quantization for each frame and an IPM network was constructed for

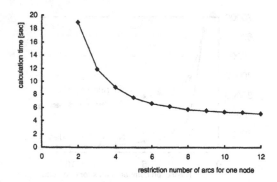

Fig. 5. Spotting search calculation time (speech to speech)

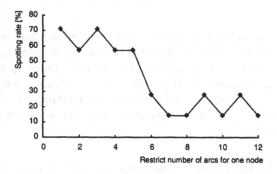

Fig. 6. Speech Spotting rate using speech query

searching. For picture feature quantities, the screen was divided into nine parts, and after calculating the low frequency part of DCT parameters of each of these parts, the video was quantized into 23,114 types of clusters. For the video query, we used 10-minutes news program containing 111 cuts from the six-hour-long sample program and proceeded with the search. This test sample 3,346 cuts include 647 cuts that were expected to be spotted by this technique.

Experimental Results Figure 8 shows the number of nodes and number of arcs in this IPM network. For video, when the restricted number of arcs is two, the total number of arcs becomes one eighth of that when the restricted number of arcs is one, and no further reduction in total number of arcs can be seen after this point. Here, if we divide total number of arcs by number of nodes, we obtain an average number of arcs per node of about 2.0. We therefore find about two arcs per node, and we can say that one of the two arcs is directed to another node, while the other arc is a self loop representing repeat of the same picture. Figure 9 shows the calculation time required for spotting. Programs for

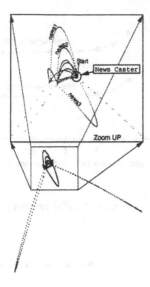

Fig. 7. Structure of sample IPM network constructed from 10-minute TV news program

experiment run on SGI indigo2(R10000 195MHz). Similar to speech spotting, calculation time is essentially proportional to the total number of arcs here. Figure 10 shows spotting rate. As can be seen, the rate drops for a restricted number of arcs between three and five. The reason for this is thought to be that the number of arcs for creating a structure other than a self loop is insufficient and that a structure cannot be adequately represented in this range.

4.3   Mutual Spotting between Speech and Video

The above two experiments target either television speech or television video, and can be referred to as "speech to speech" and "video to video" spotting. In either case, however, one is accompanied by the other in the TV program. As a consequence, when searching for speech by a speech query, the video for the retrieved speech interval can be output to achieve "video searching from speech." This is shown in Figure 11. Likewise, when searching for video by a video query as described in section 4.2, the speech for the retrieved video interval can be output to achieve "speech searching from video."

5   Summary

This paper has proposed a method for achieving compact representation of video data through the use of a self-organizing network and a high-speed mutual spotting search technique between speech and video using a continuous automaton.

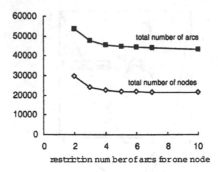

Fig. 8. Number of nodes and arcs in IPM network for TV news program

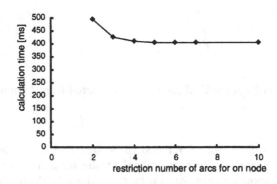

Fig. 9. Spotting search calculation time for one frame (video to video)

In a series of experiments performed to evaluate the proposed technique, a spotting rate of 70% was obtained for a search based on a speech query, and accuracy of spotting result at 83.9% was obtained for a search based on a video query in which six-hours worth of data, and it was searched in 0.5 seconds per frame.

Acknowledgments

The authors would like to extend their deep appreciation to Junichi Shimada, Director of the Real World Computing Partnership, for his valuable assistance during the course of this research. They would also like to thank everyone in the Information Basis Function Laboratory and the Multimodal Function Laboratory for their enthusiastic and enlightening discussions with regard to this research.

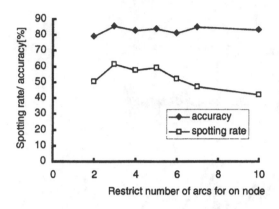

Fig. 10. Video spotting rate using video query

References

1. "Alta VistaC"http://www.altavista.digital.com/D
2. "goo,"http://www.goo.ne.jp/.
3. H. Ueda, T. Miyatake, S. Yosizawa, "A proposal of an Interactive Motion Picture Editing Scheme Assisted by Media Recognition Technology," The transactions of the Institute of Electronics, Information and Communication Engineers(D-II), Vol. J75-D-II, No.2, pp.216-225, 1992.
4. K. Otsuji, Y. Tonomura, Y. Ohba, "Cut Detecting Method by Projection Detecting Filter," The transactions of the Institute of Electronics, Information and Communication Engineers(D-II), Vol. J77-D-II, No.3, pp.519-528, 1994.
5. Y. Taniguchi, Y. Tonomura, H. Hamada, "A Method for Detecting Shot Changes and Its Application to Access Interfaces to Video," The transactions of the Institute of Electronics, Information and Communication Engineers(D-II), Vol. J79-D-II, No.4, pp.538-546, 1996.
6. A. Nagasaka, T. Miyatake, H. Ueda, "Realtime Video Scene Detection based on Shot Sequence Encoding," The transactions of the Institute of Electronics, Information and Communication Engineers(D-II), Vol. J79-D-II, No.4, pp.531-537(1996-4).
7. Y. Yaginuma, M. Sakauchi, "A Proposal of a Synchronization Method between Drama Image, Sound and Scenario Document Using DP Matching," The transactions of the Institute of Electronics, Information and Communication Engineers(D-II), Vol. J79-D-II, No. 5, pp.747-755, 1996.
8. R. Lienhart,"Indexing and Retrieval of Digital Video Sequences based on Automatic Text Recognition,"Proc. of Int. Multimedia Conf. 96, pp. 419-420, 1996.
9. M. Abdel-Mottaleb, N. Dimitrova, "CONIVAS: CONtent-based Image and Video Access System," Proc. of Int. Multimedia Conf. 96, pp. 427-428, 1996.
10. Hatano, Quian, Tanaka, "A SOM-Based Information Organizer for Text and Video Data," Proc. of the International Conference on Database Systems for Advanced Applications, 1997.
11. R. Oka, Y. Itoh, J. Kiyama, C. Zhang, "Concept spotting by image automaton," In RWC Symposium '95, pp. 45-46, 1995.

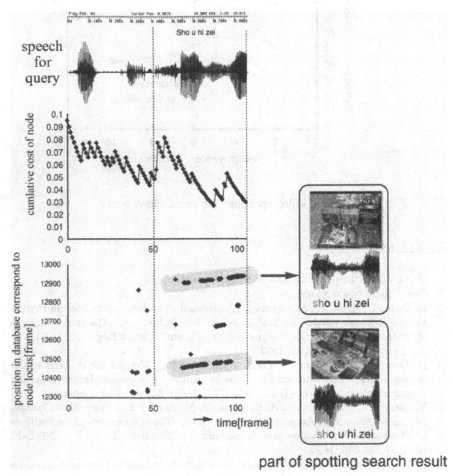

Fig. 11. Video Spotting search result using speech query

12. R. Oka, "Spotting Method Approach Towards Information Integration," Proc. of 1997 Real World Computing Symposium, pp. 175-182, 1997.
13. R. Oka, "The effect of blurred vector field in phone element spotting," Technical Report SP88(100), ISS(Japan), 1988.

Content-Based Retrieval in Multimedia Databases Based on Feature Models

Peter Apers[1] and Martin Kersten[2]

[1] University of Twente, Enschede, the Netherlands,
apers@cs.utwente.nl,
http://www.utwente.nl/~apers
[2] CWI, Amsterdam, the Netherlands
mk@cwi.nl
http://www.cwi.nl/~mk

Abstract. With the increasing popularity of WWW, the main challenge in computer science has become content-based retrieval of multimedia objects. Until now access of multimedia objects in databases was done by means of keywords. Now, with the integration of feature-detection algorithms in database systems software, content-based retrieval can be fully integrated with query processing. In this invited paper, we describe our experimentation platform under development that fully integrates traditional query processing and content-based retrieval and that is based on feature databases, making database technology available to multimedia.

1 Introduction

Large scale multimedia information retrieval is one of the major scientific challenges of this decade. This focus of attention results from significant advances in technology to capture and store raw material in databases and in files on the world-wide web [2]. As a result the research field has entered a third stage.

First generation multimedia database systems focussed on *blobs* to efficiently store the sizeable objects. Since the early 90s, database vendors provide support for these non-interpreted byte streams in their core products, leaving timing, synchronization, and quality-of-service to specialized co-processors. Video-on-demand applications slowly make their way into the homes.

The second generation concerned techniques for annotation and linking media objects. Most of this activity found itself a breeding ground in user interface research and multi-media authoring system. The database merely contains the textual annotations and search accelerators using conventional information retrieval techniques.

The third wave of multi-media database retrieval research concerns itself with developing effective techniques for indexing and retrieval by content [6][20]. The ideal searched for are algorithms to automatically index objects according to a semantic framework.

Unfortunately, this vision is not feasible in the foreseen future. And probably in general not possible either, because semantic descriptions are too tightly coupled with the frame or reference (domain) of the intended audience.

S. Nishio, F. Kishino (Eds.): AMCP'98, LNCS 1554, pp. 119–130, 1999.
© Springer-Verlag Berlin Heidelberg 1999

At best what we can hope for is to make progress in derivation of syntactic features that aid pre-selection in a large multi-media database. Progress in this area is already demonstrated for retrieving still-images based on color distribution, directionality, texture etc.

In this paper we present ongoing research in the area of content-based retrieval using a novel view on the architecture of a multi-media database search-engine. The key scientific questions driving our research are:

content-based retrieval How to effectively process a user's query by mapping concept relationships onto the basic features of multimedia objects stored in a database.

feature databases How to efficiently derive simple and complex multi-media features for widely distributed sources of raw material and making this available as an index for query resolution.

The importance of these research challenges is illustrated by the abundance of funding available worldwide. Within the Netherlands alone, the authors are involved in the following large national programs:

AMIS This national project, bringing together researchers from image processing, computer graphics, database technology, and operating systems, focuses on indexing and searching of multimedia databases. [1]

Digital Media Warehouse This project, which runs under the umbrella of the Telematics Institute, brings together academia and industry to look at the usage of multimedia database in a cooperative environment [5][1].

Work presented in this invited paper is heavily based on [13] and [9]. It is organized as follows. In Section 2 a motivating example is given. This is followed by Section 3 in which we introduce the architecture of 3rd generation multi-media database system geared at content-based multi-media information retrieval. In the two following sections, these issues are elaborated on a bit more: Section 4 discusses our research line content-based retrieval and Section 5 the feature database model and processing scheme to construct a database of index data. In Section 6, we indicate challenges ahead and secondary roads to be explored.

2   An Example

To set the stage for research, the following informal example illustrates the scope of problems to be dealt with.

Imagine a journalist looking for information for a TV news item on El Niño and its effect on weather. What he needs are some video fragments and background data to support his story. For his work he has access to a distributed, multimedia database containing news items. We will look at the way he searches depending on what is provided by the database.

---

[1] (http://www.cwi.nl/~acoi/DMW)

In the simplest case all video fragments are labeled with relevant keywords. In this case he would search for fragments labeled with the keyword El Niño. This would give a perfect *precision*, but a very low *recall*. Meaning that he would of course find all fragments with the right label, but would skip all fragments that were not explicitly labeled El Niño. A more advanced database may have subtitles included. Many programs nowadays have subtitles for deaf people. In this case it would be possible to apply *Information Retrieval* techniques to search for El Niño among the subtitles. One step further would be that video fragments are searched in which an ocean is shown and that the corresponding audio track contains spoken text regarding worms stream before the coast of Peru. The holy grail is that the multimedia retrieval engine can conclude from a video that it concerns the effects of a warm ocean stream before the coast of Peru on the weather in other parts of the world.

In the first example the search is a *boolean search* with which we are very familiar in databases. The query is formulated in terms of what is available in the database. The second example requires the integration of database query processing with *Information Retrieval techniques*. Both the query and the contents of the database (the subtitles) require some processing, e.g. words are replaced by their stems. In the third example the gap between the query formulated by the user and what is in the database becomes larger. In this case features are used. *Features* are complex functions that are applied to the raw representation of a multimedia object. Examples of features are: color distribution, directionality, circularity, texture etc. Of course a user will never be able to phrase his query in terms of feature values. For this type of search it is more common to search for a key frame from a video fragment that comes close to what one is looking for and ask the system to find other videos that have feature values close to the example provided. Via *relevance feedback* the user can indicate which of the returned video fragments are better than other ones, in this way changing the relative weight of each of the features.

The final example is of course something for the future. It requires the attachment of concepts (semantics) to video fragments and of relationships among video fragments.

3   MIR Architecture

The structure of the multi-media information retrieval (MIR) architecture pursued is illustrated in Figure 1. From the top level the system consists of four components: a (web-based) query interface, a multi-media query processor, a feature detector engine, and an extensible database server. Their role within the overall architecture is summarized as follows:

Query Interface

The query interface for multimedia databases differ considerable from the traditional straight line approach encountered in OQL or SQL. The query formulation

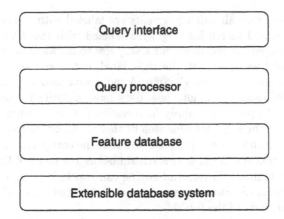

Fig. 1. Multi-media information retrieval architecture

is composed of a mix of textual descriptions, component clipping, and expression of temporal /spatial /topological relationships.

We have taken a pragmatic approach to assume that such interfaces are largely built on Java with an identifiable clean interface to a database. Except for occasional demonstrators we do not invest in the area of user-interfaces per se.

Multimedia Query Processing

In [19] it was stated that the three fundamental issues in probabilistic information retrieval are: representation of documents, query formulation, and a ranking function. These three issues come back in the three layers (see Figure 2): *concept space*, which manages the basic concepts in documents, *evidential reasoning*, which implements the ranking function, and *relevance feedback*, which implements the query facility.

The basis for these layers is a database extended with proper metrics to relate objects within feature space. *Feature clustering* is used as a first step from points in a feature space to concepts. For text retrieval the features are the words, or their stem, in a text document. These words very much correspond to concepts in real life. For multimedia documents, the computation of a feature is a point in an abstract feature space, often without semantic meaning in real life. Feature clustering algorithms have been proposed as the prime means to cluster documents, such that similarity queries are easily (and efficiently) answered. The hypothesis underlying this approach is that clustering leads to kind of concept - not necessarilly semantic meaningful- for query processing.

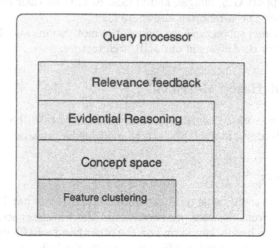

Fig. 2. Architecture multimedia query processor

Feature Databases

The query processing layer is supported by a feature database, i.e. an instantiation of a feature space model. Such models differ from traditional storage models - object-oriented and relational - in their rich provision for partial- and multi-view descriptions of the underlying object.

In many cases it is insufficient to derive a single feature value, e.g. color distribution, for a complete object, but we should detect and store discriminative features for object components. Including retention of temporal and spatial relationships. The state of the art further impose a constraint to deal with an ever-increasing set of such feature types and their costly detection.

Given the distribution and volatility of the underlying database, construction of a multi-media feature index is a continuing activity. At any time, the query processor should be aware of the fact that the feature index is incomplete to answers all requests immediately.

Database Support

Realization of a feature database calls upon the facilities offered by modern extensible databases. They provide both the facilities to deal with multi-media data items and the mechanisms to implement the necessary search accelerators for boosting the metric search in feature spaces.

The underlying database system deployed is Monet, a novel and powerful extensible DBMS. Monet provides the concept of modular extension, a technique in line with *data cartridges* and *data blades*, which encapsulate the routines and data structures for a particular data type. In particular, the system provides modules to support GIS, images, and videos. Results on their functionality and performance have been reported elsewhere [3].

In the following subsections we discuss the motivations and design considerations of the two core layers in our MIR architecture.

4   Content-Based Retrieval

The goal of our work on content-based retrieval is to take the strong points of Information Retrieval and adapt them to multimedia retrieval.

Concept Space

As mentioned, in multimedia retrieval there is an enormous gap between the real world concepts from the user perspective and the feature spaces the system is using. On the one hand a sunset and on the other hand a RGB value; the former is a concept that can be used in different contexts, the latter is a probably unique point in a large feature space.

Before using these feature points as indexing term they have to be clustered. Two techniques are used: supervised and unsupervised feature clustering. In *supervised* clustering the system is trained to learn the differences between two concepts. Similar to the way neural networks work. At GMD in Darmstadt experiments have been done to learn the system to see the difference between "type of light" (artificial or natural) [7]. The result is that from then on a user can refer to the concept articial light and the system knows how to translate that to feature values. One of the objections against this method is that manually it has to be decided which concepts are relevant.

In *unsupervised* clustering documents are clustered based on the proximity of their feature points in the feature space. Very likely the significance of this proximity reveals an underlying concept. However this concept may not be well understood in real life, and is therefore not exposed to the user. We expect that these detected concepts may be of great help in query processing. This is based on the results of the FourEyes learning agent for the Photobook image database [11]. Experiments will have to show whether we are right.

Evidential Reasoning

This part of the system has to determine which documents are relevant for a particular query. In Information Retrieval this is of course done by means of a ranking function. The relevance of a document for a particular query is based on the evidence found in the representation of a document. Matching documents to a query is based on a theory called *evidential reasoning*. The evidence is based on the presence or absence of concepts derived in the concept layer, very similar to traditional Information Retrieval.

Many models exist to implement evidential reasoning, for example, probability theory, Demster-Shafer theory, fuzzy logic [12], and Bayesian belief networks

[15]. The architecture is such that any of these can be used. In the miRRor project [10] we have chosen for Bayesian networks. In spite of the fact that Bayesian inferencing is NP-hard [4], it becomes tractable for more restricted networks. INQUERY, a text retrieval system, is based on the inference network model [16]. The additional advantage of using Bayesian networks is to handle several features, possibly coming from different media of one document. For example, evidence that a document is the right document for a particular query can come from subtitles, keyframes from a video, and the corresponding audio track. Research shows that evidence obtained from different representations is a better support and gives better results.

Relevance Feedback

In [17] we argued that the best way to handle multimedia queries is by means of a *dialogue* between the user and the database system. The idea is that during this dialogue the low-level concepts are identified that are relevant for the user.

Basically, there are two approaches to relevance feedback: query-space modification and document-space modification. In *query-space modification* [14] the relative importance of terms is adjusted. For example, a set of picture is presented to a user. Based on positive and negative feedback the weight of the various low-level concepts or features is adjusted, resulting in a follow-up query.

In *document-space modification* [8] concepts of a document are added or dropped based on a large set of queries for which this document is found relevant. The representation of a document (its attached low-level concepts) is adjusted based on the fact that this document should or should not be included in a result set of queries.

Although we regard both types of modification as important, currently we focus on query-space modification. One of the major advantages is the fact that this can be done during query execution.

5   Feature Database

Query processing a multi-media database presupposes a rich feature database. The designer of a feature database is challenged with finding a balance between flexibility, storage, and performance. Flexibility to support a broad spectrum of possibly proprietary feature detectors, to store their multi-dimensional results in a database with ease of access, and high performance to permit index construction. In this section we discuss the ingredients to built them focusing on the model requirements and construction of the feature database.

Feature Models

The model proposed is based on the observation that indexing an arbitrary multimedia object leads to a hierarchical structure that describes the components of interest for the search. Such hierarchical structures are concisely described with formal grammars. Our interpretation of parsing, however, slightly differs from conventional techniques in language processing.

To recall, we describe a language of properties using a grammar $G = (V, T, P, S)$ where V is a collection of variables, T a set of terminals, P productions of the form $V \rightarrow (V \cup T)^*$, and S the start symbol taken from V. A sentential form α is a string of terminals and variables, such that $S \xrightarrow{*} \alpha$. The collection of parse trees is denoted by PT.

A sub-language $L(G_w)$ is described with the sub-grammar $G_w = (V_w, T_w, P_w, w)$, taking a consistent subset of the corresponding components of G. It describes the structure of sub-sentences in the language $L(G)$.

The terminals T are ordinary typed lexicals. The built-in set of types encompasses the traditional programming types int $\cdots$ str. Furthermore, type extensibility of Monet provides for more complex types, such as image. The terminals are collected into token sequences or sentences $TS = [t_0(v_0), \cdots t_k(v_k)]$ where $t_i \in T$ is an atomary type name, and v_i a value in $domain(t_i)$. A token sequence ts belongs the language $L(G)$, i.e. ts is parsed against grammar G, if there exists a sequence of productions such that $S \xrightarrow{*} ts$.

Turning back to our main objective, we consider a feature database a collection of sentences with indexing values. Their parse tree denotes a hierarchical structure and provides a name space to access and manipulate components. Actually, there exists a natural mapping from sententials to complex objects. In particular, the (non-)terminals are mapped into object attributes; repetition into a list constructor; and alternatives as elements in abstract classes. When a class description is needed for application interfacing, it can readily be derived and refined with application specific behavior. This leads to the following observation:

Definition 1. For $v \in V \cup T$ the class C_v denotes the class of complex objects equivalent to the sub-language G_v.

Feature detectors fit in this framework as operations associated with non-terminals, which massage a token sequence to steer correct parsing of the corresponding sub-language. For this they may inspect the parse tree under construction (its sentential form).

Definition 2. A *feature detector* $d \in D \subset V$ is a function that maps a token sequence $w \in TS$ into $w' \in TS$ using its parse tree d_t, such that the head of w' is a sentence in the sub-language G_d.

A feature detector may involve user interaction to identify the element in F or even extend F in the classification process. For example, the detector could ask the user explicitly for the classification information using a dialogue initialized with a set of choices *ask("car","house",...)* or to let the user draw geometric structures on the screen to identify the portions of interest, e.g. *faces*.

Ordinary functions differ from the feature detectors in that the information derived is not kept permanently in the database for recall. As such, they are also total functions instead of partial functions (over the database extent).

Since detectors may be introduced long after the database has been created, the indexing process necessarily is incremental, because the source may not be available at all times. This leads to two sub-classes for any class C as follows:

Definition 3. The object class indexed by feature detector d is denoted by $\overrightarrow{C}_d$. Those not yet indexed are denoted by $\overleftarrow{C}_d$. At any time class $C_d = \overleftarrow{C}_d \cup \overrightarrow{C}_d$.

```
%ATOM      image;
%ATOM      str server, directory;
%ATOM      str basename, extension
%ATOM      int width, height;

%DETECTOR url;
%DETECTOR picture;
%DETECTOR icon(image);
%DETECTOR avatar SELECT thumnail WHERE picture.width=48
          AND picture.height=64;

mmo:        url category;
url:        server directory* basename extension;
category:   thumbnail | avatar;
thumbnail:  picture icon;
picture:    image width height;
icon:       picture;
```

Fig. 3. A Feature Grammar Example

To illustrate, consider the feature grammar defined in Figure 3. The top part defines atoms (typed terminals) and feature detectors. Detector avatar is a *white*-box detector; its behavior is defined by an expression understood by the feature detection tool kit. The other detectors are black box detectors, known by their name only. It is up to the user to supply an implementation. Their body may inspect parse tree - it provides access to contextual information- and change the token sequence to assure proper continued parsing.

The bottom part contains a grammar for a hierarchical structured feature space. An object o that is known to obey this grammar has an implied syntax tree where the edges are labeled with the names of the corresponding production rules. Components of this parse tree can be accessed with regular (path) expressions.

Unlike traditional grammars, alternation between thumbnail and avatar is not exclusive. Both productions describe alternate views on the same underlying object. The category rule succeeds when for all succesfull alternatives produce the same remaining token pool for continuation. An alternative that fails is further ignored.

Observe that semi-structured databases follow the same pattern, a document is a hierarchical composition whose structure is conveniently described by a grammar (e.g. SGML, HTML, XML, Hytime). However, in Acoi we expect an a priori geven grammar and do not derive the schema on the fly from the documents in the database.

Feature Engines

Feature extraction is a time consuming operation, because the multi-megabyte source is often stored remotely and the detection algorithm is often compute intensive. With our focus on the volatile web as the primary source for retrieving multi-media objects calls for a mechanism to schedule feature detection activities. In an ideal situation, all relevant features for a given user query have been pre-computed and consolidated in the query result. More often, though, we may have to calculate on the fly feature properties, such that the user can be satisfied with (at least) a partial answer. Unlike traditional database queries it is out of the question to wait for all objects of interest to be processed.

An informal description of how the feature grammar is used to obtain the index runs as follows (using the feature grammar in Figure 3). At some point in time, a string (e.g. "http://www.cwi.nl/~ monet/lady.gif") is inserted in the token pool from which the grammatical structure is parsed. The start node `mmo` creates a parsing context that ultimately leads to acceptance or rejection of the object as a `mmo` object. This proof is attempted by proving the right hand side of the `mmo` rule, which starts with calling the `url` detector. It searches the pool for a string and breaks it into components as follows:
`[server(www.cwi.nl),directory(~ mk),basename(lady),extension(gif)]`
and the detector returns SUCCEED. The modified token pool is consumed by the parser looking for a valid url. The `mmo` rule can then proceed with the `category` proof with two alternatives, `thumbnail` and `avatar`, both are valid continuations.

The `thumbnail` rule triggers the detector `picture`. Its body has access to the complete parse tree built so far. It uses this information to access the file being referenced and determine its type from the extension component. Upon success (it is a `gif` file) it opens the corresponding file and generates atoms `[image(cache/lady.gif), width(85), height(250)]` pushed in front of the token queue.

Subsequently the `icon` detector is called with the most recent `image` object as parameter. It derives a small icon, leaving it behind in the token stream for consumption as `[image(cache/lady.icon.gif),width(75),height(75)]`. When `thumbnail` proof has ended successfully, the category proof proceeds with the next alternative, `avatar`.

The `avatar` is an example of a predicate-based detector. The `thumbnail` argument sets the context. But there are two pictures available in the parse tree (thumbnail and icon). Therefore, the path should explicate the context to locate the correct `width` and `height`.

The `category` rule succeeds if at least `thumbnail` or `avatar` reports success. When the complete `mmo` rule has been proven the original string object has been parsed into a hierarchical structure containing classification and feature information.

This execution model gives a systematic parsing method to classify a new object. The feature detector engine uses this method to steer feature detector behavior. Basically classification is based upon the success or failure of parsing the token sequence. The detectors merely assure that the proper classification information is available just in time.

6   Conclusions

For multimedia retrieval one of the main challenges is to translate concepts in the real-world environment of users to features that can be computed from the raw data of multimedia objects. And, furthermore, fully integrate multimedia retrieval with tradational query processing thereby showing the full power of database technology.

Feature database can concisely be described with a formal grammar, which captures both the inter-component structure and provides the semantic basis for evidential reasoning. A direct mapping to either a relational or object-relational scheme makes is attractive intermediate model. As such, the approach is in line with XML, where the grammatical structure and annotations form the underlying model for the information modelled.

The novel way to look at parsing as a concerted action of multiple feature detectors, make the approach emendable for wide scale (and parallel) deployment against multi-media indexing on the web.

Currently, a large scale experimentation platform is under construction to demonstrate the technology against a database of 1M images and other multimedia objects gathered from the web.

7   Acknowledgements

We would like to thank our researchers working in our multimedia projects. They are the ones that have to realize our ambitious goals. A special word of thanks goes to Arjen de Vries, Niels Nes, and Menzo Windhouwer for their research contribution to this invited paper.

References

1. AMIS, Advanced Multimedia Indexing and Searching,
   http://www.cwi.nl/ acoi/Amis/index.html
2. P.M.G. Apers, H.M.Blanken, M.A.W. Houtsma (eds), Multimedia Databases in Perspective, Springer Verlag, ISBN 3540761098, June 1997.
3. P. Boncz and M.L. Kersten, Flattening an object algebra to provide performance, ICDE, 1998.

4. G.F. Cooper, The computational complexity of probablistic inference using Bayesian belief networks, Advances in Knowledge Discovery and Data Mining, AAAI Press, 1995.
5. DMW, Digital Media Warehouses, http://www.cwi.nl/ acoi/DMW/index.html
6. C. Faloutsos, Searching multimedia databases by content, Kluwer Academic Publishers, 1996.
7. R. Ferber, Accessing documents to knowledge discovery methods and intelligent retrieval, ERCIM-97-W001, pp 17-22, 1996.
8. N. Fuhr, and C, Buckley, A probabilistic learning approach for document indexing, ACM Transactions on Office Information Systems, Vol 9, No 3, pp. 223-248, July 1991.
9. M.L. Kersten, M.A. Windhouwer, and N.J. Nes, A Feature Database for Multimedia Objects, Proc. workshop ERCIM DBRG, May 1998, Schloss Birlinghoven, Germany.
10. MiRRor, Multimedia Information Retrieval Reducing information OveRload, http://wwwis.cs.utwente.nl:8080/DOLLS/.
11. T.P. Minka and R.W. Picard, Interactive learning using a "society of models", technical report TR-349, MIT Media Laboratory Perceptual Computing Section, 1997.
12. S. Parsons, Current approaches to handling imperfect information in data and knowledge bases, IEEE Transactions on Knowledge and Data Engineering, Vol 8. No 3, pp. 353-372, June 1996.
13. A.P. de Vries and Henk Blanken, The Relationship between IR and Multimedia Databases, accepted for publication at IRSG'98.
14. S.E. Robertson, On term selection for query expansion, Journal of documentation, Vol 46, No 4, pp. 359-364, 1990.
15. H.R. Turtle, Inference networks for document retrieval, PhD Thesis, University of Massachusetts, 1991.
16. H. Turtle and W.B. Croft, Evaluation of an inference network-based retrieval model, ACM Transactions of Information Systems, Vol 9, No 3, 1991.
17. A.P. de Vries, G.C. van der Veer, and H.M. Blanken, Let's talk about it: Dialogues with multimedia databases. Database support for human activity, Displays, 1998, 18, 4, pp. 215-220.
18. A.P. de Vries, B. Eberman, and D.E. Kovalcin, The design and implementation of an infrastructure for multimedia digital libraries, Proc 1998 Int Database Engineering & Applications Symposium, 1998, Cardiff, UK, July, pp. 103-110.
19. S.K.M. Wong and Y.Y. Yao, On modeling information retrieval with probabilistic inference, ACM Transactions on Information Systems, Vol 13, No 1, pp. 38-68, January 1995.
20. J.K. Wu, A.Desei Narasimhalu, B.M. Mehtre, C.P. Lam, and Y.J. Gao, CORE: a content-based retrieval engine for multimedia information systems, Multimedia Systems, Vol 3, pp 25-41, 1995.

An Efficient Index Structure
for High Dimensional Image Data

Jae Soo Yoo[1], Myung Keun Shin[2], Seok Hee Lee[1], Kil Seong Choi[1],
Ki Hyung Cho[1], and Dae Young Hur[3]

[1] Department of Computer & Communication Engineering,
Chungbuk National University, San 48, Gaesin Dong,
Heungduk Ku, Cheongju, Chungbuk, South Korea, 361-763
yjs@cbucc.chungbuk.ac.kr
[2] CAIS, KAIST, Taejon, South Korea, 305-701
mkshin@camis.kaist.ac.kr
[3] Database section, ETRI, Taejon, South Korea, 305-380

Abstract. The existing multi-dimensional index structures are not adequate for indexing higher-dimensional data sets. Although conceptually they can be extended to higher dimensionalities, they usually require time and space that grow exponentially with the dimensionality. In this paper, we analyze the existing index structures and derive some requirements of an index structure for content-based image retrieval. We also propose a new structure, called CIR(Content-based Image Retrieval)-tree, for indexing large amounts of point data in high dimensional space that satisfies the requirements. In order to justify the performance of the proposed structure, we compare the proposed structure with the existing index structures in the various environments. We show through experiments that our proposed structure outperforms the existing structures in terms of retrieval time and storage overhead.

1 Introduction

Many recent applications such as image databases, medical databases, GIS, CAD/CAM and so on require enhanced indexing for content-based image retrieval. Content-based image retrieval is to query large on-line databases using the images' content as the basis of the queries. Examples of the content include color, texture, and shape of image objects and regions. In the applications that need content-based retrieval, indexing of high-dimensional data has become increasingly important. For example, in multimedia databases, the multimedia objects are usually mapped to feature vectors in some high-dimensional space. The queries are processed against a database of those feature vectors. The index structures for the content-based retrieval also efficiently need to process similarity queries that are related to some measure of similarity between feature vectors.

S. Nishio, F. Kishino (Eds.): AMCP'98, LNCS 1554, pp. 131-144, 1999.

There are a few index structures for high dimensional data such as SS-tree[11], TV-tree[10] and X-tree[13]. The SS-tree was proposed as an index structure for efficiently supporting similarity search. The idea of TV-tree is based on the observation that in most high-dimensional data sets, a small number of the dimensions bears most of the information. The main idea of X-tree is to avoid overlap of bounding boxes in the directory by using a new organization of the directory that is optimized for high-dimensional space. However, they are not suitable to indexing for content-based retrieval. The reason is that although conceptually they can be extended to higher dimensionalities, they usually require time and space that grow exponentially with the dimensionality.

In this paper, we derive some design requirements of an index structure for content-based image retrieval. We also propose a new structure, called CIR(Content-based Image Retrieval)-tree, for indexing large amounts of point data in high dimensional space that satisfies the derived requirements. We perform extensive experiments with various data distributions such as uniform, normal and exponential distributions. The relationships among various performance parameters are thoroughly investigated. We show through performance comparison based on experiments that regardless of data distribution, the CIR-tree significantly improves performance in both the retrieval time and the storage overhead over SS-tree, TV-tree and X-tree.

The remainder of this paper is organized as follows. In section 2, we describe an overview of the currently existing multidimensional index structures. In section 3, we present a few requirements of an index structure for content-based retrieval. In section 4, we propose a new indexing structure that satisfies the requirements. Section 5 performs experiments to show that the proposed index structure outperforms existing index structures. Finally, conclusions are described in section 6.

2 Existing High Dimensional Index Structures

The R-tree and its most successful variant, the R*-tree have been used most often for indexing high dimensional data in the database literature. The R-tree[19] is a height-balanced tree in which each node includes multidimensional rectangles as complete objects without clipping them. Each internal node in the R-tree stores entries of the form (*I, pointer*), where *pointer* is the address of a child node in the R-tree and *I* represents the minimum bounding rectangle of all rectangles that are entries in that child node. A leaf node contains entries of the form (*I, Oid*), where *Oid* refers to a record in the database. Each leaf node includes MBRs that minimally cover spatial objects on the space.

The R*-tree[9] has two major enhancements over the R-tree. First, rather than considering the area only, it minimizes margin and overlap of each enclosing rectangle in the inner nodes. Second, the R*-tree introduces the notion of forced reinsert to make the shape of the tree less dependent to the order of insertion. However, the R-tree and the R*-tree explode exponentially with the dimensionality, eventually reducing to sequential scanning.

The TV-tree[10] is a method in the database literature that was proposed specifically for indexing high-dimensional data. The basis of the TV-tree is to use dynamically contracting and extending feature vectors. That is, it uses as little features as possible which are necessary to discriminate among the objects. An Example of TV-tree is given in Fig. 1. The points designated from A to I denote data points (only the first two dimensions are shown). In the root level, region R3 uses only one dimension for discrimination. But the other regions use two dimensions for discrimination.

However the TV-tree would not handle overlap properly. To solve overlap problem of TV tree, we will suggest improved ChooseSubtree algorithm in section 4.3.1, and adapt supernode concept in Split algorithm.

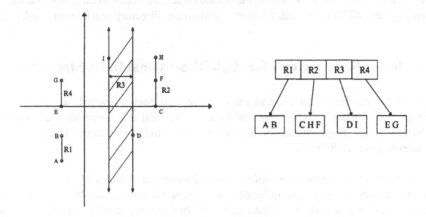

Fig. 1. Example of a TV-tree. In the root level, region R3 uses only one dimension for discrimination. But the other regions use two dimensions for discrimination.

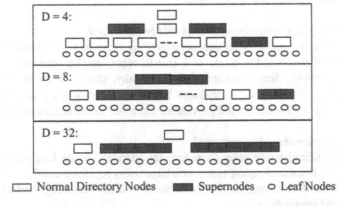

Fig. 2. Various shapes of the X-tree in different dimensions

The X-tree[13] was proposed as an index structure to avoid splits that would result in a high degree of overlap in the directory. To do this, the X-tree uses a split algorithm minimizing overlap and additionally utilizes the concept of supernodes. Supernodes are large directory nodes of variable size(a multiple of the usual block size). Supernodes are created during insertion to avoid splits in the directory that would result in highly overlapped structure. The X-tree uses the notion of maximum overlap(MaxO) value to decide to split or to make supernode. The suggested value of MaxO in [13] is 20%.

Due to the fact that the overlap is increasing with the dimension, the number and size of supernodes increase with the dimension. Fig. 2 shows three examples of X-tree with different dimensionalities.

Although the overlap was reduced in directory, the X-tree loses the efficiency of hierarchical structure. In Fig. 2, when D=32, supernodes are so large, the structure of the tree looks like linear. If we use more smaller feature vector in directory, the size of supernodes will be decreased. Of course total size of directory will be decreased.

3 Design Requirements for High Dimensional Index Structures

Under the condition that the features of image has been extracted, we analyze the properties of the previously proposed high dimensional index structures and present desired design requirements for high dimensional index structures. The design requirements are as follows.

• *The index structure should deal with high dimensional features efficiently.*
Index structures for a content-based image retrieval system should deal with high dimensional image features. Most existing multi-dimensional index structures are not adequate for handling high dimensionality. When going to higher dimensions, they become extremely inefficient because the number of nodes increases exponentially. When constructing index structure, the number of nodes should not increase exponentially as the number of dimension increases.

• *The overlap between directory regions should be minimized.*
In general, overlap means a region that is covered by more than one directory area. As the amount of data and the height of a tree increase, overlap area increases remarkably with growing dimensionality of data. Usually, since overlap increases the number of paths to be traversed, it produces bad effects on processing queries. As a result, the new index structure should provide an algorithm to minimize overlap.

• *Storage utilization should be optimized.*
Higher storage utilization will generally reduce the query cost as the height of the tree will be kept low. Eventually, query types with large query regions are influenced more since the concentration of regions in several nodes will have a stronger effect if the number of found keys is high.

- *The index structure should be appropriate to similarity retrieval.*

Unlike conventional database system, a content-based image retrieval system processes queries based on similarity since images are not atomic symbol and unformatted data. Therefore the index structure should process similarity queries efficiently.

- *The index structure should employ a similarity measure that can evaluate well similarity between high-dimensional features.*

In content-based retrieval system, image features are expressed as points on multi-dimensional space. We use the euclidean distance between two point objects as a similarity. In general, since the dimensions of image features are independent with one another and different in respect of relativity and distribution, measuring the similarity between dimensions with just euclidean distance measure suffers limits on exactness. As a result, another similarity measure should be employed.

- *The index structure should process various query types efficiently.*

An index structure has to be able to process various query types such as exact match query, partial match query, and k-nearest neighbor search query. It also should guarantee the accuracy of the results of search.

- *An index structure for content-based image retrieval system has to handle high-dimensional features dynamically.*

Though there are certain applications having archival nature, i.e., insertions are less frequent and updates/deletions are seldom necessary, the content-based image retrieval system in practice requires a dynamic information storage structure.

4 CIR(Content-Based Image Retrieval)-Tree

4.1 Characteristics

Various index structures for high dimensional data sets have been proposed. However, most of them have the dimensionality problem, as surveyed in the previous sections, eventually loosing the efficiency as an index structure. TV-tree and X-tree are the index structures proposed to support efficient query processing of high-dimensional data. It is true that they are more adequate index structure for high-dimensionality than existing index structures such as R-tree and its variants. As we mentioned in the section 2, however, they suffer from processing image data with numerous features.

We propose a new high dimensional index structure, called CIR-tree, in order to alleviate the problem. The proposed CIR-tree satisfies the design requirements mentioned in section 3. The idea of CIR-tree came from the insights of these two structures. We applied the main idea of both tree structures to CIR-tree in order to solve the dimensionality problem. For nodes that are close to the root node we use just a few dimensions so that we can store more branches and obtain a high fan out. On the other hands, we use more and more dimensions as descending tree so that we can see more discrimination. In the CIR-tree, it is assumed that feature vectors for data objects are ordered in ascending order by its importance, and the importance can be obtained by employing various conversion functions[10].

Like other index structures, CIR-tree represents data with hierarchical structure. In detail, the clustered feature vectors for objects are stored in the leaf node of the tree, and the descriptions of their minimum bounding region (MBR) are stored in parent node. Then until the root node is created, the parent nodes are clustered recursively. The CIR-tree makes up disadvantages of the index structures of R-tree group. According to experimental evaluation of overlap in the R*-tree directories, overlap increases to about 90% for high dimensionality larger than 5. The increase of overlap deteriorates the performance of index structure remarkably. The overlap is made when splitting nodes. The CIR-tree uses supernode concept of X-tree to reduce the number of node splits and a split algorithm to avoid overlap when overflow occurs. That is, the CIR-tree avoids overlap whenever it is possible without allowing the tree to degenerate. Otherwise, the CIR-tree uses extended variable size directory nodes, socalled supernodes. Therefore the structure of the CIR-tree is the mixture type of the linear array structure for representing supernode and the hierarchical structure of the R-tree.

In general, the existing index structures use euclidean distance as a similarity measure on retrieval. However, the euclidean distance is not appropriate as a distance measure for high dimensional data because of its exactness limit. To alleviate such a problem, CIR-tree uses the weighted euclidean distance such as Equation #1. The weighted euclidean distance processes various kinds of similarity queries more efficiently than the euclidean distance.

$$D(X,Y) = \sqrt{(x-y)^T diag(w)(x-y)} \ ------------ \ (Equation \ \#1)$$

In the Equation #1, x and y are feature vectors and w is a vector representing relative weight.

4.2 Structure of the CIR-Tree

The structure of the CIR-tree is similar to TV-tree except for supernodes. Each node consists of pointers to child branches, and a MBR represents each child branch. The MBR is a region containing all descendants of that branch, and has the feature vector as much as necessary for discrimination.

The data structures of MBR are as the follow:

```
struct      MBR { Feature    nonsig,
                  Feature    lower,
                  Feature    upper };
```

```
struct Feature { float      feature_value[];
                 int        no_of_dimensions };
```

where Feature denote 'feature vector'.

A directory node contains the MBRs that represent minimum bounding region of all their descendents. The data structure is as follows.

```
struct Branch_node { int     no_of_element;
                     list of(MBR) };
```

A leaf node includes actual feature vectors. The structure of the leaf node is as follows.

```
struct Leaf_node     { int     no_of_element;
                       list of(Feature) };
```

A supernode is created when splitting a directory node. We will discuss the conditions of creating supernodes in section 4.3.2, when we describe the split algorithm. The structure of supernodes is represented as a continuous array of nodes

4.3 Algorithms in CIR-Tree

4.3.1 Insertion Algorithm

To insert a new object, we should find the branch at each level that seems most suitable to hold the new object, and then insert the new object in there. If overflow occurs at this time, we can cope with it by reinserting some entries in the node or splitting the node. We select the items to be inserted from root node. It provides a possibility of eliminating dissimilar data from the node so that it accomplishes more efficient clustering. After inserting, splitting, and reinserting a node, we update the MBRs of its parent node.

The insertion algorithm calls ChooseSubtree algorithm first. ChooseSubtree is very important to make well-clustered tree structure. However, the TV-tree overlooks the clustering of data. But in the CIR-tree, the second criteria shown below clusters similar object together. Eventually, this reduces the overlap and outperforms in retrieval.

The algorithm ChooseSubtree uses the following criteria, in descending priority:

1. Select the MBR that has minimum number of new pairs of overlapping MBR within the node. An example is in Fig. 3 (a).
2. Select the MBR that uses more dimensions for discrimination. Fig. 3 (b) shows only the first two dimensions. The R1 and R2 are overlapped and R1 uses one dimension for discrimination and R2 uses two dimensions for it. The R2 may have more region in the direction of the next dimension. When inserting the point P, R1 is selected over R2 because R2 uses more dimensions. Using more dimensions means that similar object are clustered in the small region.
3. Select the MBR whose center is close to a new object.

Handling the overflow of a branch node and a leaf node is performed by the virtue of split algorithm that we will discuss in the following section and the split may propagate to upper level. If the area of overlap within the node exceeds threshold value when splitting, the node is extended to supernode. The pseudo code for insertion algorithm is as follows.

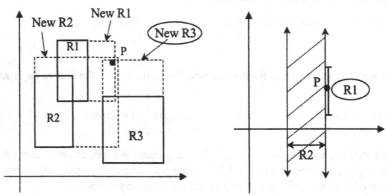

(a) R3 is selected because extending R1 or R2 will leas to a new pair of overlapping regions

(b) R1 is selected because R1 uses two dimensions for discrimination, but R2 uses only first dimension.

Fig. 3. Illustration of the criteria of ChooseSubtree algorithm

Algorithm Insertion
```
1. ChooseSubtree() // choose the best branch to follow,
                    // descend the tree until the leaf
                    // node is reached
2. Insert a new object into the leaf node.
3. If(node overflows)
4.      Call Reinsert
5.      if(Reinsert fail)
6.          Call Split
5.              If( the split routine returns supernode
                                        as a return value)
6.                  Extend the leaf node to supernode
7.              else
8.                      Insert the MBRs of two splitted node
                        into parent node
9. UpdateTree()    // update the MBRs of the
                   // parent node.
```

4.3.2 Split Algorithm

The purpose of splitting is to divide the set of MBRs of vectors into two groups in order to facilitate upcoming operations and provide high space usability. The creation and extension of a supernode occur if there is no possibility to find a suitable hierarchical structure. In other words, if the dividing of the MBRs does not result in minimal overlap split, we does not split the node, but create a super node of twice block size, or appending a block size if the current node is a supernode.

In case of the splitting of a directory node, we first find if overlap exceeds the MaxO value by trying (topological split). If the overlap exceeds the MaxO value, the directory node is extended to supernode. As mentioned in section 2, we set the MaxO value to 20%.

Algorithm Split

```
1. try split  // topological split.
2. If(overlap_ratio > 0.2) return supernode
3. else return the MBRs of two splitted nodes
```

4.3.3 Search Algorithm

In this algorithm, the search starts from root node. It examines whether there is intersection between entries in the node and search area or not. If the intersection exists, we traverse the child nodes of the entries. Because MBRs are allowed to overlap, multiple branches can be traversed. The following is the pseudo-code of search algorithm.

Algorithm Search

```
1. If(accessed node == Leaf node)
2.    Evaluate the similarity of the query and the
      entries in the node.
3.    Return the objects satisfying the query according
      to similarity.
4. else // for directory nodes
5.    Select all MBRs including the query for
      active dimensions.
6.    Call the search algorithm recursively with
      child node that the selected MBR points to.
7. end
```

4.3.4 Deletion Algorithm

This algorithm is quite simple unless underflow occurs. In this case, the remaining branches of the node will be deleted and reinserted. The underflow may propagate to upper level. The pseudo-code is as follows.

Algorithm Delete
```
1. Call Search
2. Continue Search until we arrive at leaf node including
   the target object.
3. Delete the target object from the leaf node.
4. If(underflow)
5.     Delete the corresponding MBR of parent node
6.     Repeat
7.         If (leaf node)  reinsert the remaining objects
8.         else  reinsert remaining MBRs
9.         Move to a parent node
10.    If(Underflow)
11.        Delete the corresponding MBR of parent node
11.    else  return
13. end
```

4.3.5 The Properties of CIR-Tree

The proposed CIR-tree uses a variable number of dimensions when constructing tree to support high-dimensional data efficiently. For nodes that are close to root node, we use just a few dimensions to store more data in a node. This tree provides higher fan out in the top levels, so the height of tree becomes lower. In that result, the number of disk accesses reduces, similarity retrieval becomes easier, and the efficiency of storage space increases. It processes various query types more efficiently, and facilitates deletion and insertion process as well. Also, the CIR-tree uses weighted Euclidean distance for more exactly evaluating similarity between a query and a object. Using supernode, it minimizes overlap, so it reduces the factors that deteriorate retrieval performance. However, since it employs weighted Euclidean distance, in order to give weight to each feature we need to get advice from domain experts.

5 Experiments

We will show the superiority of the proposed CIR-tree by comparing its performance with that of TV-tree and X-tree. In this experiment, we used Trigem Tgstation SDT-820 workstation with 64Megabytes of main memory and 3 Gigabytes of hard disk. All simulation programs were implemented with ANSI C++ and compiled with GNU C++ compiler. We used the TV-tree and the X-tree programs without modifying that were implemented by the authors of the papers. For the experiment, we generated 2,000,000 uniformly distributed floating point numbers, and then, grouped them with desired dimensions to make the dta points. The dimension was varied from D=4 with 500,000 data points up to D=18 with 111,111 data points. The size of each block used for our experiments is 4Kbytes.

5.1 Storage Space

Fig. 4 shows an experimental result of each index structure in terms of storage space. Due to the fact that the CIR-tree and X-tree create similar numbers of leaf nodes, the comparison of the number of leaf node is meaningless. The figure shows that the space usage of the X-tree is increase with dimension, but the space usage of the CIR-tree is kept as stable. This is because the CIR-tree stores small number of features in the directory node for all dimensions. The fact that the CIR-tree creates small number of nodes means that it uses the storage space efficiently. As a result , the performance comparison in terms of storage space shows that the storage overhead of the CIR-tree is much less than that of the X-tree.

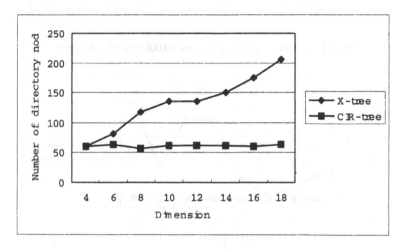

Fig. 4. Number of directory nodes depending on the dimensionality

5.1 Retrieval Performance

Fig. 5 and Fig. 6 show the result of comparison of three index structures for uniformly distributed data. We have applied 50,000 exact match query for each dimensions. Note that, to count the page accesses, the access to supernode of size s was counted as s page accesses. As shown in the Fig. 5, the CIR-tree outperforms other index trees. The decrease of page accesses of the TV-tree with dimension is due to the number of data points is decreased with dimension. The retrieval performance of the X-tree depends on the size of supernodes. For large supernodes, for example D=12, the increase of page accesses of the X-tree is significant. Because the CIR-tree maintains small size of directory, the number and size of supernodes are smaller than the X-tree. Eventually the CIR-tree provides always better performance for each dimension.

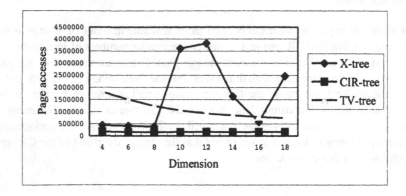

Fig. 5. The number of page accesses for 50,000 exact match query

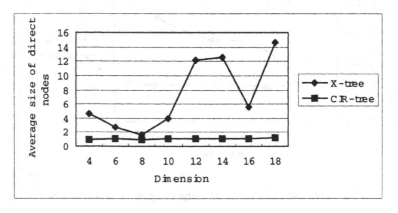

Fig. 6. Average size of directory nodes

6 Conclusion

In this paper, we have analyzed existing index structure for high dimensional data and have proposed the several desired design requirements that the new index structure should have. We also have proposed the new efficient index structure, called CIR-tree, which follows the design requirements. Since the proposed CIR-tree uses few dimensions at upper level, it can store more data at a node. This method produces high fan out at upper level. In that result, the height of the tree becomes lower, so it solves the dimensionality problems that we mentioned several times before. This supports high dimensional data more efficiently, and diminishes disk accesses and improves the disk storage utilization. It also uses weighted euclidean distance measure to overcome the exactness problem of euclidean distance and uses supernode in order to minimize overlap.

We have compared the proposed CIR-tree with TV-tree and X-tree through various experiments to manifest the superiority of our tree. The experiments show that the CIR-tree outperforms the TV-tree and X-tree in the terms of retrieval speed and space requirements.

In this paper we have carried out the performance evaluation of CIR-tree with virtual data sets. We will carry out performance test with the features extracted from real image data in the nearest future. We expect to obtain high performance when constructing content-based image retrieval system with the proposed CIR-tree.

Acknowledgment

This work was partially supported by the Institute of Information Technology Assesment.

References

1. W. E. Mackay and G.Davenport., "Virtual video editing in interactive multimedia applications," Communications of the ACM, 32:802-810, July 1989.
2. Myron Flickner and et. al., "Query by Image and Video Content : The QBIC System." IEEE Computer, 28(9), 1995.
3. Charles E. Jacobs, Adam Finkelstein, David H. Salesin., "Fast Multiresolution Image Query." Proceedings of the 1995 ACM SIGGRAPH, New York, 1995.
4. W. Niblack, R. Barber, W. Equitz, M. Flickner, E. Glasman, D. Petkovic, P. Yanker, C. Faloutsos, G. Taubin., "The QBIC project: Querying image by content using color, texture and shape." Proceedings SPIE Storage and Retrieval for Image and Video Databases, pages 173-187, February 1993.
5. Y. Alp Aslandogan, chuck Their, Clement T. Yu, Chengwen Liu, Krishnakumar R. Nair, "Design, Implementation and Evaluation of SCORE(a System for Content based Retrieval of pictures)," Proceedings of 11th international conference of Data Engineering, 1995, pp280-287.
6. P. M. Kelly, T. M. Cannon and D. R. Hush., "Query by image example: the CANDID approach.," Proc. SPIE Storage and Retrieval for Image and Video Database III, 2420:238-248, 1995.
7. C. Faloutsos, R. Barber, M. Flickner, J. Hafner, W. Niblack, D. Petkovic, W. Equiz., "Efficient and Effective Querying by Image Content," Journal of Intelligent Information System(JIIS), 3(3):231-262, July 1994.
8. J. K. Wu, A. Desai Narasimhalu, B. M. Mehtre, C. P. Lam, Y. J. Gao., "CORE: a content-based retrieval engine for multimedia systems.," ACM Multimedia Systems, 3:25-41, 1995.
9. N. Beckmann, H.P. Kriegel, R. Schneider and B. Seeger "The R*-tree: An Efficient and Robust Access Method for Points and Rectangles ", ACM SIGMOD, pp.322-331, May 1990.
10. K.I. Lin, H. Jagadish, and C. Faloutsos, "The TV-tree: An Index Structure for High Dimensional Data", VLDB Journal , Vol. 3, pp.517-542, 1994.
11. D. A. White and R. Jain, "Similarity Indexing with the SS-tree," In Proc. 12th Intl. Conf. On Data Engineering, New Orleans, pp.516-523, 1996.

12. D.A. White and R. Jain, "Similarity Indexing: Algorithms and Performance," In Proc. of the SPIE : Storage and Retrieval for Image and Video Databases IV, Vol. 2670, pp.62-75, 1996.
13. S. Berchtold, D. A. Keim, H-P. Kriegel, "The X-tree:An Index Structure for High-Dimensional Data," Proceedings of the 22nd VLDB Conference, Bombay, India, 1996
14. B.Furht, S.W.Smoliar, H.Zhang, "Video and Image Processing in Multimedia Systems," Kluwer Academic Publishers, 1994.
15. Lomet. D., "A Review of Recent Work on Multi-attribute Access Methods," ACM SIG-MOD RECORD, Vol. 21, No. 3, pp. 56-63, Sept. 1992.
16. M. J. Swain and D. H. Ballard., "Color indexing. International Journal of Computer vision," 7(1):11-32, 1991.
17. Y. Gong et al., "An image database system with content capturing and fast image indexing abilities.," In Proceedings of the International Conference on Multimedia Computing and Systems, pages 121-130, Boston, MA, May 1994. IEEE.
18. J. T. Robinson, "The K-D-B-Tree : A Search Structure for Large Multidemensional Dynamic Indexes," ACM SIGMOD, pp. 10-18, Apr. 1981.1. Baldonado, M., Chang, C.-C.K., Gravano, L., Paepcke, A.: The Stanford Digital Library Metadata Architecture. Int. J. Digit. Libr. 1 (1997) 108–121
19. Guttman A., "R-trees: A Dynamic Index Structure for spatial Searching" Proc. 7th Int. Conf. on Data Engineering, 1991, pp.520-527.

Color-Based Pseudo Object Model
for Image Retrieval with Relevance Feedback

Tat-Seng Chua and Chun-Xin Chu

School of Computing
National University of Singapore
Lower Kent Ridge Road, Singapore 119260
Tel: +65-874-2726 Fax:+65-779-4580
chuats, chuchunx@comp.nus.edu.sg

Abstract. Color has been widely used in content-based image retrieval
system. The problem with using color is that its representation is low
level and hence its retrieval effectiveness is limited. This paper examines
the issues related to improving the effectiveness of color-based image retrieval system. It explores the choice of suitable color space and color
resolutions for representation and retrieval. This work also emphasizes
the use of color coherent vector (CCV) as the basic model for retrieval.
CCV is an extension of Color Histogram method to provide low-level
representations of objects within the images. A relevance feedback (RF)
technique is developed that uses the pseudo object information from relevant images to enhance subsequent retrieval performance. The overall
system is tested on a large image database containing over 12,000 images. Tests were performed to evaluate the effectiveness of pseudo object
based retrieval method with RF at a number of color resolutions. Results
indicate that the RF method is effective and a medium color resolution
of 316 colors performs the best.

1 Introduction

Images are being generated at an ever-increasing rate by a variety of sources. A
content-based image retrieval (CBIR) system is required to retrieve these images
effectively and efficiently. Such a system helps the users (even those unfamiliar
with the database) to retrieve relevant images based on visual properties such
as colors, or pictorial entities such as the shapes of objects. In a typical CBIR
system, the representations of images' contents in the database are pre-computed
automatically. A user initiates a retrieval process by using an existing image as
a query. The system then extracts the query image's content and matches it
against those stored in the database. The result of query is a set of images that
are similar to the query image rather than an exact match.

Color is a dominant visual feature and is widely used in CBIR. The selection
of color space and color resolution is fundamental to this approach. A suitable
color space is one that is uniform, complete, compact and natural. These criteria
can partly be served by utilizing a linear color space such as the CIE *Luv* color

S. Nishio, F. Kishino (Eds.): AMCP'98, LNCS 1554, pp. 145–160, 1999.
© Springer-Verlag Berlin Heidelberg 1999

space [5]. Also of importance is the selection of colors within the space, and the choice of a suitable color resolution for image retrieval that best matches user's ability to perceive and differentiate colors in natural images.

Color histogram has been widely used for color-based CBIR [2, 8, 11, 19]. Histogram technique is a global method without any local information, so images with very different appearances can have similar histograms [24]. The color coherent vector [10, 11] (CCV) was recently proposed as a refinement to histogram with spatial information. In CCV the pixels within a given color bucket are split into two classes based on spatial coherence. A coherent pixel is part of a sizable contiguous region, while an incoherent pixel is not. The coherent pixels often correspond to object and thus CCV is a pseudo object representation scheme.

As CCV is a low-level representation of objects and color distributions within the images' contents, the retrieval accuracy of system based on CCV will inevitably be limited. Any retrieval is bound to return a list of relevant images along with many irrelevant ones. It is reasonable in such situation to enlist the services of users to identify relevant images from the list of images retrieved. This information can then be used to refine the search, and thus potentially improving the quality of retrieval. This process is known as relevance feedback (RF).

This paper describes the implementation of a simple color based retrieval method using color coherent vector method (CCV). As CCV partly models objects within its coherent components, this paper aims to investigate the effectiveness of pseudo object based RF technique. The project also investigates the choice of color space and the tradeoffs in effectiveness and efficiency between color resolution and the use of perceptually similar colors.

The rest of the paper is organized as follows. Next section reviews related works in color-based image retrieval. Section 3 discusses the choice of color space and color resolution. Section 4 describes the development of a pseudo object based image retrieval method with relevance feedback. The implementation and testing are described in Sect. 5. Finally Sect. 6 contains our concluding remarks.

2   Related Works

The use of color for CBIR has previously been explored in [2, 8, 11, 19, 21]. The fundamental problems of this approach include the selection of color space and color resolution. As color perception is highly subjective [25], there exist a large variety of color spaces in practice. For example, Swain and Ballard [21] quantized an opponent-axis (OPP) color system [1] into 2048 colors. The IBM QBIC system [8] selected the k best Munsell colors (typically, k = 64) through clustering. Pass et. al [11] simply quantized the RGB color space uniformly into 64 colors. While Gray [4] transformed the RGB into CIE-Luv, before partitioning the space into 512 colors.

A good color clustering method on a perceptually uniform color space can improve the retrieval effectiveness significantly. This is because in a linear color space, the perceptually similar colors have the higher probability of lying in the same clustered color bins. [6, 22, 23] discussed the color space clustering methods and the relationship between the retrieval effectiveness and color resolution. [23] compared a number of different clustering methods and concluded that the hierarchical clustering method performs the best. [22] examined the color distributions and clustering strategies in different color spaces, and suggested that there is a need for a thorough study to establish the relationship between the retrieval effectiveness and color resolution.

The color histogram has been shown to be efficient and effective in content-based image retrieval [8, 9, 12]. However, due to the lack of spatial information, the histogram method is liable to false positives. The problem is especially acute for very large image databases. Also, the histogram method is not robust against large appearance changes.

Several recent schemes incorporate spatial information in color representation in order to improve the effectiveness of color based method [3, 10, 11, 19, 20, 24]. Hus. et. al [24] proposed a color spatial method by extracting single-colored regions and their spatial locality in the image space. Pass and Zabih [10, 11] developed a color coherence vector (CCV) by partitioning the histogram bins into coherent and non-coherent portions. CCVs are fast to compute and appeared to perform better than histograms. The CCV method was extended in [10] to model objects within the center region of the image.

Relevance Feedback (RF) techniques were introduced in free-text IR systems over 20 years ago [16]. Price et al [13] employed free-text RF technique to enhance the retrieval effectiveness of a text-based image retrieval system. Shen [18] adopted the IR RF technique to a color-based image retrieval system by modifying the color query vector using the dominant colors extracted from relevant images. Huang et al [7] applied RF to color correlograms for content-based image retrieval. A color correlogram expresses how the spatial correlation of pairs of colors changes with distance, and RF technique was developed to give higher weights to common components of correlogram found in relevant images. A similar approach was employed in [15] on texture feature representations.

3 Choice of Color Space and Color Resolution

3.1 Color Space Selection

The ease in navigating the color space impacts the user's ability to construct color-based image queries. Our purpose is to select a color space that possesses the properties of uniformity, completeness, compactness and naturalness. Color space with such properties permits the image contents to be represented more effectively. These properties can be defined as:

Property 1 (Uniformity). The metric proximity between colors indicates the perceived similarity of colors.

Property 2 (Completeness). The color space includes all possible perceptually distinct colors.

Property 3 (Compactness). Each color in the color space is perceptually distinct from the other colors.

Property 4 (Naturalness). The color space provides for a natural breakdown of colors into the three basic perceptual attributes of color: brightness, hue and saturation.

Digital images are normally represented in RGB color space, which is the color space used by CRTs. However, RGB color space is perceptually non-uniform. To overcome this problem the HSV and Munsell systems are developed. Both systems describe a color by hue, brightness and colorfulness, which are similar to human sensation of colors. In particular, the CIE *Luv* system gives a quantitative expression of the Munsell system for color classification. It is composed of three components in which L defines the luminance, while u and v define the chrominancy. It is intended to yield a perceptually uniform spacing of colors in which any two colors that are equal in distance are perceived as equal in difference by the viewers. It is also implicit that the pixel distribution is uniform and the changing of color is smooth in terms of human perception. The CIE *Luv* space is also capable of representing completely any color from the visible frequency domain. So it satisfies the uniformity, completeness and naturalness properties. By employing a proper color space clustering method it is also possible to make the space compact. Thus, the CIE *Luv* color space is chosen for color representation and retrieval.

3.2   The Division and Extraction of Colors in LUV Color Space

Since *Luv* is linear along all three axes, it is reasonable to divide the axes uniformly and partition the space into cubical cells as illustrated in Fig. 1. This is a simplification of the hierarchical clustering method which possesses the highest ability to conserve color information [23]. The center of the cube is chosen to represent all the colors within the cube. The distance d from the center of the cube c to any vertex will be the *color distance* for the cube (see Fig. 2).

Since *Luv* can be derived from RGB through some transformations, it is possible to get the maximum and minimum values for L, u and v that correspond to some valid colors in RGB space. In our experiments we found that the L component has the range of $[0, 100]$, the u component has the range of $[-134, 220]$, and the v component has the range of $[-140, 122]$. For a given d, the width of the cube a in Fig. 2 is given by $a = \sqrt{3}/2 * d$. By dividing the axes equally by a from the minimum to the maximum points, we get three sets of values for each axis: L, U and V. A Cartesian product is taken over the sets to get all possible color combinations, $P = L \times U \times V$. From P, we can extract those *Luv* colors that correspond to valid RGB colors. This set of colors forms a subspace of *Luv* that we called $L'u'v'$. This subspace inherits the uniformity and naturalness of *Luv* color space. As every RGB color can be mapped to this subspace through a color map table, the $L'u'v'$ fulfills the completeness requirement for the color space. The space also meets the compactness property as it can be observed from Fig. 1 that the space is the smallest space that contains the color set. The generated $L'u'v'$ space can thus be used effectively for image representation and retrieval.

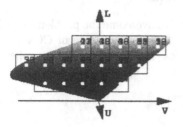   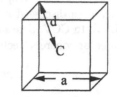

Fig. 1. Division of LUV color space   **Fig. 2.** Cube used in clustering

3.3   The Choice of Different Color Resolutions in Retrieval

The next key issue is to determine the suitable number of cubes to cover the $L'v'v'$ space. This corresponds to the number of colors used in the color representation, or color resolution. The color resolutions in our $L'u'v'$ space can be decided by the *color distance* of a cube. For a given color distance, there is a corresponding color resolution within the space. With a lower color resolution, the efficiency of retrieval can be improved significantly, but the effectiveness may be affected because it may cause distinct colors to be mapped into the same color bin, making them non-distinguishable. On the other hand, higher color resolution can provide a much finer representation of colors. However some colors may be beyond the user's ability to differentiate them, and thus may render the retrieval process less effective and efficient. It is generally observed that human has the ability to perceive a region consisting of multiple distinct colors in closed proximity as a smooth shade of colors. Such capability can partially be achieved by clustering color space into appropriate color resolution, or by incorporating perceptually similar color computation during the retrieval process. The latter approach will add to computation costs.

In this study, we aim to find a suitable color resolution that provides a good trade-offs between effectiveness and efficiency. We define two colors in $L'u'v'$ color space to be perceptually similar if they are within a distance of $2 * d$ units (see Fig. 2). The size of d determines the size of cube and hence the color resolution used. We define a perceptually similar color matrix for each color resolution in such a way that $S(i,j)$ defines the similarity between colors i and j.

4   Pseudo Object Based Image Retrieval with Relevance Feedback

4.1   Pseudo Object Based Image Retrieval

Color Coherent Vector Representation. To overcome the problems of using color histogram as discussed in Sect. 2, we investigate the use of CCV [11] in image retrieval. In CCV, pixels are classified as either coherent or non-coherent. We define a color coherent vector as:

$$\mathbf{ccv} = (\mathbf{H^c}, \mathbf{H^{nc}}) \ . \tag{1}$$

- $\mathbf{H^c}$ is normalized coherent color histogram
- $\mathbf{H^{nc}}$ is the normalized non-coherent color histogram.

In general, coherent regions tend to correspond to part of objects within an image while non-coherent pixels tend to come from image background. Thus the coherent color histogram can be regarded as a crude pseudo object representation. By using CCV, we can prevent coherent pixels in one image from matching non-coherent pixels in another. This allows fine distinctions that cannot be achieved with using color histogram alone. Also, matching of coherent histograms partly corresponds to matching of objects.

The CCV representation can be constructed as follows:

a. Image Decoding and Color Space Conversion.
   First the J-PEG files are decoded into 24 bits RGB files. Then these RGB colors are mapped into appropriate values in $L'u'v'$ space.
b. Segmentation of Contents.
   The image is segmented into 8-connection color regions based on the Luv values. For every pixel, check its 8-connection neighbors to find those pixels with the same color and mark it. The process is repeated for every neighboring pixel of resulting region until the region does not grow anymore. The perceptually similar color is not considered at this stage in order to avoid ambiguous partition.
c. Vector Construction.
   If the size of a connected region is greater than 1% of the size of the image, it is considered to be a coherent region. These pixels are accumulated into the coherent color histogram. Otherwise the pixels are accumulated into the non-coherent color histogram. Both histograms are normalized after the construction.

Retrieval Process. The similarities between the query image and all images in the database are computed based on the correlation of corresponding coherent and non-coherent histograms with contributions of perceptual similar colors. The aim of retrieval is to find those images in image database having similar objects to that of query image. We therefore give more weight to the contribution of coherent part.

The similarity is computed as follows:

a. Compute the normalized difference between query Q and image D for an exact color i as:

$$Diff_{ccv}(Q, D, i) = w_c * \frac{|Q_i^c - H_i^c|}{max(Q_i^c, H_i^c)} + w_{nc} * \frac{|Q_i^{nc} - H_i^{nc}|}{max(Q_i^{nc}, H_i^{nc})} \ . \quad (2)$$

Here we set $w_c = 0.7$ and $w_{nc} = 0.3$ to give more weights to the contributions of coherent colors.

b. Convert the difference formula to a similarity one by:

$$Sim_{ExactCol}(Q, D, i) = 1 - Diff_{ccv}(Q, D, i) \ . \quad (3)$$

c. Determine the contribution of colors perceptually similar to color i:

$$Sim_{PerCol}(Q, D, i) = \sum_{j \in s_p} [Sim_{ExactCol}(Q, D, i) * s(i, j)] \quad (4)$$

where S_p is the set of colors perceptually similar to color i. $S(i, j)$ gives the similarity between colors i and j.

d. Compute the overall similarity contributed by color i:

$$Sim_{Color}(Q, D, i) = Sim_{ExactCol}(Q, D, i) * (1 + Sim_{PerCol}(Q, D, i)) \ . (5)$$

Here the perceptually similar contribution is scaled by (3) to give less weight to the perceptually similar colors.

e. Finally, compute the similarity between query Q and image D as:

$$Sim_{Color}(Q, D) = \sum_{i=1}^{N} Sim_{Color}(Q, D, i) * (Q_i^c + Q_i^{nc}) \quad (6)$$

where N is the number of color used. Here, the contribution of color i is scaled by $(Q_i^c + Q_i^{nc})$ in order to give higher weight to more dominant query colors.

4.2 Relevance Feedback

The relevance feedback process will make use of the coherent parts of CCV representations common to all relevant images. This is equivalent to extracting common object information from the relevant image set. This information is used to modify the query so that the new query can capture more common object information. This would enable the new query to retrieve more images containing similar objects and thus lead to more accurate retrievals.

Given the above guiding principals, the object-based RF process can be carried out as follows:

a. *Process relevance feedback information and select dominant coherent colors*
Given the set of relevant images judged by the users, the first step is to
extract dominant coherent colors that are common to most relevant images.
Since the relevant images are judged based on user's perceptions, it is essential to incorporate the contribution of perceptually similar colors in the
selection of dominant coherent colors. The simplest way to do this is to incorporate the perceptually similar components into the coherent color vector
H^c directly. The revised i^{th} coherent color is given by:

$$P_i^c = \sum_{j=1}^{N}(H_j^c * s_{ij}) .$$ (7)

Here s_{ij} is the (i, j) element of the color similarity matrix defined in Sect.
3.3.
The revised coherent histogram P^c of relevant images are then used to determine dominant colors to represent the relevant image set. By definition, a
dominant color is one that appears frequently in relevant images, but rarely
in the rest of image collection. This information can be estimated using a
probabilistic 0.5 formula, which has been widely used in free-text RF processes [16]. The 0.5 formula uses the distribution of terms in relevant images
and the rest of collection to estimate the importance of terms. By treating
a coherent color as equivalent to a text term, we can adopt the 0.5 formula
to estimate the weight of color i as follows:

$$w_i = log\left(\frac{(r_i + 0.5)(N - n_i - R + r_i + 0.5)}{(R - r_i + 0.5)(n_i - r_i + 0.5)}\right)$$ (8)

where:
 − N is the number of images returned to the user.
 − R is the total number of relevant images indicated by the user.
 − n_i is the number of images among those returned whose P_i^c is above the
   noise threshold.
 − r_i is the number of relevant images with P_i^c above the noise threshold.
To illustrate the characteristic of 0.5 formula, we plot the relationships between the occurrence frequency of coherent colors and w_i's in Fig. 3 for a
specific retrieval. The color resolution used is 176. Figure 3a shows the occurrence frequency of coherent colors in the retrieved and relevant retrieved
images. Figure 3b shows the corresponding probabilistic weights (0.5 formula) of the coherent colors. From Fig. 3b, it can be observed that there are
only a few colors that dominate the whole scene. So for most colors, their
occurrence frequency is near zero. Corresponding to the low-occurrence of
colors in all images, there is a "Zero Line" as shown in Fig. 3b. Colors above
the zero line are meaningful to our feedback process, as these are colors
with comparatively big r_i and small n_i. These colors typically correspond to
meaningful objects in the relevant images and are thus chosen as dominant
colors.

We use w_i as the basis to rank the importance of a color. From extensive tests, we found that it is only necessary to select up to top k (k is set to 10) dominant coherent colors whose w_i is above the Zero Line. These dominant colors are referred to as $D_i^c, i = 1, \ldots k$, where $k \le 10$.

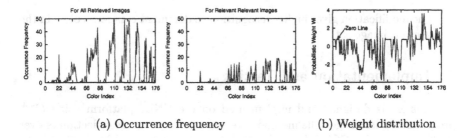

(a) Occurrence frequency   (b) Weight distribution

Fig. 3. The relationship between occurrence frequency and probabilistic weight for coherent colors

b. *Generate New Query*

In order to make the new query closer to the contents of relevant image set, we use the relevance judgment information to modify the query CCV. As we are interested only in extracting common object information from relevant image set, only the coherent component of query CCV is modified using the set of dominant coherent colors extracted. The non-coherent part of query CCV remains unchanged.

We adopted the RF process used in Vector Space Model for free text [16] to modify the coherent component of query CCV. The process is:

$$Q_i^{c(i+1)} = \begin{cases} \alpha Q_i^{c(0)} + \beta AvgR_i^c & \text{if } i \in D^c \\ Q_i^{c(0)} & \text{Otherwise} \end{cases} \tag{9}$$

$$Q^{nc(i+1)} = Q^{nc(0)} \tag{10}$$

$$\text{where} \quad AvgR_i^c = \frac{\sum_{\forall j \in D^c} P_j^c}{k}. \tag{11}$$

Here α and β are used to control the query modification process. We choose $\alpha = 0.15$ and $\beta = 0.85$ to give more weight to the relevant images.

The rationale for developing the above formula is as follows. We use the average value of dominant coherent colors to update the corresponding component in Q_i^c in order to make the query closer to relevant image set. On the other hand, we perform the modification on the original query vector to ensure that the original intent of the user is preserved as much as possible. The new query is normalized before a new retrieval is performed.

c. *Conduct a new retrieval and repeat the RF process until the user is satisfied with the retrieval result*

In the new retrieval, we modify only (5) as follows:

$$Sim_{\text{Color}}(Q, D, i) = (1 + \gamma) * Sim_{\text{ExactCol}}(Q, D, i)$$
$$* (1 + Sim_{\text{PerCol}}(Q, D, i))$$
$$\text{where} \quad \gamma = \begin{cases} w_i & \text{if } i \in \boldsymbol{D^c} \\ 0 & \text{otherwise.} \end{cases} \tag{12}$$

The modification gives higher recognition to dominant coherent colors in the relevant image set.

5   Implementation and Testing

The system is designed and implemented on the UNIX platform using C^{++} programming language. The method is tested on a large image collection of over 12,000 images. The CCV information for all images are extracted beforehand and stored in a CCV database in order to support interactive image retrieval. The images cover a wide range of categories. For testing purpose only the categories as shown in table 1 are used. For each query, a sample image is selected and used for the initial query. It is used to search for relevant images in the entire collection. The set of relevant images for each query is also predetermined as indicated in the table.

Table 1. Test queries

Query Number	Query description	Number of Relevant images
Q1	Images of apes	41
Q2	Images of horses in outdoor environment	94
Q3	Images of red and/or white birds only	74
Q4	Images of natural scenery of land, sea and sky	109
Q5	Images of birds flying	51
Q6	Images of pink flowers	87
Q7	Images of some red flowers only or some yellow flowers only	97
Q8	Images of sunset with the full sun in view	74

The purpose of this research is to evaluate the use of pseudo objects in image retrieval with RF, and the role of color resolution in retrieval effectiveness, We will test the CCV method with RF over a range of color resolutions. By choosing an appropriate value for d (see Fig. 2), we can derive different color spaces with different color resolutions. In this study, we have chosen appropriate d to obtain color space at color resolution of low (176 colors), medium (316 colors) and high (407 colors) (see Fig. 4). At each color resolution, we conduct two relevance feedback iterations. The corresponding CCV methods used are referred to as CCV176, CCV316 and CCV407 respectively.

For evaluation purpose, we compare the performance of CCV methods with that of a normal color histogram method, which is similar to that described for CCV in Sect. 4, except that we perform the color computation and RF on the whole histogram. The normal color histogram method at the color resolution of 316 colors is chosen and referred to as NH316. The result of retrieval is presented on the screen in decreasing order of similarity from left to right and top to bottom (see Fig. 5). Figure 5 gives the top 18 images retrieved using queries, Q6 and Q8, with 2 RF iterations. The relevant images are marked with red dot at the bottom of the images. The figure shows a steady improvement in retrieval effectiveness with RF operations.

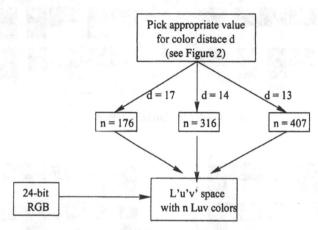

Fig. 4. Process of converting 24-bit RGB color into $L * u * v$ color

As it is not possible to display all retrieved images, only the top 150 images are presented to the user and used for relevance feedback. The normalized precision and normalized recall are used to evaluate the retrieval effectiveness of the method [14, 17].

The normalized precisions and recalls for all the queries and methods are summarized in Tables 2 and 3 respectively. For clarity, the comparative graphs of average normalized precision and recall of all methods are plotted in Fig. 6. From the figure, it can be seen that the use of RF has led to steady improvement in retrieval effectiveness for all method. This demonstrates that the RF technique developed is consistent and effective. Also, all CCV methods outperform the normal histogram method in all experiments. The improvement of CCV316 over NH316 after two RF iterations is 45.4% in normalized precision and 10.7% in normalized recall. The results demonstrate that CCV method is more effective than normal histogram method, and that the use of pseudo object relevance feedback technique is effective.

In terms of color resolution, it can be seen that CCV316 achieved the best retrieval effectiveness as compared with CCV176 and CCV407 after the second round RF. As described in Sect. 3.3, the color resolution of 176 colors is probably

Query Q6

Query Q8

(a) The sample queries

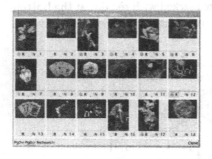

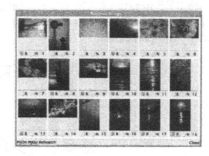

(b) Results for initial retrieval

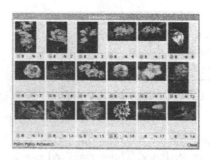

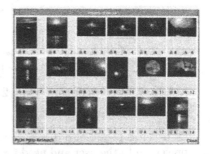

(c) Results for 1st RF

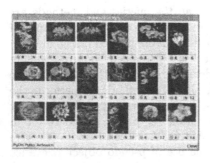

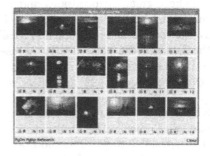

(d) Results for 2nd RF

Fig. 5. Retrieval results for queries Q6 and Q8 using CCV316

insufficient to capture all distinct colors. This has resulted in the inability of the system to distinguish many colors, thus leading to poor retrieval effectiveness.

Table 2. Normalized precision

Q (p)	NH316			CCV176			CCV316			CCV407		
	0	1	2	0	1	2	0	1	2	0	1	2
Q1	0.1911	0.1881	0.1881	0.1930	0.2504	0.2867	0.2057	0.3148	0.3492	0.2339	0.3460	0.3482
Q2	0.2194	0.2408	0.2421	0.2016	0.3377	0.3815	0.2650	0.4244	0.3871	0.2387	0.4117	0.4203
Q3	0.1869	0.2014	0.2336	0.1602	0.1782	0.1784	0.1895	0.2059	0.2144	0.1681	0.2016	0.2090
Q4	0.1541	0.1473	0.1527	0.1666	0.1869	0.1888	0.1472	0.1721	0.1738	0.1652	0.1838	0.1846
Q5	0.3966	0.3969	0.4033	0.2642	0.2693	0.2805	0.4125	0.4198	0.4193	0.2855	0.3349	0.3443
Q6	0.2218	0.2458	0.2551	0.2714	0.3377	0.3340	0.3252	0.4361	0.4486	0.2370	0.2648	0.2562
Q7	0.2736	0.2666	0.2669	0.2788	0.3096	0.2910	0.2738	0.2766	0.2990	0.2828	0.3332	0.3228
Q8	0.2163	0.2356	0.2473	0.4409	0.5260	0.5940	0.4214	0.5491	0.6000	0.4131	0.5361	0.5687
Avg	0.2325	0.2403	0.2486	0.2471	0.2995	0.3169	0.2800	0.3498	0.3614	0.2530	0.3265	0.3318
%+[a]	-	-	-	6.296	24.64	27.47	20.46	45.60	45.40	8.857	35.90	33.47

[a] % Improvement over NH316

Table 3. Normalized recall

Q (r)	NH316			CCV176			CCV316			CCV407		
	0	1	2	0	1	2	0	1	2	0	1	2
Q1	0.5295	0.5295	0.5295	0.5473	0.5718	0.5961	0.5593	0.6085	0.6210	0.5835	0.6327	0.6332
Q2	0.5498	0.5608	0.5608	0.5445	0.6096	0.6364	0.5822	0.6579	0.6365	0.5661	0.6577	0.6527
Q3	0.5322	0.5391	0.5531	0.5242	0.5244	0.5244	0.5378	0.5448	0.5516	0.5308	0.5448	0.5450
Q4	0.5081	0.5081	0.5127	0.5189	0.5237	0.5237	0.5095	0.5190	0.5190	0.5189	0.5236	0.5237
Q5	0.6546	0.6546	0.6547	0.5866	0.5867	0.5964	0.6846	0.6752	0.6752	0.6061	0.6357	0.6357
Q6	0.5549	0.5667	0.5725	0.5836	0.6128	0.6072	0.6127	0.6596	0.6654	0.5605	0.5721	0.5664
Q7	0.5796	0.5848	0.5848	0.5899	0.6005	0.5902	0.5797	0.5747	0.5954	0.5849	0.6111	0.6110
Q8	0.5594	0.5731	0.5800	0.7135	0.7354	0.7694	0.6937	0.7487	0.7694	0.6805	0.7286	0.7557
Avg	0.5585	0.5646	0.5685	0.5761	0.5956	0.6055	0.5949	0.6236	0.6292	0.5789	0.6133	0.6154
%+[a]	-	-	-	3.153	5.501	6.514	6.534	10.45	10.68	3.663	8.635	8.262

[a] % Improvement over NH316

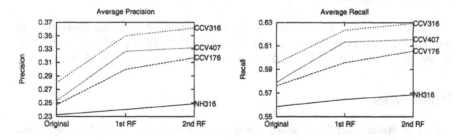

Fig. 6. Averaged normalized precision and recall over 8 Queries

On the other hand, at the high color resolution of 407 colors, it might lead to the reverse problem of too many shades of colors being captured by the system. As general users tend to perceive regions with multiple colors in closed proximity as smooth shades of colors, the system does not have such capability. Thus the use of too many colors may result in an inaccurate set of colors being chosen to represent the shades of a region, especially during RF.

The best compromise is achieved with using only 316 colors. The tests demonstrate that there is no advantage in using too high a color resolution in retrieval as advocated by others. Higher color resolution is effective only if there are good techniques to approximate colors in closed proximity by a smooth shade of colors as generally perceived by the users.

6 Conclusion

In this paper, we adopted the CCV method as the basic model for color-based image retrieval. By treating the color coherent component as a low-level representation of object within the images' contents, we develop a pseudo-object-based method for image retrieval with relevance feedback. The main idea of this method is to use pseudo object representation for feedback, thus permitting the system to retrieve more new relevant images with similar objects.

The other purpose of this research is to evaluate the role of color resolutions in image retrieval. To this end, we test the resulting CCV schemes on a range of color resolutions. The testing is carried out on an image collection containing over 12,000 images. The testing demonstrates the effectiveness of pseudo object retrieval scheme with relevance feedback, and that the most effective scheme is achieved with a moderate color resolution of 316 colors.

As images/videos are best represented with multiple attributes such as colors, texture, shapes and text annotations, we will investigate the use of RF technique for image retrieval using multiple attributes. We will also examine the use of higher level object-based model for image representation and retrieval.

Acknowledgements

The authors would like to acknowledge the support of National University of Singapore for the provision of a research grant #RP3960687 (Video Classification Based on Contents and Motions) under which this research is carried out.

References

[1] D. H. Ballard and C. M. Brown. *Computer Vision*. Prentice-Hall, Inc, Englewood Cliffs, NJ, 1982.

[2] F. Ennesser Bigün and G. Medioni. N-finding waldo and or focus of attention using local color information. *IEEE Trans. Pattern Anal.*, 17(8), 1995.

[3] T. S. Chua, Swee kiew Lim, and H. K. Pung. Content-based retrieval of segmented images. In *The second ACM International Multimedia Conference*, pages 211–218, 1994.

[4] Robert S. Gray. Content-based image retrieval: Color and edges. Technical report, Dartmouth University Department of Computer Science technical report 95-252, 1995.

[5] Roy Hall. *Illumination and Color in Computer Generated Imagery.* Springer-Verlag, New York, 1989.

[6] Christopher G. Healey and James T. Enns. A perceptual color segmentation algorithm. Technical report, Department of Computer Science, University of British Columbia, 1996.

[7] Jing Huang, S Ravi Kumar, and Mandar Mitra. Combing supervised learning with color correlograms for content-based image retrieval. In *The Fifth ACM International Multimedia Conference*, pages 325–334, 1997.

[8] W. Niblack, R. Barber, and W. Equitz. The qbic project: Querying images by content using color, texture, and shape. Technical report, IBM RJ 9203 (81511), February 1993.

[9] V. Ogle and M. Stonebraker. Chabot: Retrieval from a relational database of images. *IEEE Computer*, 28(9):40–48, 1995.

[10] Greg Pass and Ramin Zabih. Histogram refinement for content-based image retrieval. In *IEEE Workshop on Application of Computer Vision*, pages 96–102, 1996.

[11] Greg Pass, Ramin Zabih, and Justin Miller. Comparing images using color coherence vectors. In *The Fourth ACM International Multimedia Conference*, pages 65–73, 1996.

[12] A. Pentland, R. Picard, and S. Sclaroff. Photobook: Content-based manipulation of image databases. *Intl. Journal of Computer Vision*, 18(3):233–254, 1996.

[13] R. Price, T.S Chua, and S Al-Hawamdeh. Applying relevance feedback on a photo archival system. *Journal of Information Science*, 18:203–215, 1992.

[14] Rosanne J. Price. Applying relevance feedback to a photo archival system. Technical report, Department of Information Systems and Computer Science, National University of Singapore, 1991.

[15] Y. Rui, T.S. Huang, and S. Mehrotra. Content-based image retrieval with relevance feedback in mars. In *Proceedings of IEEE International Conference on Image*, 1997.

[16] Gerard Salton. *Automatic Text Processing.* Addison-Wesley Publishing Company, Cornell University, 1989.

[17] Gerard Salton and Buckley. *Introduction to Modern Information Retrieval.* McGraw-Hill Inc., 1983.

[18] M.J. Shen. Image retrieval using multiple attributes with relevance feedback. Technical report, Dept of Information Systems and Computer Science, National University of Singapore, 1996.

[19] J. R. Smith and S.-F. Chang. Tools and techniques for color image retrieval. *Symposium on Electronic Imaging: Science and Technology: Storage and Retrieval for Image and Video Databases IV*, 2670:426–437, 1996.

[20] John R. Smith. *Integrated Spatial and Feature Image Systems: Retrieval, Analysis and Compression.* PhD thesis, Columbia University, 1997.

[21] Michael J. Swain and Dana H. Ballard. Color indexing. *International Journal of Computer Vision*, 7(1):11–32, 1991.

[22] Xia Wan and C. C. Jay Kuo. Color distribution analysis and quantization for image retrieval. In *SPIE:Storage and Retrieval for Still Image and Video Databases'96(IV)*, 1996.

[23] Jia Wang, Wen jann Yang, and Raj Acharya. Color clustering techniques for color-content-based image retrieval from image databases. *Proceedings of the International Conference on Multimedia Computing and Systems*, pages 442–449, 1997.

[24] Hsu Wynne, T. S. Chua, and H. K. Pung. Integrated color-spatial approach to content-based image retrieval. In *The third ACM International Multimedia Conference*, pages 305–313, 1995.

[25] G. Wyszecki and W. S. Stiles. *Color science : concepts and methods, quantitative data and formulae. The Wiley series in pure and applied optics.* John Wiley and Sons, Inc., New York, 1982.

InvenTcl: A Fast Prototyping Environment for 3D Graphics and Multimedia Applications

Sidney Fels[1]* and Kenji Mase[2]

[1] University of British Columbia, Vancouver, BC, V6T 1Z4, Canada,
ssfels@ece.ubc.ca, http://www.ece.ubc.ca/~fels,
+1 (604) 822-5338, fax: +1 (604) 822-5949
[2] ATR MI&C Research Laboratories, Seika-cho, Soraku-gun, Kyoto, 619-02, Japan,
mase@mic.atr.co.jp, http://www.mic.atr.co.jp/~mase,
+81 774 95 1440, fax: +81 774 95 1408

Abstract. This paper describes *InvenTcl* which is an interpretive version of Open Inventor, a 3D graphics toolkit. To create InvenTcl, the Open Inventor toolkit is "wrapped" inside the interpreter Tcl/Tk and [incr Tcl]. To wrap InvenTcl the Open Inventor header files are parsed to create [incr Tcl] interpretive objects with the same names as objects in Open Inventor. Additionally, window event management, non-objects and object bindings are included and managed by InvenTcl. The advantages of InvenTcl include: script-able and direct manipulation of 3D objects in an Open Inventor scene, easy prototyping of 3D graphics and animation, low bandwidth communication of 3D scenes and animations (using scripts), and easy integration of 3D graphics with other media for fast prototyping of multimedia applications.

1 Introduction

There have been many 3D graphics packages and libraries available such as PEXlib [8], OpenGL [4], Open Inventor [9], GKS-3D [13], PHIGS [12]; however, they are usually precompiled toolkits, and thus not well-suited for fast prototyping new ideas, rapid experimentation with 3D scenes, or easy extension and integration with user-defined code. In addition, they often lack the combination of two useful modes of interaction; direct (mouse-click) mode and command line or script-based mode. Open Inventor falls into this category of 3D graphics packages.

InvenTcl extends Tcl/Tk/[incr Tcl] by providing interpretive access to Open Inventor. InvenTcl provides a window for creating, displaying, animating, and interacting with 3D objects. This is achieved by combining the Open Inventor C++ library [21, 9] with the Tcl/Tk library [18, 20] and the [incr Tcl] libraries [14]. By embedding the library of Open Inventor in Tcl/Tk, a high-level 3D scene interface is created with which objects can be manipulated on the fly via device

---

* Sidney Fels was a visiting researcher at ATR MI&C Research Laboratories when this research was done.

S. Nishio, F. Kishino (Eds.): AMCP'98, LNCS 1554, pp. 161–176, 1999.
© Springer-Verlag Berlin Heidelberg 1999

interaction as well as command line interaction. As a result, we accomplish both, 3D scene access to Tcl and interpretive access to Inventor. Prototype applications can be programmed entirely in InvenTcl; using Tk for the traditional 2D interface widgets (buttons, text widgets, pull down menus, etc.) and InvenTcl calls to Open Inventor for creating and displaying 3D scenes and implementing direct manipulation within the scene. Further, embedding other media toolkits into Tcl/Tk provides similar access flexibility and integration.

Open Inventor provides an objected oriented view of 3D graphics. The objected oriented view of 3D graphics programming is well suited to an interpreted access model. For this reason, Open Inventor was chosen over a function based model such as OpenGL for the basis of InvenTcl.

The simplicity of using InvenTcl makes it very suitable for both novice and expert users of 3D graphics toolkits and C++ programming. The potential uses for InvenTcl include: multimedia prototyping, VR prototyping, 3D graphics education, 3D GUI prototyping and scientific visualization. In section 3, we describe our use of InvenTcl to create a VR prototype of an architectural walkthrough of our laboratory. We also have used InvenTcl for teaching concepts of 3D graphics. Students without formal 3D graphics training have been able to learn to make relatively complex 3D scenes within a few hours, including, animating a sheet and exploring animating a walking robot. In another project, we were able to link a musical research project [15] with InvenTcl to display 3D representations of gesture space. Linking the two systems, developing suitable 3D graphics, and getting useful results took less than one hour (with no recompiling necessary). The flexibility of Tcl as a "glue" language makes InvenTcl a powerful toolkit for the scientist wanting to connect research code to InvenTcl for 3D visualization of their results.

Several people have created other 3D graphics extension to Tk [19, 11]. These applications are using the low-level OpenGL [17] library. Other researchers have created interpretive 3D toolkits such as Alice [5] and Obliq-3D[16]. These approaches are similar to InvenTcl, however, they use either their own 3D graphics toolkits or interpreter rather than a pre-existing one. InvenTcl is the combination of two popular systems, thus leveraging the work (and support) of them. Another approach is taken in SWIG [3] where one can wrap additional C-code around an existing C/C++ function to create a new Tcl-command. Applying this program to the Open Inventor library would result in a globalisation of all methods of each class in the object oriented library, which would be problematic. Our approach keeps the object oriented feature of Open Inventor, but also tailors some higher level commands instead of using the basic Inventor methods. Interestingly, if SWIG is used to generate Python commands the object oriented structure should be maintained since SWIG supports object oriented code for Python. An early version of InvenTcl can be found in [6]. Compared to using VRML, InvenTcl has the advantage of using Tcl/Tk as the master shell, making it easy to bind in new applications without recompilation.

Our immediate goal with InvenTcl is two-fold; one, provide an interpretive version of Open Inventor, and two, develop a complete 3D Tk canvas widget version which will behave in a similar fashion to the current 2D canvas widget. The interpretive version of Open Inventor is geared to the novice and expert Open Inventor programmer and has the following advantages over directly using C++ linked with Open Inventor's libraries:

- script-able and direct manipulation of objects in a scene
- easy prototyping of 3D graphics and animation;
- easy prototyping of GUIs for interacting with 3D scenes
- low bandwidth communication of 3D scenes and animations (using scripts).
- easy integration of 3D graphics with other software

A 3D canvas widget can be built on-top of InvenTcl for the Tk programmer to shield them from the details of Open Inventor and make access have the look and feel of Tk.

The power and flexibility of InvenTcl is a function of embedding a C++ based 3D graphics toolkit inside an extensible interpreter. InvenTcl provides a strong demonstration of the fruitfulness of this direction of research and development. This interpreter philosophy used for InvenTcl should be applied to other toolkits to allow both compiled and interpreted modes to be available for the developer. Further, the interpreter should be an extensible one so that users can integrate 3D graphics (and other toolkits, such as MET++[1]) into their own applications easily. Wrapped in this way provides highly interactive, easy to use, and extensible toolkits.

This paper describes how we converted Open Inventor and some of the key points we addressed. In all likelihood, for future conversion of graphics toolkits some of the same issues will be important, thus, in addition to describing a useful tool, this paper can also be used as a reference for wrapping other 3D graphics or multimedia toolkits.

The first section of this paper describes how the Open Inventor libraries were wrapped. A simple example is provided to show how InvenTcl is used. The second section of this paper discusses an example application created with InvenTcl. The application is an architectural walkthough (and walkthrough builder) connected to a person tracker system and database system. While the application may not be particularly interesting on its own, what is significant for this paper is that the application was written in 6 days by one programmer with only medium expertise in 3D graphics. The entire application was written in InvenTcl, thus, required no C++ code or compilation. Finally, the future directions of InvenTcl and some conclusions are made.

2   Making Open Inventor Interpretive

To create InvenTcl, the Open Inventor libraries need to be wrapped so that all the objects, their methods and public fields are accessible from the Tcl shell. Further, to make the interpreter version appear similar to the C++ version the object hierarchy and naming convention needs to be maintained. To integrate with the Tcl/Tk environment the 3D objects should be "bindable". That is, objects in the 3D scene can be bound to Tcl scripts which execute according to some user interaction such as a mouse button press. Finally, any non-object parameters accepted by the Open Inventor library functions need to have a representation in [incr Tcl] so that they can be passed as arguments. In summary, the five main requirements are:

1. Convert all Open Inventor classes to [incr Tcl] classes, including: methods, public fields, and static functions.
2. Integrate Open Inventor's event management loop into Tcl/Tk's event management loop.
3. Add support for 3D object binding so that interaction events, such as mouse and keyboard events, will call Tcl scripts.
4. Convert any non-object arguments, such as, enum types, arrays, and FILE pointers, accepted by Open Inventor object methods to [incr Tcl] objects to allow for run-time checking and value passing.

Most of the work to wrap Open Inventor is done automatically by parsing the header files of Open Inventor. The parser automatically converts all the classes and methods. Some parts are converted by hand, including some of the non-objects and event manager. A block diagram representing the necessary parts to create InvenTcl is shown in figure 1.

2.1   Converting Open Inventor Classes to [incr Tcl]

A program called Itcl++ [10] was used as the starting point for converting the Open Inventor class structure into [incr Tcl] classes. This program parses the header files of the class libraries and creates [incr Tcl] class structures. The [incr Tcl] class structure created provides the ability to instantiate objects and call methods of each object from the interpreter. The class inheritance structure is maintained in the [incr Tcl] class structure. Methods use run time checking of argument types. We have enhanced Itcl++ to also provide access to objects' public fields. When an object is created in [incr Tcl] the [incr Tcl] class constructor calls some C++ code which actually calls the *new* C++ operator to instantiate the Open Inventor object. The instantiation in C++ returns a pointer. This pointer is associated with the [incr Tcl] name assigned to the object created in [incr Tcl]. Thus, an intuitive way to think about the relationship between the interpreter and Open Inventor representations of objects is that InvenTcl provides string names as pointers to objects and Open Inventor uses integer pointers to objects. InvenTcl maintains the relationship between the two.

Overloaded Methods   Many methods in Open Inventor are overloaded. In InvenTcl we use run-time type checking of arguments to perform the necessary casting so that the appropriate method signature is used. Due to the interpretive nature of InvenTcl, type checking must be performed at run time to correctly implement the overloaded methods. In contrast, the original scheme used by Itcl++ used different method names to deal with method overloading. In this scheme, for each argument signature a unique method name was adopted. The method name was the name of the overloaded method with a version number appended to it. For example, if a method **setValue** is overloaded with an **int** or a **float** argument, two methods are created in the interpreter called **setValue1** taking an **int** argument and **setValue2** taking a **float** argument. We found this scheme very difficult to work with from a user's point of view. It was often the case that the user could not remember which version of the method to use and had to always refer to the online help to figure it out. By performing run-time checking we eliminated this problem so that, for example, only one method called **setValue** was created.

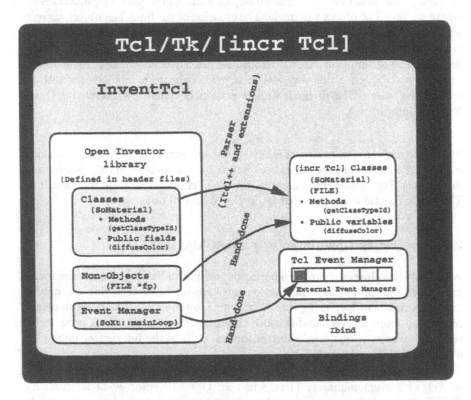

Fig. 1. Block diagram showing relationship of InvenTcl to Open Inventor and Tcl/Tk/[incr Tcl] and how InvenTcl is created from Open Inventor header files. InvenTcl is the set of interpreted classes connecting Open Inventor classes and [incr Tcl], the event manager embedding inside of Tcl/Tk and the Open Inventor object to Tcl binding mechanism.

Public Fields Some objects in Open Inventor provide public fields. We have modified Itcl++ to allow access to these public fields. For example, an *SoMaterial* node has several public fields, including *ambientColor* which is of type *SoMFColor*. When an *SoMaterial* node is created in C++ the *ambientColor* field is also created and can be accessed with a pointer reference. We provide access to this field from the interpreter by creating an object for the *ambientColor* field and associating it with a public variable in the [incr Tcl] object. Thus, the public variable in the *SoMaterial* [incr Tcl] object contains the name of the instantiated *ambientColor* [incr Tcl] object, and thus, can access the field. Access to the actual values maintained by the field are available through the methods provided.

The only modification to the parser that was needed was to make sure that for each public field of an instantiated object a call to the [incr Tcl] functions to create an [incr Tcl] object was made. The creation of the [incr Tcl] object assigns a name and associates the pointer with that name. Public fields play the role of members variables in an object. Open Inventor is consistent in only providing fields instead of simple member variables. As discussed in [21], one reason Open Inventor was designed this way was to provide consistent methods for setting and getting values. This design feature is extremely valuable when making Open Inventor interpretive.

Static Member Functions Finally, there are some public static functions associated with Open Inventor classes which have associated public procedures in the [incr Tcl] classes. For example, the *SoXt* class has a static function *init*. The static functions are provided for in the [incr Tcl] objects.

2.2   Integrating Open Inventor's Event Manager

A typical Open Inventor program performs some initialization, creates the scene graph and sets up any user interaction mechanisms before entering an infinite event management loop using the *SoXt::mainLoop* function. The event management loop loops forever handling any Open Inventor events which need to be serviced. This structure is modified in InvenTcl since the interpreter must also check for Tcl/Tk events too.

To solve this problem we embedded Open Inventor's event management inside of Tcl/Tk's event manager. Thus, when an Open Inventor event is found by Tcl/Tk in the main event queue, Open Inventor is called to manage the single event and return control to Tcl/Tk's event manager. In this way, events for Tcl/Tk and Open Inventor are handled appropriately.

From the user's point of view, nothing has changed. That is, in InvenTcl when they want to start Open Inventor's event handling they issue the *SoXt::mainLoop* command. But, rather than starting the infinite loop, InvenTcl installs the embedded version of the event handler and returns control to the interpreter. This is illustrated in the example below in section 2.5.

2.3   3D Object Binding

One of the powerful user interaction capabilities offered in Tcl/Tk is the ability to dynamically bind Tcl scripts to Tk widgets[1] which are triggered by events (which typically come from the user). We have implemented the same binding mechanism in InvenTcl for 3D objects. That is, in InvenTcl the user can bind Tcl scripts to 3D objects which trigger on user events such as mouse button clicks, mouse movements or keyboard presses.

The function created for performing the binding is called *Ibind*. The Ibind command expects four arguments: name of the object being bound, the name of the head of the scene graph which contains the object, the user event to watch for and finally the Tcl script to execute. For example, if we have an object called *cone* which is in the scene graph with *root* as the top we can do the following:

```
Ibind $cone $root <Button1> {puts "InvenTcl is great!"}
```

This command will make it so that when the user presses mouse button 1 the string "InvenTcl is great!" is printed.

To add this functionality, we add a generic call back node to the top of the scene graph that is triggered on any user generated event. This callback node retrieves the user event and the path where the event occurred and checks a list of all bindings to see if any of them should trigger given this information. The Ibind command also allows the current mouse position and the name of the [incr Tcl] object where the event occurred to be passed to the Tcl script. When the *Ibind* command is executed the appropriate binding is added to the list.

This binding mechanism is extremely useful. Prototyping different user interfaces can be achieved very quickly since developers can immediately try different types of user interactions. Remember, the Tcl scripts can be *any* Tcl scripts, thus, one can control: 3D graphics, all the Tk widgets, the Tcl interpreter or any "glued" in application such as video, audio or text toolkits. This mechanisms for implementing this event based binding provides the necessary infrastructure for developing the high-level synchronization mechanisms needed in a multimedia toolkit, such as found in [1].

2.4   Converting Non-object Structures

Open Inventor is mostly consistent in its objected oriented approach, however, there are a number of non-object structures which can be used in Open Inventor. Generally speaking, these structures are used as input parameters to some methods. The main structures are: enumerated types, arrays (1D, 2D, 3D and nD), FILE pointers, and function pointers. For each of these some appropriate [incr Tcl] object (or access method) was created[2].

---

[1] Tk widgets are graphical elements such as buttons, sliders, canvases, etc. which are used for creating GUIs

[2] Function pointers have not been implemented yet.

Enumerated Types For enumerated types, we provide a static procedure of the same name to access the value. For example, the class *SoNormalBinding* has an enumerated type called PER_FACE. In C++ this is accessed using

```
x = SoNormalBinding::PER_FACE;
```

where as, in InvenTcl there is a static procedure called PER_FACE which returns the enum value. The interpreter version of the above C++ code is:

```
set x [SoNormalBinding::PER_FACE]
```

Arrays For arrays, the original version of Itcl++ provides a 1D array object with values of basic types: **int**, **float**, **char**, and *SbBool* and methods to set and get their values. The types can be short or long, signed or unsigned where appropriate. We have extended this support for InvenTcl to include 2D, 3D and 4D arrays. Currently, we are still implementing support for arrays of complex object types as well as n-dimensional arrays.

File Pointers We have created a *FILE* [incr Tcl] class which associates an [incr Tcl] name with a file pointer. An object of this class can be used as the input argument for methods which require a file pointer argument. During initialization, we create FILE objects for the standard I/O file pointers: stdin, stdout and stderr. These *FILE* objects are referenced by the Tcl variables **stdin**, **stdout** and **stderr**.

The next section shows a simple example which demonstrates how InvenTcl works.

2.5   A Simple InvenTcl Example

This example shows some of the most basic features of InvenTcl. The example covers the main aspects necessary to create 3D graphics and setting up user interaction from Tcl to Open Inventor and from Open Inventor to Tcl using InvenTcl.

The example consists of creating an active 3D cone as shown in figure 2. If the user clicks button 1 on the mouse the message "running a Tcl script" appears. Also, a simple GUI is created with Tk for changing the colour of the cone. Remember, all this code is typed in directly (or sourced from a file) within the Tcl shell. The full source code is shown in figure 3.

Referring to numbered parts in figure 3, here is an explanation of some of the important features of the example. The numbered parts are explained below.

1. This piece of code calls the static procedure to initialize the scene data base.
2. This piece of code creates a root node to head the scene graph.
3. This piece of code creates a material node to control the material properties of the cone. This node is used below to create a GUI to allow the user to dynamically change the properties of the cone.
4. This code creates the cone and puts it in the scene.

5. This code starts the Open Inventor event manager. The embedding of Open
   Inventor's event manager inside of Tcl/Tk's event manager is discussed in
   section 2.2. The critical point here is that control is returned to the Tcl
   interpreter so that further manipulation of the scene graph can occur while
   the user sees the current scene graph.

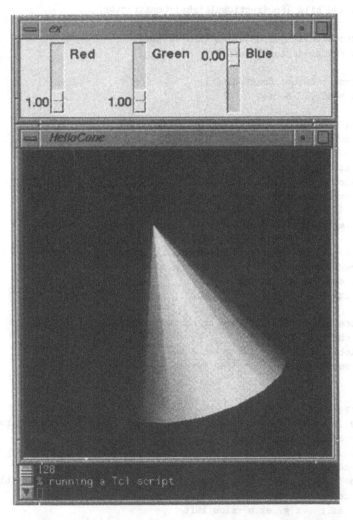

Fig. 2. 3D cone example with Tk sliders to manipulate colours. The top widget
shows the GUI for manipulating the colours of the cone. The middle window
shows the cone and the bottom part of the figure shows the Tcl shell. Notice,
that we are viewing this right after the user clicked mouse button 1 on the cone.

```
1. set window [SoXt::init "HelloCone" "HelloCone"]

2. set root [SoSeparator::Constructor]
   $root ref

   # make camera and light to see the cone
   set camera [SoOrthographicCamera::Constructor]
   $root addChild $camera
   $root addChild [SoDirectionalLight::Constructor]

3. set coneMaterial [SoMaterial::Constructor]
   $root addChild $coneMaterial

4. set cone [SoCone::Constructor]
   $root addChild $cone

   set ra [SoXtRenderArea::Constructor $window]
   $ra setTitle "HelloCone"
   $camera viewAll $root [$ra getViewportRegion] 1
   $ra setSceneGraph $root
   $ra show
   SoXt::show $window

5. SoXt::mainLoop

6. set coneColour [lindex [$coneMaterial configure -diffuseColor] 2]
   set red 0.5
   set green 0.5
   set blue 0.5

   # procedure to connect cone colour with sliders
   proc changeColor {val} {
      global coneColour red green blue
      $coneColour setValue $red $green $blue
   }

7. toplevel .ex
   scale .ex.r -from 0 -to 1 -resolution 0.01 -label R -variable red \
      -command changeColor
   scale .ex.g -from 0 -to 1 -resolution 0.01 -label G -variable green \
      -command changeColor
   scale .ex.b -from 0 -to 1 -resolution 0.01 -label B -variable blue \
      -command changeColor
   pack .ex.r .ex.g .ex.b -side left

8. lbind $root $cone <1> {puts "running a Tcl script"}
```

Fig. 3. Code to draw a cone with Tk sliders to adjust its colour. The binding is set so that if button 1 is pressed when the cursor is on the cone the message "running a Tcl script" is displayed.

6. This set of code defines a procedure which manipulates the material node in the scene graph. The material node's colours are changed by setting the *coneMaterial*'s public field *diffuseColor* values to the current values of the global variables: red, green and blue. The red, green and blue values are manipulated by the Tk sliders and the changeColor procedure is called whenever a change in one of the values occurs.

7. This section of the code creates the Tk sliders and connects changes to the sliders to the *changeColor* procedure and the values of the red, green and blue variables.

8. Finally, this piece of code shows the Ibind command being used. The command makes it so that when the user clicks mouse button 1 and the cursor is on the cone the Tcl script, {puts "running a Tcl script"} is run which prints the message. The result of the click can be seen at the very bottom of figure 2. This piece of code demonstrates how Open Inventor objects can be connected to arbitrary Tcl scripts. The scripts could execute more graphics commands, Tcl commands, Tk commands, operating system commands or any application code that has a Tcl front end.

This simple example demonstrates some of the main features of InvenTcl. However, what may not be obvious from the example is the high level of flexibility that InvenTcl provides. InvenTcl code can be modified *on-the-fly* to suit the developer's needs and can be integrated with other applications easily. The next section briefly describes an example application created completely using InvenTcl which takes advantage of many of the features of InvenTcl.

3   Example of Using InvenTcl for Prototype Walkthrough

To illustrate the power and ease of using InvenTcl we created an example application for demonstration during our laboratory's open house. The application we created was an architectural walkthrough and walkthrough builder. The walkthrough was also connected to an active badge system from Olivetti. The active badge system tracked visitors to our laboratory using infrared badges and badge sensors. The walkthrough would show the current location of all the visitors in the laboratory. The laboratory set up was not going to completed until the night before the actual open house so the map of the floor would not be available until just before the open house started. For this reason, the walkthrough had to have a walkthrough builder which would allow the developer to change the layout of the walls very quickly.

The entire system was programmed in six days by one programmer with only medium expertise in graphics programming and no experience with walkthroughs. The programmer did not use any C++ code or compilation[3]. The entire application is a set of Tcl files which are sourced to run the application.

---

[3] Of course, an application developer can use C++ in combination with InvenTcl for creating applications.

The system consisted of two main parts, a 2D map layout and editor and a 3D view of the floor. The 2D map editor was built using Tk widgets and is shown in figure 4. The 3D view used the InvenTcl objects and is shown in figure 5. When a line was drawn on the 2D map a corresponding wall was displayed in the 3D window. The wall in the 3D window was active so that the developer could click on it to popup a Tk based wall editor. This editor allowed the developer to specify properties of the wall including colour and texture. The 2D and 3D views of the floor were always kept in sync by using the binding mechanisms in Tk and InvenTcl.

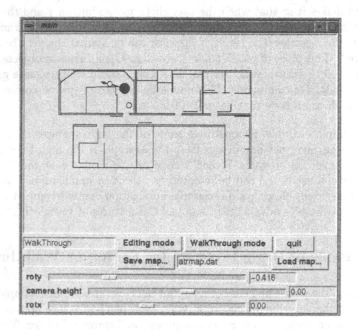

Fig. 4. 2D GUI for laying out, editing, saving and reading maps. There are also controls on the GUI for manipulating the viewpoint of the scene. The large red dot is the current viewing position and the arrow indicates the direction of view. The small yellow dots indicate visitors in the lab.

The 2D map editor allowed the system to be reconfigured easily to suit the dynamic environment of the preparations for our open house. On the night before open house, when the final layout of the floor of the laboratory was finalized, the final map was drawn in, photographs taken of the walls (for texture mapping) and the complete walkthrough was ready for the next morning.

The other aspect of the system that was required was to include a representation of all the visitors in the lab detected by the active badge system. The system used both a 2D and 3D representation of each visitor. The 3D representation was

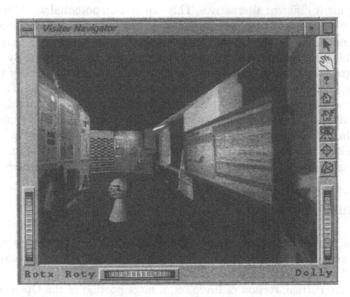

Fig. 5. 3D Open Inventor window displaying the lab floor.

a doll whose position could be dictated by the active badge system. One doll is shown in figure 5. The position of the visitor was determined by the active badge system. Position information was exchanged as Tcl scripts using a client/server relationship with a Tcl based position server. By making InvenTcl a Tcl server any application can connect to it and send InvenTcl scripts to control graphics. In this way, it is easy to connect Tcl extended application running on a different machine to InvenTcl to take advantage of the 3D graphics capabilities.

A late breaking requirement was that each visitor was going to be assigned a cartoon character representation as a representation of their own personal tour guide agent. This was done as part of the C-Map project [7]. Each visitor's 3D doll was required to display the character on its face. The assignment of the character was done at a registration desk. This information was communicated using the same Tcl client/server mechanism used with the active badge system; that is, once the assignment is made, a Tcl command is sent from the registration server (written in Java) to the Tcl client which then updates the 3D graphics. Because of the flexibility and ease of modifying InvenTcl code this change was integrated in less than 2 hours.

One aspect of InvenTcl which was very useful for this application is the ease which different user interaction techniques can be tested. The developer of the system was not too familiar with walkthrough environments. Thus, he was unsure of how to allow a user to move through the 3D floor space. He was able to test out many different techniques of moving the user's viewpoint around using 1D sliders and 2D widgets. He was able to experiment "on-the-fly" with new ideas as he discovered them. This fast testing cycle led to him exploring many different alternatives. This exploration potential would be nearly impossible using a standard program/compile/debug cycle of development.

The conclusion reached from this example application is that InvenTcl is a powerful paradigm for creating and manipulating 3D graphics. By making the 3D graphics library interpretive development time is drastically reduced in creating complex worlds. The binding mechanisms provided by Tk for 2D widgets and the one provided by InvenTcl for 3D objects make creating GUIs simple and quick. These properties allow many different GUIs to be explored for determining which is best. This feature becomes even more important as different media toolkits are embedded since the interaction paradigms are less understood and GUI experimentation becomes critical (in contrast to the approach in [2]).

4   Future Work

There are two main areas for the future of InvenTcl. The first deals with improvements to the current system. The second area deals with expanding the concept of InvenTcl to include other communities.

With the current version of InvenTcl, a large portion of the Open Inventor libraries is accessible from the interpreter. The main areas of improvement are:

- implementing function and procedure callbacks,
- integrating Open Inventor's draggers and manipulators,
- using other interpreters such as Python[4],
- improving speed, and
- integrating Tk widgets and the Open Inventor window so they appear in one window.

To improve speed we have implemented an interactive mode switch which allows users to turn the interpreter on and off on demand to allow the Open Inventor event manager to run at full speed. However, when the switch is on the Tcl event manager is off, thus, preventing access to the interpreter.

There are other 3D toolkits available. It is hoped that this paper motivates 3D graphics developers to see the merits of having an interpretive version of other toolkits and create them. Further, the interpreter used should be extensible and matched to the structure in the toolkit, i.e., if the toolkit is object oriented the interpreter should support object oriented code. Such efforts[5] have been pursued, in particular, World Tool Kit (WTK) by Sense8 has been "wrapped" in Python. In extending InvenTcl, we plan to wrap Cosmo3D[6].

The approach of wrapping toolkits is general in nature. For example, by wrapping a toolkit such as MET++, a scriptable interface is achieved. Further, as Tk provides a GUI builder, various multimedia interaction techniques can be built and experimented with, including a visual programming environment [2].

---

[4] Python web address: http://www.python.org/

[5] PyWTK: http://www.mic.atr.co.jp/~gulliver/PyWTK/www/

[6] Cosmo3D: http://www.sgi.com/Products/cosmo/cosmo3D/

The versatility of InvenTcl demonstrates the advantages of embedding toolkits inside of interpreters. Multimedia researchers and developers are in an excellent position to take advantage of the merits of extensible interpreters.

5   Conclusions

We had two main goals when creating InvenTcl; one, provide an interpretive version of Open Inventor, and two, develop a complete 3D Tk canvas widget version which will behave in a similar fashion to the current 2D canvas widget. We have achieved the first goal. This paper has discussed the implementation of the interpretive version of Open Inventor. Most of the objects and methods available in the Open Inventor library have been wrapped in Tcl/[incr Tcl]. Additionally, mechanisms have been created to allow 2D GUIs to directly control the 3D environment and for 3D user interaction in the Open Inventor window to call back to the Tcl interpreter.

InvenTcl leverages all the advantages of interpretive languages and brings them to bear on the Open Inventor toolkit. Thus, using InvenTcl it is possible to have:

- script-able and direct manipulation of objects in a scene
- easy prototyping of 3D graphics and animation;
- easy prototyping of GUIs for interacting with 3D scenes
- low bandwidth communication of 3D scenes and animations (using scripts) and
- easy integration of 3D graphics with other applications and toolkits.

The current version of InvenTcl[7] is available.

6   Acknowledgements

We are grateful for the contributions made by Kazuhiro Kawagoe, Tameyuki Etani, Silvio Esser, Armin Bruderlin, and Ryohei Nakatsu for their assistance with creating InvenTcl. We also thank other members of the C-MAP team, Yasuyuki Sumi, Nicolas Simonet, and Kaoru Kobayashi.

References

[1] Phillipp Ackermann. *Developing Object-Oriented Multimedia Software - Based on the MET++ Application Framework*. dpunkt Verlag, 1996.

[2] Phillipp Ackermann, Dominik Eichelberg, and Bernhard Wagner. Visual programming in an object-oriented framework. In *Proceedings of Swiss Computer Science Conference*, Zurich, Switzerland, Oct. 1996.

[3] D. M. Beazley. Swig: An easy to use tool for integrating scripting languages with C and C++. In *Proceedings of Tcl/Tk Workshop, Monterey, CA*, July 6-10, 1996.

---

[7] InvenTcl: **http://www.mic.atr.co.jp/organization/dept2/inventcl/**

[4] OpenGL Architecture Review Board. *OpenGL Reference Manual*. Addison-Wesley, 1992.

[5] Randy Pausch et al. Alice: A Rapid Prototyping System for 3D Graphics. *IEEE CG&A*, 15(3):8–11, May 1995.

[6] S. S. Fels, A. Bruderlin, S. Esser, and K. Mase. Inventcl: Making open inventor interpretive with tcl/[incr tcl]. In *Visual Proceedings of SIGGRAPH'97*, page p. 91, Aug 1997.

[7] S. S. Fels, Y. Sumi, T. Etani, N. Simonet, K. Kobayashi, and K. Mase. Progress of c-map: a context-aware mobile assistant. In *Proceedings of the AAAI Spring Symposium on Intelligent Environments*, pages pp. 60–67, Mar 1998.

[8] T. Gaskins. *PEXlib Programming Manual*. O'Reilly & Associates, Inc., 1992.

[9] Open Inventor Architecture Group. *The Inventor Reference Manual*. Addison-Wesley, New York, 1994.

[10] W. Heidrich and P. Slusallek. Automatic generation of Tcl bindings for C and C++ libraries. In *Proc. of the Tcl/Tk Workshop*, July 1995.

[11] I. Hsu. Tksm a mesa/opengl 3d modeling widget extension for tcl 7.[45]/tk. In *http://www.isr.umd.edu/%7Eihsu/tksm.html*.

[12] American National Standards Institute. *American National Standard for Information Processing Systems – Programmer's Hierarchical Interactive Graphical System (PHIGS) Functional Description, Archive File Format, Clear-Text Encoding of Archive File, X3.144-1988*. ANSI, New York, NY, 1988.

[13] American National Standards Institute. *International Standard Information Processing Systems – Computer Graphics – Graphical Kernel System for Three Dimensions (GKS-3D) Functional Description, ISO 8805:1988(E)*. ANSI, New York, NY, 1988.

[14] M. McLennan. [incr Tcl]: Object-oriented programming in Tcl. In *Proc. 1st Tcl/Tk Workshop*, University of Berkeley, CA, USA, 1993.

[15] A. Mulder, S. S. Fels, and K. Mase. Empty-handed gesture analysis in Max/FTS. In *Proceedings of Kansei - The Technology of Emotion, AIMI International Workshop*, pages pp. 87–91, Oct 1997.

[16] Marc A. Najork and Marc Brown. Obliq-3D: A high-level, fast-turnaround 3D animation system. *IEEE Trans. on Visualization and Computer Graphics*, pages 175–193, June 1995.

[17] J. Neider, T. Davis, and M. Woo. *OpenGL Programming Guide*. Addison-Wesley, New York, 1993.

[18] J. K. Ousterhout. *Tcl and the Tk Toolkit*. Addison-Wesley, New York, 1994.

[19] B. Paul. Togl: Togl allows opengl or mesa to render graphics into a special tk canvas. In *http://www.ssec.wisc.edu/%7Ebrianp/Togl.html*.

[20] B. B. Welsh. *Practical Progamming in Tcl and Tk*. Prentice Hall, New Jersey, 1995.

[21] J. Wernecke. *The Inventor Mentor*. Addison-Wesley, New York, 1994.

The NAVL Distributed Virtual Reality System*

Martine Wedlake, Kin F. Li, and Fayez El Guibaly

Department of Electrical and Computer Engineering,
University of Victoria
Box 3055
Victoria, BC CANADA V8W 3P6
{martine, kinli, fayez}@engr.uvic.ca

Abstract. Current distributed virtual reality systems suffer from many inherent difficulties: inadequate network architecture, poor object distribution and coherency, poor system resource management, and poor overall performance. This paper outlines the NAVL DVR system, designed to overcome these problems with a distributed client/server network architecture and master/slave autonomous object distribution mechanism using ForceLet simulation.

1 Introduction

Distributed virtual reality (DVR) is a form of distributed multimedia (DM). Unlike many DM applications, which have minimal interaction with the participants, DVR systems have a very high degree of interaction. While there are some DM applications that exhibit degrees of interaction, they are often limited by poor quality voice and slow refresh video (e.g., video conferencing).

DVR participants need quick access to virtual objects[1] containing multimedia content (e.g., computer graphics, sound and text). A major challenge for DVR system developers is to find ways to support highly interactive environments with limited system resources.

Distributed VR systems can be viewed as a collection of five service layers, listed from top to bottom:

1. The **Application Layer** represents the DVR applications such as virtual conferencing.
2. The **Language/Application Programming Interface (API) Layer** supports DVR Applications through a structured interface between applications and the PS Layer.
3. The **Programming System (PS) Layer** provides the infrastructure for managing DVR environments.

---

* This work was supported by the Natural Sciences and Engineering Research Council of Canada.

[1] In this paper, the term virtual object refers to the object in the virtual world. This removes confusion arising due to object oriented programming.

S. Nishio, F. Kishino (Eds.): AMCP'98, LNCS 1554, pp. 177–193, 1999.
© Springer-Verlag Berlin Heidelberg 1999

4. The **Operating System (OS) Layer** contains machine services such as CPU scheduling and memory allocation.
5. The **Protocol Layer** defines the structure and mechanisms of communication between remote hosts in a DVR system.

1.1 Case Studies

This section outlines a summary of some of the more prominent DVR systems found in the literature and how they fit into the DVR layer structure; the DVR layers involved are listed parenthetically (notice how many DVR systems are focused on a small subset of the DVR layers).

VRML (2, 5)[11] was created to provide a standard framework for developing VR languages over the Internet. It provides a language element to describe scenes, and a protocol for sending the scene descriptions over the Internet (similar to HTTP).

DIVE (2, 3)[1] uses a fully distributed shared database model. Applications change node-local copies of virtual objects, which are replicated by the DIVE infrastructure to all other participants. Mutual exclusion operators are provided to serialize simultaneous updates.

Alpha World (1)[2] is an application available over the Internet that allows multiple people to socialize in a three-dimensional setting - essentially a DVR chat program. The architecture is client/server with special gateways that one can travel through to switch servers.

MR Toolkit with Peers Package (2, 3)[9] is a collection of libraries that can be used to build small DVR applications (less than five participants) using a fully connected network model.

DIS (5)[4] describes a protocol for managing warfare simulators over a wide range of communication media. To reduce network load, DIS uses a dead reckoning protocol. Each simulator maintains two states: the internal (e.g., real) state, and the approximate (e.g., dead reckoning) state. Only the dead reckoning state is transmitted to receiving applications. When the dead reckoning state diverges from the internal state, the dead reckoning state is updated to match the internal state. Messages are only transmitted to receiving applications when the dead reckoning state changes.

1.2 Shortcomings with Existing DVR Systems

Many of the existing DVR systems exhibit shortcomings that limit their potential in practical DVR applications. The following four areas needing improvement have been identified:

1. **Network Efficiency**: This involves managing both bandwidth and latency of the network. These factors can be controlled by the design of the network architecture and efficient use of the available resources.
2. **Object Distribution and Coherency Models**: Proper management of virtual objects in a DVR system is critical to ensure good scaling performance as the system grows.

3. **Inadequate System Resource Management**: As with OS developers, DVR developers need to control many diverse kinds of resources (e.g., network, computation, graphics, and memory). Indeed, many DVR developers leverage OS services directly to manage system resources. However, while OS developers are typically interested in providing fairness[10], DVR system developers are concerned with providing immersion and interaction to the participant.
4. **Overall Performance**: The authors define a high performance DVR system as high synchronization fidelity, low environment lag, low display lag, large number of participants, and high world complexity.

1.3   Description and Motivation of NAVL

Many current DVR systems tightly couple the application layer through to the programming system layer. Often to the extent that the entire DVR system must be re-compiled in order to extend the existing application. The Newtonian Architecture for Virtual Landscapes (NAVL), takes a different approach. Instead of bonding the application to the operating environment, NAVL tries to separate the two. Application Programming Interfaces (APIs) provide a general purpose environment to support DVR applications. The NAVL architecture incorporates the following four elements:

1. **Autonomous virtual objects**: allows an encapsulation of the virtual object that may be moved as needed to balance the CPU and networking loads of the DVR system.
2. **Distributed client/server network architecture**: reduces bottlenecks within the network by spreading the bandwidth throughout the network, while simultaneously reducing the load to the network as compared with a peer-to-peer architecture.
3. **Master/slave virtual object distribution and coherency model**: splits the virtual object into a single master object and several slave objects. The master object is responsible for the state of the object while the slave objects represent the local manifestation of the object. The slave objects execute high-order commands called ForceLets that describe a path through space for the object to follow.
4. **Local simulation of virtual objects by slave objects**: reduces network bandwidth requirements by executing much of the virtual objects actions local to the user rather than at a remote node.

Unlike traditional DVR systems, NAVL supports both applications and autonomous virtual objects. This has the effect of splitting the DVR layers into dual stacks (see Fig. 1) for the applications and the virtual objects.

An application in the NAVL model requires both the application code and one or more virtual objects (represented as master objects). Applications communicate their requests to the programming system via an API, while the autonomous virtual object communicates via an object programming interface (OPI). In general, the virtual objects may be regarded as "toolkits" for application developers.

(1)	Application			Master Object
(2)	Rendering API	NAVL API		NAVL OPI
(3)	Renderer	Programming System		
(4)	Window System	Operating System		
(5)	Window System Protocols	Protocols		

Fig. 1. NAVL DVR layers. Notice how the DVR layers are split to accommodate applications and master objects

2   NAVL DVR Part I: Design Architecture

NAVL was designed as an answer to the observed problems in DVR systems. Many of the problems are handled by NAVL by the use of the master/slave coherency model. Figure 2 shows the high-level design architecture with the virtual object represented as a master object and two slave objects.

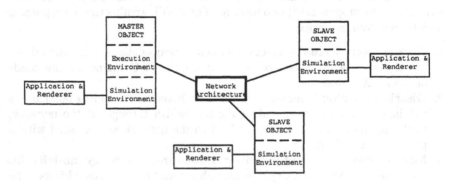

Fig. 2. Design architecture elements. The execution environment and simulation environment represent the set of services provided by the NAVL programming system

The *master objects* are responsible for maintaining the virtual object's state, and *slave objects* replicate the virtual object throughout all the NAVL hosts[2]. Unlike client/server paradigms, applications make requests of the objects themselves. When the master object resides on the same host, it is invoked directly. When the master object resides on a remote host, the application communicates to the local slave object. The slave will forward the request to the master object automatically.

---

[2] NAVL hosts are machine nodes that participate in the NAVL DVR system.

The NAVL DVR system supports distributed interactive multimedia applications (e.g., virtual reality), which typically have high demands for 3D computer graphics. NAVL encapsulates these services into a rendering engine library linked directly in the application.

The rendering engine is linked in the application (instead of being provided as a standard NAVL service) because most applications are likely to use more than a simple 3D viewspace. The use of window system primitives (e.g., menus, buttons, and dialog boxes) are left up to the application developer. Placing the renderer in the application allows the greatest degree of flexibility for the application designer.

2.1 Network Architecture

Network architectures describe how information will move throughout the DVR system, three example architectures are shown in Fig. 3.

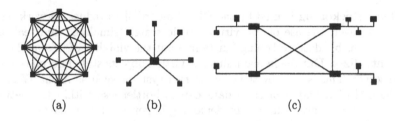

(a)   (b)   (c)

Fig. 3. Network architectures: (a) peer-to-peer, (b) client/server and (c) distributed client/server

Peer-to-Peer network architectures have interconnections between every host, distributing the network load evenly throughout the network, at the expense of more message traffic.

Client/Server network architectures have a special node designated the server that services requests from a collection of client nodes. The client/server concentrates the network into the server node, thereby reducing the number of messages sent over the network as a whole, but increases the load at the central point. The server may opt to buffer messages for a specified duration in order to reduce network bandwidth (at the expense of latency), this is deemed the buffered client/server network architecture.

Distributed Client/Server is a combination of the client/server and peer-to-peer architectures. The network is divided into regions, several local area networks (LANs) and the single wide area network (WAN). The LANs are broadcast driven, while the WANs are peer-to-peer. The larger boxes in Fig. 3(c) are special message servers called message managers, very similar to those in WAVES and HIDRA[7, 8]. Message managers serve four functions:

1. Act as a gateway between the peer-to-peer WAN and the broadcast LAN.
2. Perform directory services (e.g., host and object lookup).
3. Filter messages coming into and leaving the LAN to reduce bandwidth over the WAN.
4. Regulate the quality of service (QoS) services of the WAN, if any.

Analysis of network capacity has shown that the distributed client/server network architecture fairs very well against the peer-to-peer, and simple client server network architectures[13]. For this reason, NAVL uses the distributed client/server network architecture.

2.2   Object Distribution and Coherency Architectures

Distributed virtual reality applications need assurances that actions taken by different participants can and will have an impact on the other participants. To meet this need, an efficient distribution and coherency model is necessary. The literature has shown many examples of distribution and coherency models, for example:

Dead Reckoning is used by the DIS standard[4] to reduce network bandwidth. However, the use of two virtual world states (simulated and real) and lack of extensible dead reckoning functions limit the viability of this method.

Centralized Database maintains the virtual objects state information in a single server. This is used by the client/server systems such as Alpha World[2]. The lack of distribution of state data causes bottlenecks within the network fabric, which are only addressed by deploying larger and faster machines.

Distributed Database Locking enforces consistency by restricting access to the database itself. The DIVE[1] DVR system locks the database with mutual exclusions (MUTEX). Locking provides a good mechanism for ensuring that transactions are carried out sequentially, however using a locking mechanism for distributed applications involves a lot of network traffic resulting in long time lags. Typical distributed locking schemes[3] require messages to be sent repeatedly to all hosts; making it nearly impossible at current network latencies to have real-time interaction.

Autonomous Objects with Master/Slave Replication distributes a virtual object throughout the network, designating one copy as the master object and the rest as the slave objects. The master object decides what actions to take and is responsible for replicating that task to all the slave objects, using a high-level simulation protocol. The NAVL DVR system[13, 14] and HIDRA DVR system[8] use autonomous objects.

2.3   Autonomous Objects with Master/Slave Replication

Master objects are responsible for executing special behavior routines upon receiving an event, resulting in either changing the virtual objects state or issuing an event of its own. Changes to the virtual objects state affects the visual manifestation of the object (e.g., color or shape) or location. Figure 4 shows the block diagram for NAVL virtual objects. The master and slave components are shaded for clarity.

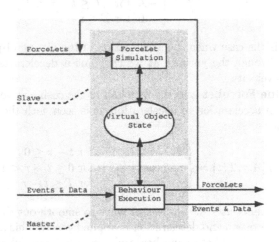

Fig. 4. Block diagram of virtual object

A special simulation environment is used when moving the virtual object through space. The simulation environment must be sufficiently complex to allow for a large range of actions, must be predictive, and must use little network bandwidth resources.

The current model for the simulation environment is based on a modification of Newtonian mechanics. ForceLets are network messages sent from the master object to the slave objects to define the path virtual objects take through space and time. ForceLets contains the following:

- Recipient and Sender Object Identifier
- Type of ForceLet
- Parameters for the ForceLet

The ForceLet notation is $F_D(A, \tau, T; t)$ where D specifies the domain for the definition (f for force, v for velocity, or d for displacement), and the parameters are: A for amplitude, τ for the start time, and T for the duration. The point in time is specified by t and is synchronized to all hosts using the network time protocol.

The authors have initially defined three ForceLets: the $\frac{1}{2}$Cosine, Sine, and Step. As research progresses, it is expected that additional ForceLet types will become identified and added to the family of ForceLets. Naturally, the list of ForceLet parameters may increase during this process as well.

The **Step ForceLet** is defined as follows:

$$F_f(A, \tau, T; t) = \begin{cases} 0 & \text{for } t - \tau \leq 0 \\ A & \text{for } 0 \leq t - \tau < \frac{1}{2}T \\ -A & \text{for } \frac{1}{2}T \leq t - \tau < T \\ 0 & \text{for } T \leq t - \tau \end{cases} \tag{1}$$

This models the case when the acceleration of the virtual object is constant until halfway through the journey, at which point it decelerates at a constant rate until zero velocity.

The $\frac{1}{2}$**Cosine ForceLet** uses the first half of the cosine wave to smooth out the transition of acceleration to deceleration as is seen with the step ForceLet. It is modeled as:

$$F_f(A, \tau, T; t) = \begin{cases} 0 & \text{for } t - \tau \leq 0 \\ mA \cos\left(\pi \frac{t-\tau}{T}\right) & \text{for } 0 \leq t - \tau < T \\ 0 & \text{for } T \leq t - \tau \end{cases} \tag{2}$$

The Sine ForceLet uses the entire sine wave to smooth out the start and end points. The acceleration and deceleration is applied gently. This allows a smooth transition from one ForceLet to another, and is especially useful when combining ForceLets. It is modeled as:

$$F_f(A, \tau, T; t) = \begin{cases} 0 & \text{for } t - \tau \leq 0 \\ mA \sin\left(2\pi \frac{t-\tau}{T}\right) & \text{for } 0 \leq t - \tau < T \\ 0 & \text{for } T \leq t - \tau \end{cases} \tag{3}$$

The Step ForceLet uses the least computation, while the Sine ForceLet uses the most[14]. Conversely, the step ForceLet is the least smooth when combining ForceLets, while the sine ForceLet is the smoothest. The $\frac{1}{2}$Cosine ForceLet fits between the two in complexity and smoothness.

2.4   The Application and Rendering Components

The application provides a semantic structure to the virtual objects. It is responsible for creating, manipulating, displaying and destroying virtual objects. Rendering is the process of displaying the virtual object on the video screen or through VR helmet. Since the display is tightly integrated with the application, it is shown together in Fig. 2.

3   NAVL DVR Part II: Implementation Architecture

Figure 1 shows three distinct parts of the NAVL system: (1) the application code *(top left of the figure)*, (2) the master object *(top right of the figure)*, and (3) the NAVL programming environment *(middle of the figure - the programming system and renderer layers)*.

These parts of the DVR model are implemented with distinct subsystems as seen in Fig. 5(a). This figure shows the implementation architecture for a single node within the NAVL system (Fig. 5(b) is discussed in the next section). The shaded regions show the separation of the process groups. The upper box represents the NAVL system, and the lower box represents the application process.

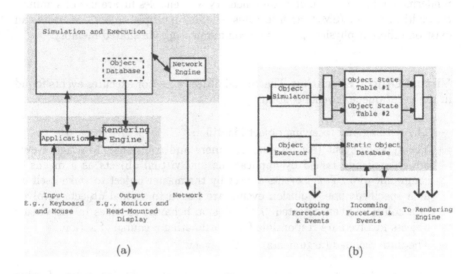

(a)   (b)

Fig. 5. Block diagram of (a) implementation architecture, and (b) Simulation and Execution Architecture. The shaded regions in (a) represent the groupings of user processes and NAVL processes, and in (b) represent shared memory

3.1 Simulation and Execution Engine

The largest and most critical element of the implementation architecture is the Simulation and Execution Engine (SEA). Its purpose is to replicate virtual objects according to the master/slave paradigm. The block diagram of the SEA is shown in Fig. 5(b). This figure is the dual to the virtual object diagram (Fig. 4), which shows object execution from the system perspective instead of the virtual object perspective.

The object simulator continually applies ForceLets to slave objects to obtain the object state table. The object state table represents the point-in-time state of all the slave objects in the NAVL host. There are two object state tables to store the previous frame while constructing the next frame. The rendering engine (e.g., application) will always retrieve data from the previous frame, not the frame being constructed. This preserves integrity during the rendering process and allows them to execute in parallel without blocking on a MUTEX.

The object executor executes behavior execution routines contained in master objects upon receipt of events (not shown). The result of the behavior execution can be a series of one or more issued ForceLets, or other state changes (e.g., shape, color or orientation).

The shaded area of Fig. 5(b) show the data structures stored in shared memory for simultaneous access from the simulation and execution engine and the rendering engine. The use of shared memory also enables future use of symmetric multi-processor (SMP) architectures whereby multiple object executors can exist on different physical processors communicating via shared memory.

Virtual Object Execution The virtual object executor executes events found in the event queue. Events contain:

- The sending and receiving object identifiers.
- The type of event (e.g., messages, timers and collisions). Message events are created and issued by applications and virtual objects as a means of communication. Timer events are set by the master object to wake itself up at a specified time. Collision events are issued to virtual objects involved in a collision when detected. The collision behavior routines in the virtual objects involved are responsible for coordinating a course of action.
- The data payload (arguments) for the event.

The virtual object executor runs the master object's behavior routine. During execution time, the behavior routine can issue ForceLets, change the virtual objects visual manifestation, or send events to itself or other virtual objects. However, the behavior execution routine must return within a specified time. This is very much like a soft real-time system. Long duration tasks can be scheduled by the user of a repeatable timer that periodically wakes up the master object. The soft real-time scheduling paradigm builds the foundation for quality of service (QoS) mechanism that can be embedded into the system at a later time.

Virtual Object Simulation The object simulator calculates the position of virtual objects based on the list of ForceLets, and the global simulation time. The simulator runs through the list of slave objects at the base simulation rate (usually about 10–20 Hz), placing the result into the active object state table.

The base simulation rate establishes the environment lag for the NAVL system, and should therefore be kept constant whenever possible. The base simulation rate is currently hard coded into the NAVL DVR prototype, however there is no reason it can not be made to adjust to the current hardware available on the host. The simulation environment decouples the NAVL hosts, allowing them to have different base simulation rates.

SEA Support Infrastructure The SEA supports the application and master objects through interfaces called the application programming interface (API) and the object programming interface (OPI).

The API uses shared memory to communicate between the NAVL system and the application. This is dramatically faster than other mechanisms such as named pipes or sockets, however it does require the application to execute on the same node as a NAVL host. Since the NAVL model extends deeply into the DVR layers, it is expected that the application will live on the same node.

Master object behavior routines are loaded into NAVL using shared library load operations, therefore the OPI is invoked through direct function calls. This allows for a more efficient interface than using the available shared memory, however there is an increased risk to the NAVL system's integrity. Security has not been addressed at this time.

3.2   Rendering Engine

The rendering engine performs two specific tasks: (1) draws scenes representing the visual manifestation of the virtual objects, and (2) detects collisions between virtual objects.

Objects are usually represented using either surface or solid modeling[5]. NAVL uses solid modeling of known shapes (e.g., cube, sphere, cone, and torus) for simplicity. While not as visually pleasing as can be obtained via surface modeling, it can be easier to use for defining shapes. The OpenGL graphics library has been chosen to provide the low level graphics rendering. This reduces the burden of the application developer from writing the rendering code itself.

To provide complex shapes, the NAVL prototype allows for shapes to be arranged in a hierarchy. The children objects live within the coordinate system of the parent. This arrangement allows for quick and easy translations and rotations of large components of the composite virtual object. The origin shape does not have any visual components, and is used to provide a reference object for coordinate transformations.

Because the rendering engine models the shape of virtual objects, it is the prime location for detecting collisions. Detecting collisions between virtual objects in a virtual environment requires an efficient algorithm for recognizing collisions between objects of arbitrary shapes and also an efficient algorithm for searching through the list of virtual objects looking for possible collisions.

The first part, recognizing collisions, depends on the shape model chosen. For example, solid modeling is typically easy to detect collisions for since they are often based on conics or other easily managed objects. Surface modeling, however causes difficulties due to the large number of polygons and arbitrary shape.

The second part, organizing the search space, is often handled in one of the following ways:

1. **Nested Loop** simply loops through all combinations of object pairs to determine if a collision occurred. Naturally, this mechanism is the most costly in terms of CPU.

2. **Locale or Segmented** divides the virtual world into regions of nearby space. The advantage is twofold: first, locales that do not touch each other can be searched in parallel; and second, the number of pairs is reduced since virtual objects that are not in the same or touching locales can not collide.

3. **Predictive** examines the virtual objects' paths through space and determines the earliest time a collision could occur[6]. When the virtual environment uses predictable paths, this can be determined fairly easily. When the virtual environment does not use predictable paths, then other assumptions must be made (e.g., fastest closure rate) to determine earliest collision time.

To detect collisions, the NAVL prototype recognizes collisions on the solid models with a simple nested loop search algorithm. It is recognized that the simulation environment within NAVL should provide a good platform for predictable collision detection. The authors are currently looking into this approach.

3.3   Network Engine

The network engine communicates ForceLets and events over the LAN using a broadcast datagram. All NAVL hosts on the LAN listen for ForceLets and events. If the ForceLet or event refers to a virtual object in the host, it handles the request immediately (via the same broadcast facility).

The message manager can filter incoming and outgoing messages to reduce network load on the WAN. In addition, it can request QoS levels commensurate with the expected load and importance of the traffic.

The NAVL prototype implements an imperative-based network protocol for communicating between the NAVL hosts on the LAN. The message manager's role has been reduced to a name service for simplicity and lack of resources for evaluating larger WAN-based implementations.

The messages in the protocol are phrased as commands to execute on the remote host, broken into five categories:

1. **Object Management** messages govern the creation, and destruction of virtual objects. Creating a slave object is called "attaching" to the master object. This emphasizes the master/slave relationship. These messages are sent to the host managing the master object.

2. **Object State Change** messages are sent by the master object whenever the virtual object changes its state. For example, ForceLets, and changing shape or orientation, require an object state change message to be broadcast.

3. **Name Service** messages enable applications to "publish" master objects so that remote hosts can attach to them. These messages are only sent between the message manager and the requesting host or application.

4. **Host Initialization** messages inform the new NAVL host of the current status of the NAVL virtual world. These messages are sent between the message manager and the new NAVL host.

5. **Simple** messages allow communication between applications and master objects. These messages are sent directly to the host with the master object.

4 NAVL DVR Part III: Evaluation

The NAVL DVR system is evaluated based on the following components: the distributed client/server network architecture, the ForceLet-based coherency model, and the OpenGL-based rendering engine with collision detection.

4.1 Network Architecture

The NAVL network architecture tries to reduce the network load by combining the peer-to-peer and client/server network paradigms. An evaluation of the capacity of the network has been made in [13]. The analysis determines the total number of packets transmitted in the network, and the peak number of packets transmitted through a node when varying number of messages (distributed throughout the network) are transmitted to all other hosts.

This evaluation concludes that the NAVL network architecture is the best compromise for minimizing both total network traffic (best) and peak network traffic (second best). To show these results visually, Fig. 6 shows the total number of messages and the peak number of messages for varying number of messages inserted into the system (we assume that no host will send more than one message). The parameters for these graphs are as follows: 500 hosts, 10 message managers, and an even distribution of message source locations.

Latency of the NAVL network architecture, however, will be slightly longer than competing architectures due to the additional hops imposed by the message managers. However, the authors feel that the distribution of the network load, QoS services, and message filtering more than compensate for the additional hops. In addition, it is expected that the network architecture will be deployed such that applications accessing common objects will be contained in the same LAN. For many cases, this will remove the hops imposed by the message managers entirely.

4.2 ForceLets

ForceLets were developed to improve the positional coherency of virtual objects following a path through space-time while using the minimum amount of network traffic. To test the capabilities of the ForceLet algorithm, an experiment was conducted with three candidates:

1. The *stream of data* algorithm that simply samples the source position at regular intervals (10 Hz), and transmits it to the destination.
2. The *dead reckoning* algorithm that uses velocity information to predict future positions in time. Only when the predicted position differs from the true position by a threshold does an update occur.
3. *ForceLets* as described in Section 2.3 using a special ForceLet type for sinusoidal paths (as one would expect for orbits).

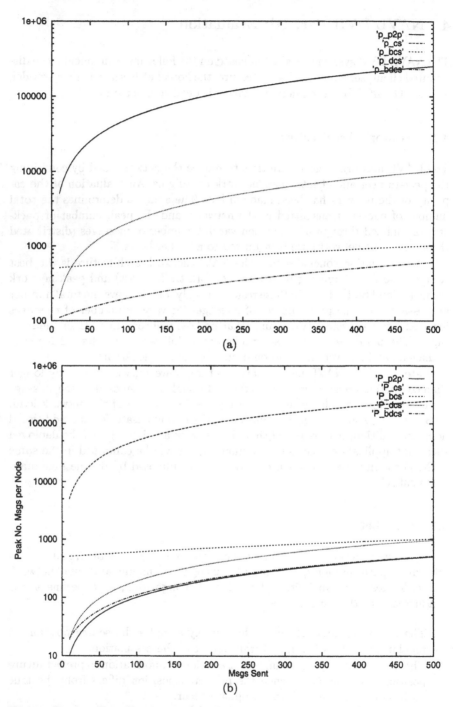

Fig. 6. Evaluation of network: (a) shows the total number of packets in the network, and (b) shows the largest number of packets for a node in the network

The algorithms were required to follow a source path through one-dimensional space. Figure 7 shows the source curve, and resulting paths for the algorithms. For this simulation, the network delay was set at 20 ms, the stream of data sampling frequency was 10 Hz, and the dead reckoning threshold was 0.5.

The *stream of data* algorithm shows quantization effects due to the sampling frequency and a time offset due to the network delay. This causes a compounded lag and jitter effect. The *dead reckoning* algorithm fairs better, however still shows a jitter effect when the threshold is exceeded. The *ForceLet algorithm* shows a small delay at the start of the path due to network lag, after which the curve follows the source curve exactly.

Numerical analysis of the results are summarized in Table 1. Notice that the ForceLet algorithm has the least average error, least maximum error, and lowest number of updates.

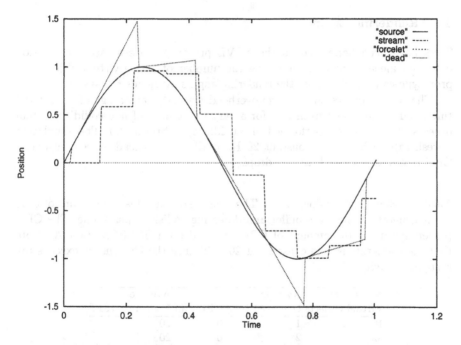

Fig. 7. Graph of coherency algorithms. The network delay was set at 20 ms, the sample frequency for stream of data was set at 10 Hz, and the threshold for the dead reckoning algorithm is set at 0.5

An additional experiment was conducted to determine the capacity of the simulator for three test cases[12]: (1) a varying number of ForceLets attributed to a single object, (2) a single ForceLet attributed to a varying number of objects, and (3) a varying number of objects with eight ForceLets attributed to each. Up to 4096 ForceLets were simulated and timed on an idle SPARCStation 20.

Table 1. Comparison of Coherency Algorithms

Algorithm	No. of Updates	Max. Error	Avg. Error
Stream	15	0.727	0.258
Dead Reckoning	6	0.609	0.146
ForceLet	1	0.125	0.00155

The results of the experiment show that the computational complexity of virtual object simulation scales linearly with the number of ForceLets at a rate of 615 thousand ForceLets/Second up to the 4096 ForceLets. Using a 10 Hz refresh rate, this translates to a capacity for 61.5 thousand ForceLets on an idle node.

4.3   Rendering Engine

Tests have been performed on the NAVL prototype to evaluate the collision detection mechanism, and determine the number of active virtual objects the prototype can handle before the rendering engine overloads the system.

The test examines the average overhead (in ms), percentage of total CPU time, and simulation refresh rate for a varying number of non-colliding virtual objects. This test was performed on an idle SparcStation 20. The simulation refresh rate is held at a constant 20 Hz until the overhead exceeds available CPU resources, at which time it drops.

Table 2. Rendering engine results. The average overhead is the amount of time spent generating the state buffer and detecting collisions per frame. The CPU percentage notes the amount of CPU required on a SPARCStation 20. Note that the framerate is held constant at 20 FPS until the CPU usage exceeds the systems capacity

Number of Virtual Objects	Average Overhead (ms)	CPU (%)	Average Framerate (FPS)
16	1	0	20
32	2	0	20
64	7	0	20
128	28	40	20
256	105	100	9.4

Notice that the overhead increases dramatically between 128 and 256 virtual objects. This is expected given the poor performance of nested-loop collision detection algorithm. While reasonable for small virtual worlds, clearly the rendering algorithm must be improved to support larger environments.

5   Summary and Future Work

This paper shows the design and implementation for a distributed virtual reality system, built from the ground up to improve network efficiency, object distribution and coherency, system resource management, and overall performance.

A prototype of the NAVL concepts has been recently completed, and the evaluation of its capabilities are underway. Preliminary results show that the fundamental systems provide a good network architecture and simulation environment, however the collision detection mechanism limits the system to small virtual worlds (less than 256 virtual objects).

References

[1] C. Carlsson and O. Hagsand. DIVE – a multi user virtual reality system. In *IEEE VRAIS*, pages 394–400. IEEE, 1993.

[2] Circle of Fire Studios, Inc. http://www.activeworlds.com.

[3] G. Couris and J. Dollimore. *Distributed Systems: Concepts and Design*. Addison–Wesley, Reading, Massachusetts, 1988.

[4] IEEE standard for distributed interactive simulation — application protocols. IEEE std 1278.1–1995, IEEE, 1995.

[5] James D. Foley, Andries van Dam, Steven K. Feiner, and John F. Hughes. *Computer Graphics: Principles and Practice*. Addison–Wesley, Reading, Massachusetts, 2nd edition, 1990.

[6] Philip M. Hubbard. *Collision Detection for Interactive Graphics Applications*. PhD thesis, Department of Computer Science, Brown University, October 1993.

[7] Rick Kazman. Making WAVES: On the design of architectures for low-end distributed environments. In *IEEE VRAIS*, pages 443–449. IEEE, 1993.

[8] Rick Kazman. HIDRA: An architecture for highly dynamic physically based multi–agent simulations. *IJCS*, pages 149–164, May 1995.

[9] Chris Shaw and Mark Green. The MR toolkit peers package and experiment. In *IEEE VRAIS*, pages 463–469. IEEE, 1993.

[10] Abraham Silberschatz and James L. Peterson. *Operating System Concepts*. Addison–Wesley, Reading, Massachusetts, 1988.

[11] The virtual reality modeling language 2.0. ISO/IEC cd14772, VRML Architecture Group.

[12] Martine Wedlake. NAVL phase 1 results: Simulation engine. Technical report, Department of Electrical and Computer Engineering, University of Victoria, February 1998.

[13] Martine Wedlake and Kin F. Li. Sailing the high seas with the NAVL virtual reality system. In *IEEE WESCANEX*, pages 208–213, Winnipeg, M.N., May 1997. IEEE.

[14] Martine Wedlake and Kin F. Li. The simulaiton and execution architecture for the NAVL DVR system. In *IEEE PACRIM*, pages 93–96, Victoria, B.C., August 1997. IEEE.

Research in Data Broadcast and Dissemination

Demet Aksoy[2], Mehmet Altinel[2], Rahul Bose[1], Ugur Cetintemel[2],
Michael Franklin[2], Jane Wang[1], and Stan Zdonik[1]

[1] Department of Computer Science, Brown University, Providence, RI 02912
[2] Department of Computer Science, University of Maryland, College Park, MD 20742

1 Introduction

The proliferation of the Internet and intranets, the development of wireless and satellite networks, and the availability of asymmetric, high-bandwidth links to the home, have fueled the development of a wide range of new "dissemination-based" applications. These applications involve the timely distribution of data to a large set of consumers, and include stock and sports tickers, traffic information systems, electronic personalized newspapers, and entertainment delivery. Dissemination-oriented applications have special characteristics that render traditional client-server data management approaches ineffective. These include:

- tremendous scale.
- a high-degree of overlap in user data needs.
- asymmetric data flow from sources to consumers.

For example, consider a dissemination-oriented application such as an election result server. Typically, such applications are implemented by simply posting information and updates on a World Wide Web server. Such servers, however, can and often do become overloaded, resulting in the inability for users to access the information in a timely fashion. We argue that such scalability problems are the result of a mismatch between the data access characteristics of the application and the technology (in this case, HTTP) used to implement the application. HTTP is based on a request-response or RPC, unicast (i.e., point-to-point) method of data delivery, which is simply the wrong approach for this type of application.

Using request-response, each user sends requests for data to the server. The large audience for a popular event can generate huge spikes in the load at servers, resulting in long delays and server crashes. Compounding the situation is that users must continually *poll* the server to obtain the most current data, resulting in multiple requests for the same data items from each user. In an application such as an election server, where the interests of a large part of the population are known *a priori*, most of these requests are unnecessary.

The use of unicast data delivery likewise causes problems in the opposite direction (from servers to clients). With unicast the server is required to respond individually to each request, often transmitting identical data. For an application with many users, the costs of this repetition in terms of network bandwidth and server cycles can be devastating.

S. Nishio, F. Kishino (Eds.): AMCP'98, LNCS 1554, pp. 194–207, 1999.

To address the particular needs of dissemination-based applications, we are developing a general framework for describing and constructing Dissemination-Based Information Systems (DBIS). The framework incorporates a number of data delivery mechanisms and an architecture for deploying them in a networked environment. The goal is to support a wide range of applications across many varied environments, such as mobile networks, satellite-based systems, and wide-area networks. By combining the various data delivery techniques in a way that matches the characteristics of the application and achieves the most efficient use of the available server and communication resources, the scalability and performance of dissemination-oriented applications can be greatly enhanced.

In this paper, we provide an overview of the current status of our DBIS research efforts. We first explain the framework and then describe our initial prototype of a DBIS toolkit. We then focus on several research results that have arisen from this effort.

2   The DBIS Framework

There are two major aspects of the DBIS framework.[1] First, the framework incorporates a number of different options for data delivery. A taxonomy of these options is presented in Section 2.1 and the methods are further discussed in Section 2.2. Secondly, the framework exploits the notion of *network transparency*, which allows data delivery mechanisms to be mixed-and-matched within a single application. This latter aspect of the framework is described in Section 2.3.

2.1   Options for Data Delivery

We identify three main characteristics that can be used to describe data delivery mechanisms: (1) push vs. pull; (2) periodic vs. aperiodic; and (3) unicast vs. 1-to-N. Figure 1 shows these characteristics and how several common mechanisms relate to them.

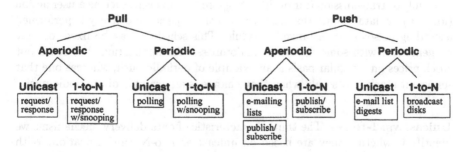

Fig. 1. Data Delivery Characteristics

---

[1] Parts of this section have been adapted from an earlier paper, which appeared in the 1997 ACM OOPSLA Conference [Fran97].

Client Pull vs. Server Push - The first distinction we make among data delivery styles is that of "pull vs. push". Current database servers and object repositories support clients that explicitly send requests for data items when they require them. When a request is received at a server, the server locates the information of interest and returns it to the client. This *request-response* style of operation is *pull-based* — the transfer of information from servers to clients is initiated by a client pull. In contrast, push-based data delivery involves sending information to a client population in advance of any specific request. With push-based delivery, the server initiates the transfer.

The tradeoffs between push and pull revolve around the costs of initiating the transfer of data. A pull-based approach requires the use of a backchannel for each request. Furthermore, as described in the Introduction, the server must be interrupted continuously to deal with such requests and has limited flexibility in scheduling the order of data delivery. Also, the information that clients can obtain from a server is limited to that which the clients know to ask for. Thus, new data items or updates to existing data items may go unnoticed at clients unless those clients periodically poll the server.

Push-based approaches, in contrast, avoid the issues identified for client-pull, but have the problem of deciding which data to send to clients in the absence of specific requests. Clearly, sending irrelevant data to clients is a waste of resources. A more serious problem, however, is that in the absence of requests it is possible that the servers will not deliver the specific data that are needed by clients in a timely fashion (if ever). Thus, the usefulness of server push is dependent on the ability of a server to accurately predict the needs of clients. One solution to this problem is to allow the clients to provide a *profile* of their interests to the servers. *Publish/subscribe* protocols are one popular mechanism for providing such profiles.

Aperiodic vs. Periodic - Both push and pull can be performed in either an aperiodic or periodic fashion. Aperiodic delivery is *event-driven* — a data request (for pull) or transmission (for push) is triggered by an event such as a user action (for pull) or data update (for push). In contrast, periodic delivery is performed according to some pre-arranged schedule. This schedule may be fixed, or may be generated with some degree of randomness.[2] An application that sends out stock prices on a regular basis is an example of periodic push, whereas one that sends out stock prices only when they change is an example of aperiodic push.

Unicast vs. 1-to-N - The third characteristic of data delivery mechanisms we identify is whether they are based on unicast or 1-to-N communication. With

---

[2] For the purposes of this discussion, we do not distinguish between fixed and randomized schedules. Such a distinction is important in certain applications. For example, algorithms for conserving energy in mobile environments proposed by Imielinski et al. [Imie94b] depend on a strict schedule to allow mobile clients to "doze" during periods when no data of interest to them will be broadcast.

unicast communication, data items are sent from a data source (e.g., a single server) to one other machine, while 1-to-N communication allows multiple machines to receive the data sent by a data source. Two types of 1-to-N data delivery can be distinguished: multicast and broadcast. With multicast, data is sent to a specific subset of clients. In some systems multicast is implemented by sending a message to a router that maintains the list of recipients. The router reroutes the message to each member of the list. Since the list of recipients is known, it is possible to make multicast reliable; that is, network protocols can be developed that guarantee the eventual delivery of the message to all clients that should receive it. In contrast, broadcasting sends information over a medium on which an unidentified and unbounded set of clients can listen. This differs from multicast in that the clients who may receive the data are not known *a priori*.

The tradeoffs between these approaches depend upon the commonality of interest of the clients. Using broadcast or multicast, scalability can be improved by allowing multiple clients to receive data sent using a single server message. Such benefits can be obtained, however, only if multiple clients are interested in the same items. If not, then scalability may actually be harmed, as clients may be continually interrupted to filter data that is not of interest to them.

2.2   Classification of Delivery Mechanisms

It is possible to classify many existing data delivery mechanisms using the characteristics described above. Such a classification is shown in Figure 1. We discuss several of the mechanisms below.

Aperiodic Pull - Traditional request/response mechanisms use aperiodic pull over a unicast connection. If instead, a 1-to-N connection is used, then clients can "snoop" on the requests made by other clients, and obtain data that they haven't explicitly asked for (e.g, see [Acha97, Akso98]).

Periodic Pull - In some applications, such as remote sensing, a system may periodically send requests to other sites to obtain status information or to detect changed values. If the information is returned over a 1-to-N link, then as with request/response, other clients can snoop to obtain data items as they go by. Most existing Web or Internet-based "push" systems are actually implemented using Periodic Pull between the client machines and the data source(s) [Fran98].

Aperiodic Push - Publish/subscribe protocols are becoming a popular way to disseminate information in a network [Oki93, Yan95, Glan96]. In a publish/subscribe system, users provide information (sometimes in the form of a profile) indicating the types of information they wish to receive. Publish/subscribe is push-based; data flow is initiated by the data sources, and is aperiodic, as there is no predefined schedule for sending data. Publish/subscribe protocols are inherently 1-to-N in nature, but due to limitations in current Internet technology, they are often implemented using individual unicast messages to multiple clients. Examples of such systems include Internet e-mail lists and some existing "push" systems on the Internet. True 1-to-N delivery is possible through technologies such as IP-Multicast, but such solutions are not universally available across the Internet.

Periodic Push - Periodic push has been used for data dissemination in many systems. An example of Periodic Push using unicast is Internet mailing lists that send out "digests" on a regular schedule. For example, the Majordomo system allows a list manager to set up a schedule (e.g., weekly) for sending digests. Such digests allow users to follow a mailing list without being continually interrupted by individual messages. There have also been many systems that use Periodic Push over a broadcast or multicast link. These include TeleText [Amma85, Wong88], DataCycle [Herm87], Broadcast Disks [Acha95a, Acha95b] and mobile databases [Imie94b].

2.3 Network Transparency

The previous discussion has focused primarily on different modes of data delivery. The second aspect of the DBIS framework addresses how those delivery modes are used to facilitate the efficient transfer of data through the nodes of a DBIS network. The DBIS framework defines three types of nodes:

1. *Data Sources*, which provide the base data to be disseminated.
2. *Clients*, which are net consumers of information.
3. *Information Brokers*, (or agents, mediators, etc.), which acquire information from other sources, possibly add value to that information (e.g., some additional computation or organizational structure), and then distribute this information to other consumers.

Brokers are the glue that bind the DBIS together. Brokers are middlemen; a broker acts as a client to some number of data sources, collects and possibly repackages the data it obtains, and then functions as a data source to other nodes of the system. By creating hierarchies of brokers, information delivery can be tailored to the needs of many different users.

The ability of brokers to function as both clients and data sources provides the basis for the notion of Network Transparency. Receivers of information cannot detect the details of interconnections any further upstream than their immediate predecessor. Because of this transparency, the data delivery mechanism used between two or more nodes can be changed without requiring changes to the data delivery mechanisms used for other communication in the DBIS. For example, suppose that node B is pulling data values from node A on demand. Further, suppose that node C is listening to a periodic broadcast from node B which includes values that B has pulled from A. Node C will not have to change it's data gathering strategy if A begins to push values to B. Changes in links are of interest only to the nodes that are directly involved. Likewise, this transparency allows the "appearance" of the data delivery at any node to differ from the way the data is actually delivered earlier in the network. This in turn, allows the data delivery mechanisms to be tailored for a given set of nodes. For example, a broker that typically is very heavily loaded with requests could be an excellent candidate for a push-based delivery mechanism to its clients.

Current Internet "push" technology, such as that provided by Point-Cast [Rama98] provide an excellent example of network transparency in action.

To the user sitting at the screen, the system gives the impression of using aperiodic push over a broadcast channel. Due to current limitations of the Internet, however, that data is actually brought over to the client machine using a stream of periodic pull requests, delivered in a unicast fashion. Thus, the data delivery between the client and the PointCast server is actually the exact opposite of the view that is presented to the user *in all three dimensions* of the hierarchy of Figure 1. This situation is not unique to PointCast; in fact, it is true for virtually all of the Internet-based push solutions, and stems from the fact that current IP and HTTP protocols do not adequately support push or 1-to-N communication.

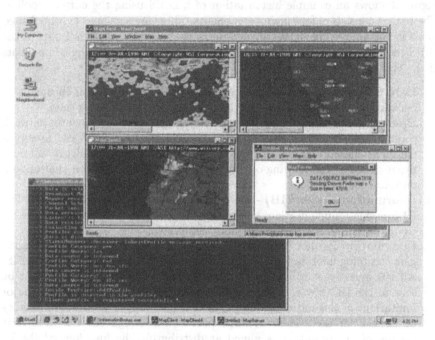

Fig. 2. The Map Dissemination Application

3   An Initial Prototype

As stated in the introduction, our ultimate goal is to build a toolkit of components that can be used to create a DBIS tailored to support a particular set of dissemination-based applications. In order to better understand the requirements and desired properties of such a toolkit, we have constructed an initial prototype toolkit and have used it to implement a weather map dissemination application.

Figure 2 shows an example screen from this application. In this application one or more "map servers" sends out updated maps of different types (i.e., radar, satellite image, etc.) for different regions of the United States. Clients can subscribe to updates for specific types of maps for specific regions. They can also pose queries to obtain the most recent versions of specific maps. The DBIS components route such queries to the appropriate server(s). In the current prototype, all maps are multicast to all clients — the clients perform additional filtering to avoid displaying unrequested results to the user. In the remainder of this section, we briefly describe the implementation of the prototype toolkit.

3.1   Toolkit Description

Figure 3 shows an example instantiation of a DBIS using the current toolkit. The toolkit consists of four main components. These are shown as lightly-shaded items in the figure. The darker shaded items are software that is not part of the DBIS toolkit, namely, the data sources and clients themselves. The components of the current prototype are:

1. **Data Source (DS) Library** - a wrapper for data sources that encapsulates network communication and provides conversion functions for data.
2. **Client Library** - a wrapper for client programs that encapsulates network communication and provides conversion functions for queries and user profiles. The client library is also responsible for monitoring broadcast and multicast channels and filtering out the data items of local interest that appear on those channels.
3. **Information Broker (IB)** - the main component of the DBIS toolkit. The IB contains communication, buffering, scheduling, and catalog management components and is described in more detail below.
4. **Information Broker Master** - The IB Master is responsible for managing global catalog information about data and about the topology of the DBIS. All IBs must register with the IB Master and all catalog updates must be sent to the IB Master. The presence of the IB Master is one of the major limitations of this initial prototype, as it is obviously a potential scalability bottleneck for the system. A large part of the design effort for the next version of the prototype is aimed at distributing the functions of the IB Master.

3.2   Data Modeling Considerations

The DBIS prototype currently uses a simple data model: the catalog consists of a set of category definitions. Categories are application-specific, that is, each application provides its own set of category definitions. Each data item is associated with a single category. In addition, a set of *keywords* can be associated with each data item. Categories and keywords are used in the specification of *queries* and *profiles*. Queries are *pull* requests that are transmitted from a client to a data source. Queries consist of a category and optional keywords. Queries are processed at a data source (or an IB); all data items that match the category (and at least one of the keywords if specified) are sent to the client from

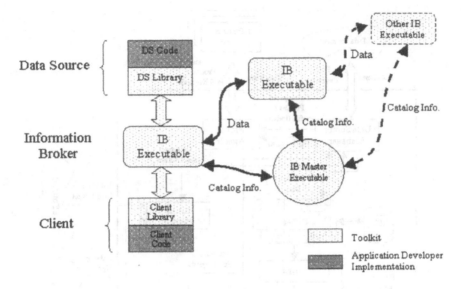

Fig. 3. An Instantiation of a DBIS

which the query originated. In contrast, profiles are used to support *push*-based delivery. When a new data item arrives at an IB, its category and keywords are compared with the user profiles registered at that IB and the item is sent to any clients whose profile indicates an interest in the item. Thus, profiles can be viewed as a form of continually executing query.

Clearly, this simple approach to data modeling must be extended to support more sophisticated applications. We are currently exploring database and WWW-based (e.g., XML) approaches for semi-structured data modeling for use in subsequent versions of the toolkit.

3.3   Information Broker Architecture

As stated above, the Information Broker module contains most of the functionality of the DBIS toolkit. The architecture of an IB is illustrated in Figure 4. Basic components of the IB are the following:

- **Catalog Manager -** This component manages local copies of catalog information for use by the processes running at the broker. Recall that the primary copy of the catalog is managed by the IB Master. All requested changes to the catalog information are sent to the IB Master, which then propagates them to the catalog managers of all other IBs.
- **Data Source Manager -** This component is in charge of receiving and filtering data items obtained from the data sources. It manages a separate listener thread for each data source directly connected to the IB.

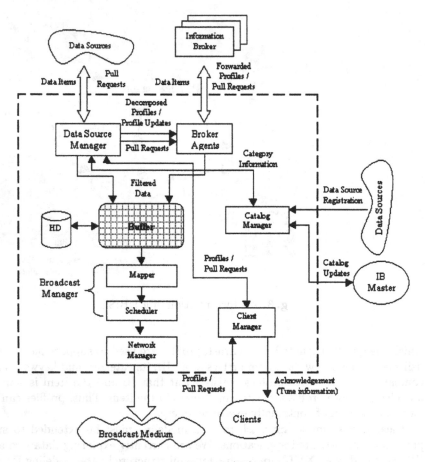

Fig. 4. Information Broker (IB) Architecture

- **Broker Agent -** This component is responsible for IB-to-IB interaction, that is, when an IB receives data from another IB rather than directly from a data source.
- **Broadcast Manager -** Once data has been filtered through the data source manager or the broker agent, it is passed to the Broadcast Manager, which has two main components. The *Mapper* assigns the data item to one or more physical communication channels. The *Scheduler* makes decisions about the order in which data items should be placed on those channels.
- **Network Manager -** This is the lowest level of the communication component of the IB. It sends data packets to the network according to the information provided by the broadcast manager.
- **Client Manager -** This module handles all requests that arrive from the clients of the IB. It forwards these requests to the proper modules within the IB and maintains communication sessions with the clients.

4 Example Research Topics

Having described our general approach to building Dissemination-Based Information Systems, we now focus on two examples of the many research issues that arise in the development of such systems.

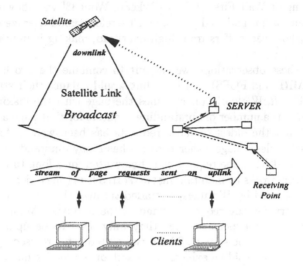

Fig. 5. Example Data Broadcasting Scenario

4.1 Topic 1: On Demand Broadcast Scheduling

As described in Section 2.1, one of the many possible mechanisms for data dissemination uses on-demand (i.e., aperiodic pull) broadcast of data. An example scenario using such data delivery is shown in Figure 5. In this scenario, two independent networks are used: a terrestrial network for sending pull requests to the server, and a "listen only" satellite downlink over which the server broadcasts data to all of the clients. When a client needs a data item (e.g., a web page or database object) that it cannot find locally, it sends a request for the item to the server. Client requests are queued up (if necessary) at the server upon arrival. The server repeatedly chooses an item from among these requests, broadcasts it over the satellite link, and removes the associated request(s) from the queue. Clients monitor the broadcast and receive the item(s) that they require.

In a large-scale implementation of such a system, an important consideration is the scheduling algorithm that the server uses to choose which request to service from its queue of waiting requests. We have developed a novel on-demand broadcast scheduling algorithm, called RxW [Akso98], which is a practical, low-overhead and scalable approach that provides excellent performance across a range of scenarios.

The intuition behind the RxW scheduling algorithm is to provide a balanced performance for hot (popular) and cold (not so popular) pages. This intuition is based on our observations of previously proposed algorithms. We have observed that two low overhead algorithms, Most Requests First (MRF) and First Come First Served (FCFS) [Dyke86, Wong88], have poor average case performance because they favor the broadcasting of hot or cold pages respectively. A third algorithm, Longest Wait First (LWF) [Dyke86, Wong88] was shown to provide fairer treatment of hot and cold pages, and therefore, good average case performance. LWF, however, suffers from high overhead, making it impractical for a large system.

Based on these observations, we set out to combine the two low-overhead approaches (MRF and FCFS) in a way that would balance their strengths and weaknesses. The RxW algorithm schedules the page with the maximal $R \times W$ value where R is the number of outstanding requests for that page and W is the amount time that the oldest of those requests has been waiting for the page. Thus, RxW schedules a page either because has many outstanding requests or because there is at least one request that has waited for a long time.

The algorithm works by maintaining two sorted lists (one ordered by R values and the other ordered by W values) threaded through the service queue, which has a single entry for any requested page of the database. Maintaining these sorted lists is fairly inexpensive since they only need to be updated when a new request arrives at the server[3]. These two sorted lists are used by a pruning technique in order to avoid an exhaustive search of the service queue to find the maximal $R \times W$ value. This technique is depicted in Figure 6

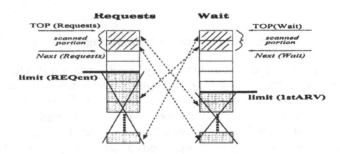

Fig. 6. Pruning the Search Space

The search starts with the pages at the top of the R list. The corresponding W value for that page is then used to compute a limit for possible W values. That is, after reading the top page in the R list, it is known that the maximum RxW-valued page cannot have a W value below this limit. Next, the entry for the page at the top of the W list is accessed and used to place a limit on the

---

[3] In contrast, for LWF the ordering can change over time, even in the absence of new requests.

R value. The algorithm alternates between the two queues and stops when the limit is reached on one of them. This technique prunes the search space while still guaranteeing that the search will return the page with the maximum RxW value.

In our experiments [Akso98], the pruning technique was shown to indeed be effective – reducing the number of entries searched by 72%. While such a substantial savings is helpful, it is probably not sufficient to keep the scheduling overhead from ultimately becoming a limiting factor as the system is scaled to the huge applications that will be enabled by the national and global broadcasting systems currently being deployed.

In order to achieve even greater reductions in the search space we developed an approximation-based version of the algorithm. By varying a single parameter α, this algorithm can be tuned from having the same behavior as the RxW algorithm described so far, to being a *constant time* approach. The approximate algorithm selects the *first* page it encounters whose RxW value is greater than or equal to $\alpha \times$ *threshold*, where *threshold* is the running average of the RxW value of the last page that was broadcast and the *threshold* at that time.

The setting of α determines the performance tradeoffs between average waiting time, worst case waiting time, and scheduling overhead. The smaller the value of the parameter, the fewer entries are likely to be scanned. At an extreme value of 0, the algorithm simply compares the top entry from both the R list and the W list and chooses the one with the highest RxW value. In this case, the complexity of making a scheduling decision is reduced to $O(1)$, ensuring that broadcast *scheduling* will not become a bottleneck regardless of the broadcast bandwidth, database size, or workload intensity. We demonstrated the performance, scalability, and robustness of the different RxW variants through an extensive set of performance experiments described in [Akso98].

4.2 Topic 2: Learning User Profiles

User profiles, which encode the data needs and interests of users, are key components of push-based systems. From the user's viewpoint, a profile provides a means of *passively* retrieving relevant information. A user can submit a profile to a push-based system once, and then continuously receive data that are (supposedly) relevant to him or her in a timely fashion without the need for submitting the same query over and over again. This automatic flow of relevant information helps the user keep pace with the ever-increasing rate of information generation. From the system point of view, profiles fulfill a role similar to that of queries in database or information retrieval systems. In fact, profiles are a form of continuously executing query. In a large publish subscribe system, the storage and access of user profiles can be be resource-intensive. Additionally, given the fact that user interests are changing over time, the profiles must be updated accordingly to reflect up to date information needs.

We have developed an algorithm called *Multi-Modal* (MM), for incrementally constructing and maintaining user profiles for filtering text-based data items [Ceti98]. MM can be tuned to tradeoff effectiveness (i.e., accuracy of the filtered data items), and efficiency of profile management. The algorithm receives relevance feedback information from the users about the documents that they have seen (i.e., a binary indication of whether or not the document was considered useful), and uses this information to improve the current profile. One important aspect of MM is that it represents a user profile as multiple keyword vectors whose size and elements change dynamically based on user feedback.

In fact, it is this *multi-modal* representation of profiles which allows MM to tradeoff effectiveness and efficiency. More specifically, the algorithm can be tuned using a threshold parameter to produce profiles with different sizes. Let us consider the two boundary values of this threshold parameter to illustrate this tradeoff: When the threshold is set to 0, a user profile is represented by a single keyword vector, achieving an extremely low overhead for profile management, but seriously limiting the effectiveness of the profile. At the other extreme, if the threshold is set to 1, we achieve an extremely fine granularity user model, however the profile size equals the number of relevant documents observed by the user, making it impractical to store and maintain profiles. Therefore, it is more desirable to consider intermediate threshold values which will provide an optimal effectiveness/efficiency tradeoff for a given application.

We evaluated the utility of MM by experimentally investigating its ability to categorize pages from the World Wide Web. We used non-interpolated average precision as our primary effectiveness metric and focused on the profile size for quantifying the efficiency of our approach. We demonstrated that we can achieve significantly higher precision values with modest increase in profile sizes. Additionally, we were able to achieve precision values with small profiles that were comparable to, or in some cases even better than those obtained with maximum-sized profiles. The details of the algorithm, experimental setting, and the results are discussed in [Ceti98].

5   Summary

The increasing ability to interconnect computers through internetworking, mobile and wireless networks, and high-bandwidth content delivery to the home, has resulted in a proliferation of dissemination-oriented applications. These applications present new challenges for data management throughout all components of a distributed information system. We have proposed the notion of a Dissemination-Based Information System (DBIS) that integrates many different data delivery mechanisms and described some of the unique aspects of such systems. We described our initial prototype of a DBIS Toolkit, which provides a platform for experimenting with different implementations of the DBIS Components. Finally we described our work on two of the many research issues that arise in the design of DBIS architectures.

Data Dissemination and data broadcasting are very fertile and important areas for continued research and development. In fact, we see a migration of data management concerns from the traditional disk-oriented architectures of existing database systems, to the more general notion of *Network Data Management*, in which the movement of data throughout a complex and heterogeneous distributed environment is of paramount concern. Our ongoing research efforts are aimed at better understanding the challenges and tradeoffs that arise in the development of such systems.

References

[Acha95a]   S. Acharya, R. Alonso, M. Franklin, S. Zdonik, "Broadcast Disks: Data Management for Asymmetric Communication Environments", *Proc. ACM SIGMOD Conf.*, San Jose, CA, May, 1995.

[Acha95b]   S. Acharya, M. Franklin, S. Zdonik, "Dissemination-based Data Delivery Using Broadcast Disks", *IEEE Personal Communications*, 2(6), December, 1995.

[Acha97]   S. Acharya, M. Franklin, S. Zdonik, "Balancing Push and Pull for Broadcast Data", *Proc. ACM SIGMOD Conf.*, Tucson, AZ, May, 1997.

[Akso98]   D. Aksoy, M. Franklin "Scheduling for Large-Scale On-Demand Data Broadcasting" *IEEE INFOCOM '98*, San Francisco, March, 1998.

[Amma85]   M. Ammar, J. Wong, "The Design of Teletext Broadcast Cycles", *Perf. Evaluation*, 5 (1985).

[Ceti98]   U. Cetintemel, M. Franklin, and C. Giles, "Constructing User Web Access Profiles Incrementally: A Multi-Modal Approach", *In Preparation*, October, 1998.

[Dyke86]   H.D. Dykeman, M. Ammar, J.W. Wong, "Scheduling Algorithms for Videotex Systems Under Broadcast Delivery", IEEE International Conference on Communications, Toronto, Canada, 1986.

[Fran97]   M. Franklin, S. Zdonik, "A Framework for Scalable Dissemination-Based Systems", Proc. ACM OOPSLA Conference, Atlanta, October, 1997.

[Fran98]   M. Franklin, S. Zdonik. "Data in Your Face: Push Technology in Perspective", *Proc. ACM SIGMOD Int'l Conf. on Management of Data (SIGMOD 98)*, Seattle, WA, June, 1998, pp 516-519.

[Giff90]   D. Gifford, "Polychannel Systems for Mass Digital Communication", *CACM*, 33(2), February, 1990.

[Glan96]   D. Glance, "Multicast Support for Data Dissemination in OrbixTalk", *IEEE Data Engineering Bulletin*, 19(3), September, 1996.

[Herm87]   G. Herman, G. Gopal, K. Lee, A. Weinrib, "The Datacycle Architecture for Very High Throughput Database Systems", *Proc. ACM SIGMOD Conf.*, San Francisco, CA, May, 1987.

[Imie94b]   T. Imielinski, S. Viswanathan, B. Badrinath, "Energy Efficient Indexing on Air", *Proc. ACM SIGMOD Conf.*, Minneapolis, MN, May, 1994.

[Oki93]   B. Oki, M. Pfluegl, A. Siegel, D. Skeen, "The Information Bus - An Architecture for Extensible Distributed Systems", *Proc. 14th SOSP*, Ashville, NC, December, 1993.

[Rama98]   S. Ramakrishnan, V. Dayal, "The PointCast Network" *Proc. ACM SIGMOD Int'l Conf. on Management of Data (SIGMOD 98)*, Seattle, WA, June, 1998, p 520.

[Wong88]   J. Wong, "Broadcast Delivery", *Proceedings of the IEEE*, 76(12), December, 1988.

[Vish94]   S. Viswanathan, "Publishing in Wireless and Wireline Environments", *Ph.D Thesis*, Rutgers Univ. Tech. Report, November, 1994.

[Yan95]   T. Yan, H. Garcia-Molina, "SIFT - A Tool for Wide-area Information Dissemination", *Proc. 1995 USENIX Technical Conference*, 1995.

Multimedia Database System for TV Newscasts and Newspapers

Yasuhiko Watanabe, Yoshihiro Okada, Kengo Kaneji, and Yoshitaka Sakamoto

Ryukoku University,
Seta, Otsu, Shiga, Japan
watanabe@rins.ryukoku.ac.jp

Abstract. It is important to use pattern information (e.g. TV newscasts) and textual information (e.g. newspapers) together. For this purpose, we describe a method for aligning articles in TV newscasts and newspapers. Also, we describe a method for extracting a newspaper article and its follow-ups. In order to align articles, the alignment system uses words extracted from telops in TV newscasts. The recall and the precision of the alignment process are 97% and 89%, respectively. On the other hand, in order to obtain a newspaper and its follow-ups, the system uses typical expressions which give signs of subsequent articles. The recall and precision are 80% and 85%, respectively. Using the results of these processes, we develop a browsing and retrieval system for articles in TV newscasts and newspapers.

1 Introduction

It is important to use TV newscasts and newspapers together. It is because TV newscasts and newspapers can complement and reinforce each other to enable more effective communication than can either medium alone. To put it another way, in a TV newscast, events are reported clearly and intuitively with speech and image information. On the other hand, in a newspaper, the same events are reported by text information more precisely than in the corresponding TV newscast. Figure 1 and Figure 2 are examples of articles in TV newscasts and newspapers, respectively, and report the same accident, that is, the airplane crash in which the Commerce Secretary was killed. However, it is difficult to use newspapers and TV newscasts together without aligning articles in the newspapers with those in the TV newscasts. To solve this problem, we propose a method for aligning articles in newspapers and TV newscasts. Also, we propose a method for extracting a newspaper articles and its follow-ups. Using the results of these processes, we develop a multiple media database system for newspapers and TV newscasts. The important points for this system are as follows:

- TV news articles and newspaper articles are cross-referenced,
- articles and their follow-ups in newspapers and TV newscasts are cross-referenced,

S. Nishio, F. Kishino (Eds.): AMCP'98, LNCS 1554, pp. 208–220, 1999.
© Springer-Verlag Berlin Heidelberg 1999

Telops in these TV news images

top left:	All the passengers, including Commerce Secy Brown, were killed
top right:	crush point, the forth day
middle left:	[Croatian Minister of Domestic Affairs] "All passengers were killed"
middle right:	[Pentagon] The plane was off course. "accident under bad weather condition".
bottom left:	Commerce Secy Brown, Tuzla, the third day

Fig. 1. An example of TV news articles (NHK evening TV newscasts; April, 4, 1996)

- the content based retrieval for TV newscasts is available by means of full text retrieval for newspaper articles which are aligned with the TV news articles, and
- A user can consult articles in TV newscasts and newspapers by using WWW browser.

米商務長官ら全員の死亡確認

クロアチア最南部のドブロブニク付近で3日午後、
旧ユーゴ各国を視察中のブラウン米商務長官ら乗員・
乗客計33人が乗った米空軍機が墜落した事故で、クロ
アチア政府は4日、ブラウン長官を含む乗客ら全員の
死亡を確認したと言明した。墜落当時、現場は強い風
雨に見舞われていた。国防総省スポークスマンは、砲
撃や爆弾テロの可能性は考えられない、と述べた。
　クリントン大統領は商務省で「バルカン半島に平和
を根付かせるため、米国の経済力の生かし方を探る
視察で、長官はたいへん意気込んでいた。長官は私に
とって最も有能なアドバイザーの1人だった」と語っ
た。ブラウン長官は今月中旬のクリントン大統領の訪
日に同行する予定だった。
　今回の事故にからみ、商務省は、メアリー・グッド
次官（技術担当）を長官代行に任命した。
　乗客は27人で、商務省職員や、旧ユーゴの復興に関
心を寄せる米企業幹部、ニューヨーク・タイムズ紙記
者らが含まれていた。

《写真》ボスニア・ヘルツェゴビナのツズラに
ある空軍基地に到着、軍用のボーイング737
型機から降りて兵士たちの出迎えを受けたブ
ラウン米商務長官。この後、同じ飛行機に再び
乗ってドブロブニクに向かう途中に事故が起き

　米国人はボスニアで、和平協議を推進した外交官3人が昨年夏、事故で死亡した。今年1月には、米
兵2人がやはり事故で死亡した。

Summary of this article: On Apr 4, the Croatian Government confirmed that Commerce Secretary Ronald H. Brown and 32 other people were all killed in the crash of a US Air Force plane near the Dubrovnik airport in the Balkans on Apr 3, 1996. It was raining hard near the airport at that time. A Pentagon spokesman said there are no signs of terrorist act in this crash. The passengers included members of Brown's staff, private business leaders, and a correspondent for the New York Times. President Clinton, speaking at the Commerce Department, praised Brown as 'one of the best advisers and ablest people I ever knew.' On account of this accident, Vice Secretary Mary Good was appointed to the acting Secretary. In the Balkans, three U.S. officials on a peace mission and two U.S. soldiers were killed in Aug 1995 and Jan 1996, respectively.
(Photo) Commerce Secy Brown got off a military plane Boeing 737 and met soldiers at the Tuzla airport in Bosnia. The plane crashed and killed Commerce Secy Brown when it went down to Dubrovnik.

Fig. 2. An example of newspaper articles (Asahi Newspaper; April, 4, 1996)

2   TV Newscasts and Newspapers

In TV newscasts, the image and the speech are the main modalities. However, it is difficult to obtain the precise information from these kinds of modalities. Telops, on the other hand, are secondary modalities in TV newscasts. However, they give us explanations of images and summaries of speeches, that is, the contents of the TV report. Moreover, it is not difficult to extract text information from TV newscasts. It is because a lots of works has been done on character recognition and layout analysis [Sakai 93] [Mino 96] [Sato 98]. Consequently, we

Fig. 3. An example of texts in a TV newscast: "*Okinawa ken Ohta chiji* (Ohta, Governor of Okinawa Prefecture)"

Fig. 4. An example of title texts: "*zantei yosanan asu shu-in tsuka he* (The House of Rep. will pass the provisional budget tomorrow)"

use these telops for aligning the TV newscasts with the corresponding newspaper articles.

On the other hand, a text in a newspaper article may be divided into four parts: headline, picture explanation, first paragraph, and the rest. In the text of a newspaper article, several kinds of information are generally given in an important order. In other words, significant words (keywords) are more frequent in the headline, the picture explanation, and the first paragraph, than in the rest. Moreover, these keywords are shared by the newspaper article with TV newscasts. For these reasons, we align articles in TV newscasts and newspapers using the following clues: location and frequency of keywords in each article and their length.

3 Aligning Articles in TV Newscasts and Newspapers

3.1 Extracting Nouns from Telops

An article in the TV newscast generally shares many words, especially nouns, with the newspaper article which reports the same event. Making use of these nouns, we align articles in the TV newscast and in the newspaper. For this purpose, we extract nouns from the telops as follows:

Step 1 Extract texts from the TV images by hands. For example, we extract "*Okinawa ken Ohta chiji*" from the TV image of Figure 3. When the text is a title, we distinguish it from the others. It is not difficult to distinguish title texts from the others because they have specific expression patterns, for example, an underline (Figure 4).

Step 2 Segment the extracted texts by the morphological analyzer JUMAN [Kurohashi 97].

---

1. explanation of contents of a TV image
   (a) explanation of a scene
   (b) explanation of an element
       i. person
       ii. group and organization
       iii. thing
   (c) bibliographic information
       i. time of photographing
       ii. place of photographing
       iii. reference image data
2. quotation of a speech
3. explanation of a fact
   (a) titles of TV news
   (b) diagram and table
   (c) other
4. information which is not concerned with a report
   (a) current time
   (b) broadcasting style
   (c) names of an announcer and reporters

Fig. 5. Information explained by telops in TV Newscasts

Step 3 Analyze telops in TV images. Figure 5 shows several kinds of information which are explained by telops in TV Newscasts [Watanabe 96]. In [Watanabe 96], a method of semantic analysis of telops was proposed and the correct recognition of the method was 92 %. We use this method and obtain the semantic interpretation of each telop.

Step 4 Extract nouns from the following kinds of telops.

- telops which explain the contents of TV images (except "time of photographing" and "reference image data")
- telops which explain a fact

It is because these kinds of telops may contain adequate words for aligning articles. On the contrary, we do not extract nouns from the other kinds of telops for aligning articles. For example, we do not extract nouns from telops which are categorized into a quotation of a speech in Step 3. It is because a quotation of a speech is used as the additional information and may contain inadequate words for aligning articles. Figure 6 shows an example of a quotation of a speech.

3.2   Extraction of Layout Information in Newspaper Articles

For aligning with articles in TV newscasts, we use newspaper articles which are distributed in the Internet. The reasons are as follows:

Fig. 6. An example of a quotation of a speech: "*kono kuni wo zenshin saseru chansu wo atae te hoshii* (Give me a chance to develop our country)"

Table 1. The weight $w(i, j)$

		newspaper			
		title	pict. expl.	first par.	the rest
T	title	8	4	4	2
V	the rest	4	2	2	1

- articles are created in the electronic form, and
- articles are created by authors using HTML which offers embedded codes (tags) to designate headlines, paragraph breaks, and so on.

Taking advantage of the HTML tags, we divide newspaper articles into four parts: (1) headline, (2) picture explanation, (3) first paragraph, and (4) the rest.

3.3 Procedure for Aligning Articles

Before aligning articles in TV newscasts and newspapers, we chose corresponding TV newscasts and newspapers. For example, an evening TV newscast is aligned with the evening paper of the same day and with the morning paper of the next day. We aligned articles within these pairs of TV newscasts and newspapers.

The alignment process consists of two steps. First, we calculate reliability scores for an article in the TV newscasts with each article in the corresponding newspapers. Then, we select the newspaper article with the maximum reliability score as the corresponding one. If the maximum score is less than the given threshold, the articles are not aligned.

If we are given a TV news article x and a newspaper article y, we obtain the reliability score by using the words $k(k = 1 \cdots N)$ which are extracted from the TV news article x:

the number of the articles in the TV newscasts	143
the number of the corresponding article pairs	100
the number of the pairs of aligned articles	109
the number of the correct pairs of aligned articles	97

Fig. 7. The results of the alignment

$$SCORE(x,y) = \sum_{k=1}^{N}\sum_{i=1}^{4}\sum_{j=1}^{2} w(i,j) \cdot f_{paper}(i,k) \cdot f_{TV}(j,k) \cdot length(k)$$

where $w(i,j)$ is the weight which is given to according to the location of word k in each article. We fixed the values of $w(i,j)$ as shown in Table 1. As shown in Table 1, we divided a newspaper article into four parts: (1) title, (2) picture explanation, (3) first paragraph, and (4) the rest. Also, we divided texts in a TV newscasts into two: (1) title, and (2) the rest. It is because keywords are distributed unevenly in articles of newspapers and TV newscasts. $f_{paper}(i,k)$ and $f_{TV}(j,k)$ are the frequencies of the word k in the location i of the newspaper and in the location j of the TV news, respectively. $length(k)$ is the length of the word k.

3.4   Experimental Results

To evaluate our approach, we aligned articles in the following TV newscasts and newspapers:

- NHK evening TV newscast, and
- Asahi newspaper (distributed in the Internet).

We used 143 articles of the evening TV newscasts in this experiment. As mentioned previously, articles in the evening TV newscasts were aligned with articles in the evening paper of the same day and in the morning paper of the next day. Figure 7 shows the results of the alignment. In this experiment, the threshold was set to 100. We used two measures for evaluating the results: recall and precision. The recall and the precision are 97% and 89%, respectively.

One cause of the failures is abbreviation of words. For example, "_shinyo-kinko_ (credit association)" is abbreviated to "_shinkin_". In our method, these words lower the reliability scores. To solve this problem, we would like to improve the alignment performance by using dynamic programming matching method for string matching. [Tsunoda 96] has reported that the results of the alignment were improved by using dynamic programming matching method.

In this experiment, we did not align the TV news articles of sports, weather, stock prices, and foreign exchange. It is because the styles of these kinds of TV news articles are fixed and quite different from those of the others. From this, we concluded that we had better align these kinds of TV news articles by the

Fig. 8. An example of a sports news article: "*senbatsu kaimaku* (Inter-high school baseball games start)"

different method from ours. As a result of this, we omitted TV news articles the title text of which had the special underline for these kinds of TV news articles. For example, Figure 8 shows a special underline for a sports news.

4   Extraction of Follow-Ups in Newspapers

As mentioned, in a newspaper article, headlines and first paragraphs contain many keywords. Consequently, a newspaper article and its follow-ups share many keywords in each headline and first paragraph. In addition, the first paragraph of a follow-up often include these kinds of clue expressions:

- *jiken / jiko / mondai + de*   (on this case/accident/subject of)

   (**S–1**) *beigunki ga tsuiraku shita jiken de*   (on the case of USAF airplane crash)

- *ni + tsuite*   (about)

   (**S–2**) *Kobe no satsujin jiken ni tsuite*   (about the murder case in Kobe)

where "*de*" and "*ni*" are Japanese postpositions. These expressions are good signs of follow-ups. For these reasons, we extract a newspaper article and its follow-ups using these kinds of clue information:

- typical expressions which give signs of subsequent articles, and
- shared keywords in each article.

The extraction method consists of four steps:

Step 1 Segment headlines and first paragraphs by the morphological analyzer JUMAN [Kurohashi 97].

Step 2 Extract an article as a follow-up when the first paragraph include the following expressions:
- *jiken / jiko / mondai + de*
- *ni + tsuite*

Step 3 Calculate relevant scores for an article x, which is extracted as a follow-up in Step 2, with each previous article in newspapers.

The relevant score between newspaper articles x and y, $SCORE(x, y)$, is calculated in this way:

$$SCORE(x, y) = w_{cn}N_{cn}(x, y) + w_{pl}N_{pl}(x, y) + w_{pn}N_{pn}(x, y)$$

where N_{cn}, N_{pl}, N_{pn} are the number of common noun, place name, and person's name, respectively. Also, w_{cn}, w_{pl}, w_{pn} are the weight for common noun, place name, and person's name, respectively. If the $SCORE(x, y)$ is more than the given threshold, the article x is determined as a follow-up of the article y.

Step 4 Merge the results in Step 3. Suppose that an article x is determined as a follow-up of an article y, and the article y is also determined as a follow-up of an article z. If the article x is not determined as a follow-up of the article z, the articles x and y are determined as follow-ups of the article z.

In this study, w_{cn}, w_{pl}, w_{pn} are set to $3, 5, 5$, respectively. We used 525 newspaper articles in this experiment. The precision and recall are 85% and 80% when the threshold is set to 40.

The following is a good example of this process. In this experiment, there were five follow-ups about the amendment of the health insurance. Two of them were extracted in Step 2, and two other articles were extracted in Step 3. These four articles are merged as follow-ups about the amendment of the health insurance in Step 4. However, the article entitled "Leaders of the Government party are in agreement on the passage of the revised Health Insurance in this session" could not be extracted. It was because this article shared only a few words with the other articles which reported the progress of deliberation in the Diet.

There is one other thing to note. We can extract TV news articles and their follow-ups when we extract newspaper articles and their follow-ups. It is because these TV news articles are aligned with newspaper articles.

5 Multimedia Database System for TV Newscasts and Newspapers

5.1 System Overview

The alignment process has a capability for information retrieval, that is, browsing and retrieving articles in TV newscasts and newspapers. As a result, using the

Interface   **Retrieval**   **Database**

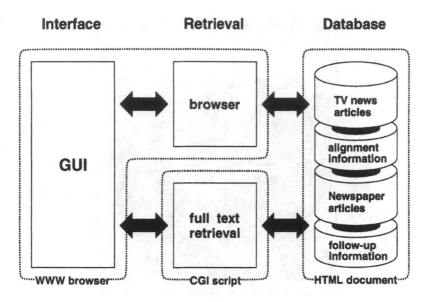

Fig. 9. System overview

results of the alignment process, we developed a browsing and retrieval system for TV newscasts and newspapers. Figure 9 shows the overview of the system. The important points for this system are as follows:

- Newspaper articles and TV news articles are cross-referenced.
- A user can consult articles in TV newscasts and newspapers by means of the dates of broadcasting or publishing.
- A user can browse newspaper articles and their follow-ups articles. In the same way, the user can consult TV news articles and their follow-ups which are aligned with the retrieved follow-ups in newspapers.
- A user can consult newspaper articles by full text retrieval. In the same way, the user can consult TV newscasts which are aligned with retrieved newspaper articles. In other words, content based retrieval for TV newscasts is available.
- Newspaper articles are written in HTML. In addition to this, the results of the alignment process and the extraction process of follow-ups are embedded in the HTML texts. As a result, we can use a WWW browser (e.g. Netscape, Internet Explorer, etc) for browsing and retrieving articles in TV newscasts and newspapers.

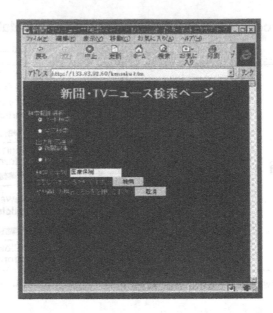

Fig. 10. The retrieval interface: the query word is "*iryo hoken* (health insurance)"

Fig. 11. The retrieval results of Figure 10: 12 newspaper articles and 4 TV news articles are obtained

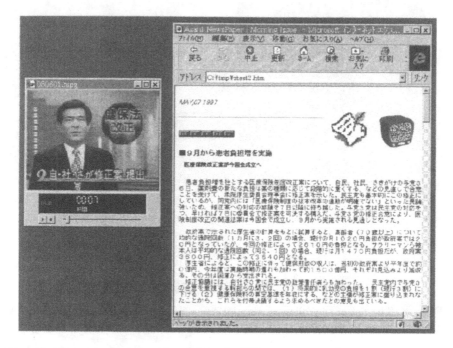

Fig. 12. The system shows the articles in TV newscasts and newspapers which a user wants to see

5.2   An Example of Browsing and Retrieval

A user can consult articles in newspapers and TV newscasts by full text retrieval in this way: when the user gives a query word to the system, the system shows the titles and the dates of the newspaper articles which contain the given word. At the same time, the system shows the titles of TV news articles which are linked to the retrieved newspaper articles. For example, in Figure 10, a user gives "*iryo hoken* (health insurance)" as a query word to the system. As a result, the user obtains 12 newspaper articles and 4 TV news articles (Figure 11). One of them, entitled "The new health insurance will be into force in September", is shown in Figure 12.

By the way, the newspaper article in Figure 12 has two icons in the above right. The right icon, looks like a TV monitor, shows this article is linked to the TV news article. When the user select this icon, the system shows the TV news article (the top left window in Figure 12). On the other hand, the next icon of "TV monitor" shows there are follow-ups of this article. When the user select this icon, the system shows the list of the follow-ups about "*iryo hoken* (health insurance)", as shown in Figure 13. The results of the full text retrieval include these five newspaper articles and three TV news articles. In other words, the results of the full text retrieval include many articles which have no relation to

220   Yasuhiko Watanabe et al.

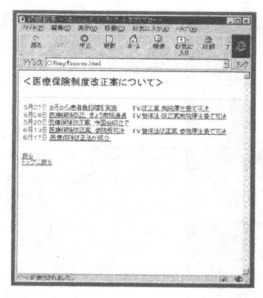

Fig. 13. The list of the follow-ups about *"iryo hoken* (health insurance)" in TV newscasts and newspapers

the amendment of the health insurance. This example makes it clear that follow-up information is good for narrowing down the articles which are obtained by full text retrieval to one subject.

References

[Kurohashi 97] Kurohashi, Nagao: JUMAN Manual version 3.4 (in Japanese), Nagao Lab., Kyoto University, (1997) [1].
[Mino 96] Mino: Intelligent Retrieval for Video Media (in Japanese), Journal of Japan Society for Artificial Intelligence Vol.11 No.1, (1996).
[Sakai 93] Sakai: A History and Evolution of Document Information Processing, 2nd International Conference on Document Analysis and Recognition, (1993).
[Sato 98] Sato, Hughes, and Kanade: Video OCR for Digital News Archive, IEEE International Workshop on Content-based Access of Image and Video Databases, (1998).
[Tsunoda 96] Tsunoda, Ooishi, Watanabe, Nagao: Automatic Alignment between TV News and Newspaper Articles by Maximum Length String between Captions and Article Texts (in Japanese), IPSJ–WGNL 96–NL–115, (1996).
[Watanabe 96] Watanabe, Okada, Nagao: Semantic Analysis of Telops in TV Newscasts (in Japanese). IPSJ–WGNL 96–NL–116, (1996).

---

[1] The source file and the explanation (in Japanese) of Japanese morphological analyzer JUMAN can be obtained using anonymous FTP from
ftp://pine.kuee.kyoto-u.ac.jp/pub/juman/juman3.4.tar.gz

A TV News Recommendation System with Automatic Recomposition

Junzo Kamahara[1], Yuji Nomura[2], Kazunori Ueda[3], Keishi Kandori[4],
Shinji Shimojo[5], and Hideo Miyahara[3]

[1] Kobe University of Mercantile Marine,
5-1-1 Fukaeminami-cho, Higashinada-Ku, Kobe, Japan,
kamahara@cc.kshosen.ac.jp,
http://www.rd.cc.kshosen.ac.jp/~kamahara/
[2] FFC Limited, 1 Fuji-machi, Hino-shi, Tokyo, Japan,
[3] Department of Informatics and Mathematical Science, Graduate School of
Engineering Science, Osaka University, 1-3 Machikaneyama, Toyonaka, Osaka, Japan
[4] Asahi Broadcasting Corporation,
2-48 2-chome, Oyodo-Minami, Kita-ku, Osaka, Japan
[5] Computer Center, Osaka University, 5-1 Mihogaoka, Ibaraki, Osaka, Japan

Abstract. In this paper, we propose a new recommendation system for
a TV news with automatic recomposition. For the time consuming browsing of the TV news articles, we propose three modes of presentation, the
digest mode, the relaxed mode, and the normal mode, where each presentation length is different. To make these presentation, TV news articles
are decomposed, analyzed, and stored in the database scene by scene.
Then, the system selects desired items and synthesizes these scenes into
a presentation based on a user's profile. For the profile of the user, we
use a keyword vector and a category vector of news articles. The system is designed so that user's control to the system becomes minimum.
Therefore, a user only plays, skips, plays previous, and rewinds news articles in the system as same as an ordinary TV. However, different from
an ordinary TV, the system collects user's behavior while he uses the
system. Based on this information, the system updates the user's profile.
We also show preliminary experimental results.

1 Introduction

Innovation of the digital television and the digital broadcasting satellite will
introduce a new age of television, that is, a digital multi-channel era. However,
two fundamental problems will be issued in the digital multi-channel era; the lack
of contents to fill up all the schedule of multi-channel and the lack of effective
way to select one from huge variety of contents. The former problem arises the
huge production cost for a TV program. Usually, a TV broadcasting company
makes a program which aim at a wide variety of audiences because selling time
slots for CM to the companies which have a large audience in return is the only
way to earn a huge amount of production cost. In the digital multi-channel era, a
TV broadcasting company is required to make a wide variety of programs which

S. Nishio, F. Kishino (Eds.): AMCP'98, LNCS 1554, pp. 221–235, 1999.
© Springer-Verlag Berlin Heidelberg 1999

aim at a smaller variety of audiences in a smaller a production cost. The latter problem is for the users. In the digital multi-channel era, a user is required to select one from a huge variety of the contents. There needs some effective way to select contents.

As a next generation TV, we are developing a recommendation system for TV news. Although many recommendation systems for news papers, web, etc. have been investigated, items for a recommendation are mostly based on text. On the contrary, TV news has video and audio portions and this makes a recommendation for TV news different. First, to browse several TV news takes a long time because we should consider time dependent media, therefore the presentation of a recommended data is very important. Second, we could not expect that users are active for the recommendation system. In the text based recommendation system, users are expected to be active for using the system. But for the TV news, users behave very passively. Usually, they are just watching the news and sometimes, zapping it by a remote controller. Therefore, we should design user's control of the system becomes minimum. Although the architecture of the TV news recommendation system consists of the server and clients, multimedia data treated by the server is huge as compared to the text or image for other recommendation systems. Therefore, there must be some mechanism to deal with such amount of data.

In this paper, we propose a new recommendation system for TV news with automatic recomposition. For the presentation of the TV news articles, we propose three modes of presentation, a digest mode, a relaxed mode, and a normal mode, where each presentation length is different. To make these presentation modes, TV news articles are decomposed, analyzed, and stored in the database scene by scene. Then, based on a user's profile the system selects desired scenes and synthesizes these scenes into presentation. As the profile of the user, we use keyword vector and category vector of news articles.

The system is designed so that user's control to the system becomes minimum. Therefore, a user only plays, skips, plays previous, and rewinds news articles in the system as same as ordinary TV. However, different from the ordinary TV, the system collects user's behavior while he uses the system. Based on this information, the system improves the user's profile.

This paper organized as follows. Firstly, we introduce the system architecture in the following section. In section 3, we describe the system how to learn the user's preference from his operation. In section 4, we propose three modes of presentation, a digest mode, a relaxed mode, and a normal mode and analysis and synthesis of the news article. And In section 5, we show the detail of the experiment in 1998.

2   The TV News Recommendation System

Firstly, we show the overview of our TV news recommendation system (Fig. 1).
This system consists of the news server, the profile server and clients. An update of TV news on the TV news recommendation system is performed at

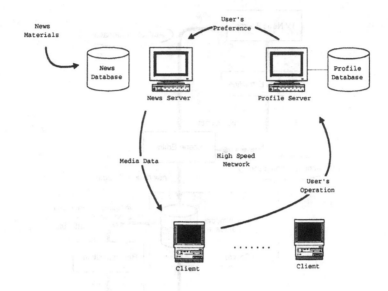

Fig. 1. The news recommendation system

any time. The server consists of two parts, the part of storing the news data into
the database and the part of recomposing and distributing the data requested.
There are two paths to add attributes to a news item(Fig. 2). First path is that
information for the news is written by a creator through the scene editor(Fig. 3),
and the other path is that it is extracted by image processing. In this process,
we use various media analysis techniques listed below, which are existed in other
study.

- Cut detection
- Closed caption detection
- keyword extraction(in Japanese).

The multimedia database manages the news items and their attributes. The
news stored in the database is distributed to the clients with some attributes. The
user's operation of the client is stored and used for updating the user's profile.
In the section 3, we propose the improving algorithm using above information
for the recommendation.

2.1   News Structure

Our basic idea is that we analyze TV news, give it a structure and synthesize it
by the preference of the user. In the synthesis shown in the previous section, we
utilize attributes attached to an item. In this section, we show how to analyze
TV news program automatically. Our method uses an audio and video segments

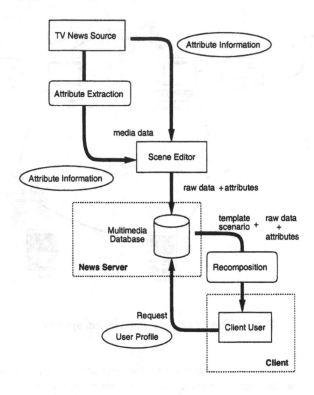

Fig. 2. A General Architecture

of TV news as well as a text segment of a closed caption or a closed caption. Based on information derived from these segments, we fragment TV news, give it a structure and store them into a news DB for later use of synthesis.

Because TV news programs have a purpose to report events to viewers, it is provided in the form which is easily and quickly understood by many viewers. In other words, they are strongly structured. We define a structure of a news program broadcasted by several companies as follows(Fig. 4):

- Company
- Program
- News Item
- Scene
- Cut

A news program consists of several news items. To introduce a semantic unit in a news item, we consider a notion of a scene as a continuous set of cuts which have a same meaning. In most cases, a TV news program has typical scenes in the following order with their types in the parentheses.

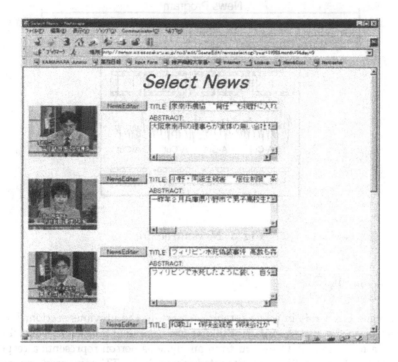

Fig. 3. Scene Editor (These images from ASAHI Broadcasting Corporation "Wide ABC DE-su")

- short description(anchor person),
- detail description(spot or reference video),
- response description(interviews or comments),
- summary or comment(anchor person).

The most news programs follow this pattern with some variations. We consider that each of the above elements corresponds to a type of a scene. By analyzing video segments in a news program, we can find the above structure. If we analyze video segments in a news program by a cut detection algorithm proposed in the literature, we can find several cuts in the news program. In addition, we find shots which contain a closed caption comparing other typical closed caption shot. A scene is found by clustering continuous cuts with similar contents automatically or manually.

Additional information could be retrieved from texts in a closed caption and audio and stored with a news structure. We use an object oriented database for management of several attributes and media data in our TV news recommendation system.

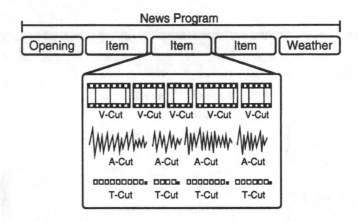

Fig. 4. News Structure

2.2   Scene Type

Each scene has a role in a news item as shown in the previous section. For example, an anchor person shot appeared at the beginning tells a short description for the a news item. A Comment from anonymous person represents a response in town for the news. Through the interview with the TV news crew, we divide scenes into the following five types which we call scene types. The scene types are used in the relax mode for selecting the remaining scenes within the presentation. The most important scene is used while the other scenes are dropped.

- Anchor Person
- Spot
- Reference Video
- Interview
- Comments

Most of these scene types are detected by image processing. We apply the closed caption detecting algorithm which is a similar algorithm for detecting an anchor person shot shown in the paper[5]. A scene type of **Comments** is found by detecting a shot viewing a person at large size. If a particular name is found in the closed caption, the scene should be classified as an **Interview**. A **Reference Video** is considered to be a shot with a still image or in slow motion. We develop a heuristic algorithm to classify scene types. A scene type is the most important and the most semantic attribute in our system and it is very effective for composing a new news program.

Table 1. Notation of vectors

	News i	Profile j
Keyword	K_i^n	K_j^u
Category	C_i^n	C_j^u

3   Improving Profile from the User's Behavior

We propose the system feedback from the watching time of the news items which is logged on the client program. The system will be able to calculate the user's interest of the operation time.

There are some studies of the filtering news using information of user's behavior. Those studies deal with the Internet news and show the relationship of the user interests and the reading time of the news article[3]. In the case of multimedia news, the system can detect the duration of viewing of the news because the duration is the time from beginning through stop. So, we consider to make the news arranged in filtered order. The system performs filtering using the weighted sum of keyword vector[6] and category vector.

We categorize news items as follows which are similar to the ones in the news papers:

- Social
- Politics
- Economy
- Sports
- Life

Because each TV news item belongs to several categories, we have developed a category vector which shows the degree of fitness to a category by analyzing the text segment of the news. The Categories classify a news to rude selection. But the detail selection depend on the keyword vector.

The score S_{ij} of user's interests is calculated by equation. Each i and j is a user and a news item, so S_{ij} is the score of a news item j which a user i watched. R_k and R_c is a weight which controls of the importance for a user.

$$S_{ij} = (K_i^n \cdot K_j^u) \times R_k + (C_i^n \cdot C_j^u) \times R_c \qquad (1)$$

Here,

$$-1 \le K_i^n \cdot K_j^u \le 1 \quad -1 \le C_i^n \cdot C_j^u \le 1 \qquad (2)$$

As we set the maxim value of S_{ij} to one when the maxim value of product of each vectors is one, following relation must be satisfied.

$$R_k + R_c = 1 \qquad (3)$$

Our system recomposes the news sequence along with the score calculated by the above equations (1) − (3).

Table 2. Play sequence and time for each modes

Mode	Sequence	Presentation time
Digest	by recommendation score	only abstract is shown
Relax	by recommendation score	by recommendation score
Normal	as original	as original

3.1 Feedback from User's Operation

The system gets a feedback from a user's operation. Since the user's operation reflects his evaluation of the news, to make this evaluation feed to the user's preference, we use the following simple feedback algorithm. For each news items, a user can perform one of following five operations:

1. a news item ends without user operation:
2. a user operates a skip.
3. a user operates a replay.
4. a user operates a previous.
5. a user operates a stop.

In all cases, the system puts a feedback on the vector by the time of user's browsing. The values of user's profile vector are increased by the rates of browsing time to increase the score of watching news item. If the user watch some news for a long time, the value of vector which are contained in the news is increased. On the other hand, if he skips a news, the keyword vector on the news is decreased.

$$K_j^{u'} = K_j^u + K_i^n \times V_{ij} \qquad C_j^{u'} = C_j^u + C_i^n \times V_{ij} \qquad (4)$$

Here, $K_j^{u'}$ is a keyword vector after feedback, and $C_j^{u'}$ is a category vector after feedback. V_{ij} and V_{ij} are weights whether user focuses on the keyword vector or the category vector.

4   Presentation Styles

To make our recommendation system same as an ordinary TV, we propose the following three mode of the presentation styles (Table 2):

- **Digest Mode**: Users browse the short news for quick understanding,
- **Relax Mode:**The system automatically adjust the presentation time for each news according to user's preference,
- **Normal Mode:**TV news are played in a ordinary way.

Digest Mode TV news has some patterns of a structure described in section 2.1. In a news item, 'who,' 'what,' 'when,' and 'where,' are firstly shown, and then 'why' and 'how' are presented. Since the former 4Ws are presented in the beginning of each news, it can be used as a digest. Therefore, we can build a quick browsing by extracting an audio segment of the very first portion of the news and composing the audio segment and other video segments into one segment(See. Fig. 5). To do this, firstly we pick up an audio segment of a scene whose type is anchor-person and then, combine it to the video segments picked from the rest of the scenes. While presenting the news digest, a user can interact with the client programs to watch the news in detail or to browse the related news. We showed the prototype system and the demonstration of news digest in [7].

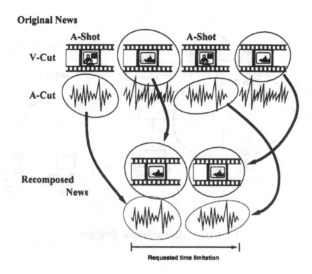

Fig. 5. The Method for Recomposing a Digest of a Day.

Relax Mode In the relax mode, the system decides the presentation time and its order in the presentation for each news item based on the user's preference. In this mode, the client interacts with the server to decide which scenes are dropped(Sec. 2.2).

Normal Mode In the normal mode, the original TV news videos are played in sequence. It is as same as the ordinary TV news.

5   An Experiment of the Proposed System

We perform the experiment of our proposed system. The purpose of this experiment is to examine a usability of our client and the efficiency of our proposed feedback algorithm by collecting user profiles of a campus wide network.

5.1   Abstract of Our Experiment

The system configuration is shown in Fig. 6. For the server, we use two PCs, which are an Intel Pentium II 266MHz machine and AMD K6 200MHz machine with MPEG capture boards. The former one creates normal size media files as Video CD format and the latter one creates small size media files with another parameter.

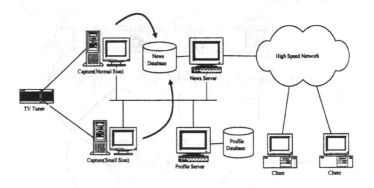

Fig. 6. Experimental System

The output from a TV tuner is distributed signals to above two PCs and those machines captures TV news at a specified time of a day. When capture finishes, the attribute extraction server detects attributes and transfer the captured movie files to the server. In the next step the server executes the detection for scene cut and news caption detection with using the attribute extraction mechanism. The extracted attributes are stored into the attribute database on the news server.

The small size movies are split into audio and video and are separated to the news items. Currently, we plan to input the speech text by hand with our program "News Editor" because in Japan the closed caption is not obligated to Japanese broadcast TV company.

We use the object-oriented database as a news database, NEC's "PERCIO," on Windows NT. The news server creates the scenario file, which specify the order of the news item by calculating the score(Sec.3) from each registered user profile after finishing pre-processing news. The scenario file currently lists the news items in a sequential order which are played on clients and have headline text and a flag whether that news plays in streaming or pre-loaded into local hard disk as a cache.

An user client always runs on user's PC and periodically maintains above the scenario file. If that file has the flag for pre-loaded news, the client retrieves listed news media files via high speed network as a cache.

The client caches such news media. When a user pushes the play button on a client, it can play the skimming news as quick viewing for this small size media data at any time. The normal size media files are determined by calculating user scores whether pre-loaded or streaming. Although, the client itself can decide to store the media data files in the own local storage in order to limit disk size.

Fig. 7. Interface of experimental client (This image from ASAHI Broadcasting Corporation "Wide ABC DE-su")

The client has some buttons of commands (Fig. 7). When the user pushes the DETAIL button while he is playing some news, the client requests to the server for streaming if that news doesn't have a normal size media file on the client. When the server accepts the request from clients, it retrieves the normal size media data and starts transferring it to the requested client in streaming. A user can control the client with following commands:

1. PLAY button plays a news item from the beginning or previously paused position,

2. PAUSE button pauses the currently playing news,
3. SKIP button skips a current news and play the next one,
4. REPLAY button stops a current news and play it from the beginning,
5. DETAIL button changes the currently playing news from small size to normal size which shows the news completely,
6. DIGEST button changes currently playing news from normal size to small size which can show news items quickly,
7. TOP button plays the top score news from begin and continue to play in score order.

The history of these commands is sent to the server and stored into the profile database by the client. The system learns the user's interests from these commands sequence. By analyzing the command sequences, the system changes the value of user's keyword vector.

5.2   Play Styles on the Experiment

For the digest mode, small size movies are made by capturing and recomposition described below. Each news items are stored in the news database and automatically transfer to the user clients in background. The client periodically checks the new news on the news server and if there are some, it downloads them into the clients. These steps are automatic. When the user pushes the play button, the client plays the digested news in the preference order which is calculated in the server.

5.3   Preliminary Results

We have some preliminary results of the experiment. For the validation of our learning algorithm, we surveyed user's evaluation of news items. This survey is based on the six level evaluations in parallel to the learning of each news item.

- 6: very interesting
- 5: interesting
- 4: bit interesting
- 3: less interesting
- 2: no interesting
- 1: time waste

Fig. 8 shows the co-relation of user's evaluation and watching time of the news taken from the command history. This graph shows that the user who watched a news for a long time gave the high evaluation for that news.

6   Evaluation

Our cut detection algorithm sometimes fail for the scene changes. But the "Scene Editor" shows the all cuts in a news items and the modification of scene boundary is easy at a glance.

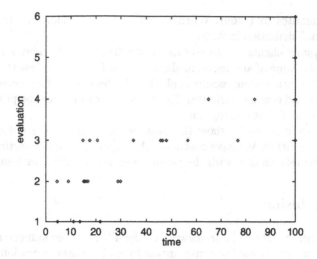

Fig. 8. The result of pre-experiment

Table 3. required disk capacity

period	MPEG file	entire
1 day	248MB	740MB
1 week	1.2GB	3.6GB
1 month	5.4GB	16.2GB

Our system can detect the special closed captions perfectly (not recognize) for the experiment news programs. The special closed captions are only appeared in the beginning of each news item. So we can detect the news item perfectly for the current news program. For other news programs we may detect their special closed captions with the region matching algorithm.

Our system require about 248 MB byte for disk capacity while capturing MPEG movie in the 20 minutes news program. After pre-processing, a mount of required disk size about 740MB. because the TV news program which we have use for our system broadcasts in week day, our system require about 16GB disk capacity per a month(Fig. 3). We are using 30GB RAID disk drive but the system requires the disk spaces for the database, so we only have about one month program in this system.

Our system may consume about 4 hour half for total processing time. The target TV news program will be captured in real time about 30 minutes. After captured, the captured MPEG file will transfer to the news server in a few minutes. For media processing of media split, cut detection and closed caption detection need 2 hour half on Sun Ultra 167MHz/Solaris2.6). Then based on such result the system produce the news items and digest movie for quick viewing in form 30 minutes to 1 hour. When we will start in 18:30, the system will can prepare to distribution in 23:00.

The system clients can receive the media files from the news server in night.

The total time of automatic media synthesis is not more fast than producing by hand. But our system requires only 30 minutes editing by a person which are text inputs and editing attributes. So our system can reduce the production cost for a digest of TV news program.

The system seems to show the news items in appropriate order with our feedback algorithm. We have continued the experiment for collecting more data to show the relationship with the result sequence and the user interest.

7   Conclusion

In this paper, we show the framework of the TV news recommendation system. This system learns from the consumption time of the user operation on his client program.

Before implementing the system, we interviewed the TV news crews in the japanese TV company in order to make the structure of news clear and to determine the scene types.

Various research for exploiting the TV news is going on. The followings are some examples. However, there are few researches treated on the analysis and synthesis of the news at the same time. Zhang et. al.[5] describe how to recognize an anchor person shot in TV news programs, and shows the basic idea of automatic parsing and indexing of news video. Our proposed system utilizes this method for finding an anchor person shot and the segmentation of the news. In the Informedia Project at CMU[2], they developed the News-on-demand system as a research of Digital Library. The skimming video which is extracted with the significant part from the video by image processing and speech recognition is very similar to one function of our system. However, they do not treat synthesis of the news. The similar project is the News in the future project[1] which suggests various application of news.

We perform the experiment of field study for our proposed system. The purpose of this experiment is to examine a usability of our client and the efficiency of our proposed learning algorithm by collecting user's profiles of campus wide various users.

We have made to the prototype systems in previous studies. We expect that the appearance of the local bus standard such as IEEE 1394 accelerates more integration of audio and video and exchanging them as digital information between the computers and videos. However, the data is made standard, the process of making standards to use the such data in various applications doesn't proceed. On the point of view for the advanced news system, we can use audio cassette tapes, floppy diskettes and video tapes in general digital format as well as a hard disks. And various information from video cam-corder or video tape can be extracted. Based on this situation, we settle the advanced news system as a multimedia database in the integrated digital video environment.

And the presentation of the multimedia contents will be made automatically by computer. We think that the significance of presentation by filtering using user's profile is increasing.

Acknowledgement The news video frames shown in this paper were taken from news programs broadcast by Asahi Broadcasting Corporation (ABC) with permission.

This research was supported by "Research for the Future" Program of Japan Society for the Promotion of Science under the Project "Advanced Multimedia Content Processing" (Project No. JSPS-RFTF97P00501).

References

1. MIT Media Lab.: News in the Future, http://nif.www.media.mit.edu/
2. Carnegie Mellon Univ.: Informedia: News-on-Demand,
   http://informedia.cs.cmu.edu/info/overview/index.html
3. Masahiro Morita,Yoichi Shinoda: Information filtering based on user behavior analysis and best match text retrieval, SIGIR '94. Proceedings of the seventeenth annual international ACM-SIGIR conference on Research and development in information retrieval (1994) 272–281
4. Y. Taniguchi, Y. Tonomura, and H. Hamada: A Method for Detecting Shot Changes and Its Application to Access Interfaces to Video(in Japanese), Trans. IEICE D-II (1996)
5. H.J. Zhang, S.Y. Tan, S.W. Smoliar, and G. Yihong: Automatic Parsing and Indexing of News Video, ACM Multimedia Systems, Vol.2, No.6 (1995) 256–265
6. Shimojo S. et Al: Design of the News database for The News on-Demand System, TECHNICAL REPORT OF IEICE DE95-50, Vol.95, No.287 (1995) 1–8
7. Kamahara, J., Shimojo, S., Sugano, A., Kaneda, T., Miyahara, H., Nishio, S., "A News On Demand System with Automatic Program Composition and QOS Control Mechanism," International Journal of Information Technology, Vol.2, No. 1 (1996) 1–22

Extended Digital Video Broadcasting with Time-Lined Hypermedia

Shinji Nabeshima, Kazuo Okamura, Takashi Kakiuchi, Kazutoshi Sumiya,
Naoya Takao, and Yoshiyuki Miyabe

Multimedia Development Center
Matsushita Electric Industrial, Co., Ltd.
1006 Kadoma, Kadoma, Osaka, 571-8501, Japan.
E-mail: {nabesima, okamura, kakiuchi, sumiya, ntp, miyabe}@isl.mei.co.jp

Abstract. In recent years, digital broadcasting services have started up and spread all over the world. Digital broadcasting is one of the most promising infrastructures to provide interactive services aimed at consumers. However, the broadcast bandwidth has an upper bound limitation. This paper describes our proposition, a hypermedia model, called Time-lined Hypermedia, which is designed to provide a large hypermedia navigation space in a limited broadcast bandwidth. We also describe an implementation for time-lined hypermedia called Digital Video eXtension system.

1 Introduction

Digital broadcasting is one of the most promising infrastructures to provide interactive services aimed at consumers. Digital broadcasting could realize a cost-conscious interactive service for every consumer without expensive bi-directional network infrastructures. Using its bandwidth which is much broader than other infrastructures, information could be broadcast with audio-visual program contents to TV terminals. TV viewers can navigate through the information space with a remote control just like they do to watch TV conventionally. In this type of interactivity, however, the size of navigation space depends on its bandwidth because of its mono-directional nature. This could be a serious problem when rich information is required with a limited bandwidth.

Issues of digital video broadcasting are discussed in DAVIC(Digital Audio-Video Council)[1]. Their topics include digital TVs and bi-directional interactive services. The DAVIC reference model is MHEG5 (Multimedia and Hypermedia Expert Group) proposed as ISO SC29/WG12[2]. Some systems based on this reference model have been proposed[3]. Our model is also based on the MHEG5 model. MHEG5 has the concept of neither renewing information in the hypermedia along the time line nor dividing the hypermedia space. Our model is extended on a real-time information renewal and delivery mechanism for time-lined hypermedia. In [4], for asymmetric

S. Nishio, F. Kishino (Eds.): AMCP'98, LNCS 1554, pp. 236-251, 1999.
© Springer-Verlag Berlin Heidelberg 1999

communication environments, an improvement of response time in the carousel type data transmitting system is discussed. However, transmission bandwidth is left out of consideration.

In this paper, we propose a hypermedia model, called Time-lined Hypermedia. It is designed to deliver a large hypermedia navigation space in a limited broadcast bandwidth. Based on the idea that a viewer's interest is constantly changed by the contents during the TV program, information to be sent is semantically divided into smaller segments that have time dependent attributes. We have implemented a commercial interactive service system called Digital Video eXtension and developed a dozen real interactive contents.

This paper is organized as follows: In Section 2, we present a definition of interactive services in digital video broadcasting and the requirements for them. In Section 3, we propose a hypermedia model, called Time-lined Hypermedia. In Section 4, we introduce our implementation based on Time-lined Hypermedia. In Section 5, we discuss the lessons we learned from the real examples we created.

2 Interactive Service in Digital Video Broadcasting

2.1 Interactive Service in Digital Video Broadcasting

We define interactive service in digital video broadcasting as follows:
- A set of data is delivered to viewers in mono-directional media.
- Viewers can navigate through the information space composed of the set of data.

This is not an "on demand" or "pull" type of interactive service where data can be retrieved from the server by viewer's request. This is a *push* type service[5] where information requested by the viewers is retrieved from the data already sent or on the air.

2.2 Requirements of Interactive Service in Digital Video Broadcasting

We have made the following requirement for interactive services in digital broadcasting:
1. Information should be associated with the contents of audio and video.
   Viewers can navigate through the information space sent via broadcast.

2. Interactive services should be naturally incorporated into the current convention of digital TV broadcasting[6].
   Viewers do not have to learn an unknown metaphor or a gadget for watching TV. An interactive program should be a TV program which belongs to a TV channel in broadcasting. A program would consist of audio, video and data. It has a start time and an end time which specify the program's lifetime.

3. Systems for interactive services should respect practicality of implementation. Efficient use of channel bandwidth in broadcasting is one of the most important

factors because the total bandwidth is not an unlimited resource. Moderate decoding cost at receivers is highly desirable.

In the following section, we propose our approach Broadcast Based Time-lined Hypermedia.

3 Our Approach

3.1 Program Related Information and Hypermedia

Our approach is based on the following ideas about information related to the contents of a broadcast program; the information can be represented as "hypermedia" so that viewers can navigate thorough it; the information can be aligned with the time line of the audio and video of the program.

The information closely related to the contents of a broadcast program is called *PRI* (Program Related Information). PRI can usually be divided into several portions. Each portion is a unit which viewers can choose and view on the TV screen. Hypermedia which represents the PRI is defined as follows:

- *PRI hypermedia* consists of scene nodes and links between scene nodes.
- A *scene node*[1] which represents a portion of the PRI, has a group of mono-media and screen layout information so that it can be displayed on the TV screen.
- A *link* which connects two nodes represents a mono-directional transition between two scenes represented by the nodes.

Suppose we have PRI hypermedia H for the PRI of a broadcast program. P is illustrated in Fig.1.

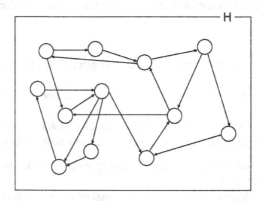

Fig. 1. PRI Hypermedia H

---

[1] Scene node is equivalent to Scene object defined in MHEG5 standards [2] which contain a group of objects used to present information (graphics, sound, video, etc.) for a screen.

Our goal is to deliver PRI hypermedia to a viewer's TV terminal along with the audio and video streams. Tuning to a program can happen anytime during the program hour, the whole data for PRI hypermedia should be sent repeatedly. However, sending the whole data all the time is not practical because it is not efficient in utilizing the limited bandwidth of broadcast waves and the TV receiver memory. Thus, we make the assumptions about the characteristics of PRI and interests of TV viewers in the next section.

3.2 Assumptions

Assumption 1: All of the scene nodes may not be necessary all the time because most of the nodes usually have a strong relevancy to specific parts of the contents of a broadcast program.

For example, consider a broadcast program of tour information for sight seeing in Paris. The program has three parts; "Welcome to the Eiffel Tower", "Invitation to the Louvre", "Shopping in Champs-Elysees". Each part may be related to its own specific information such as the height of the Tower, the list of masterpieces in the museum, or a list of shops on the street.

Assumption 2: Because the audio/visual parts are broadcast sequentially along with the time, it is very likely that viewers will become most interested in the information relevant to each part when it is broadcast.

This assumption is supported by the fact that a viewer's attitude to broadcast programs on TV is usually much more passive than their attitude towards a PC. TV viewers are not active information seekers but acceptors. In the above example, our assumption implies that the viewer's greatest attention may be changing from the structure of the tower to the famous pictures, and then finally the shop lists. Most viewers may not pay much attention to the shop list information during the museum part of the program.

The popularity of the push type interactive service shows that it is not always necessary to provide full on demand service if viewers are properly guided to a specific interest. Based on this observation on the characteristics of PRI and viewer's typical behavior, we propose a concept of Time-lined Hypermedia.

3.3 Broadcast Oriented Time-Lined Hypermedia

The contents of the broadcast program change as time passes. In a specific time period of the program, only a certain portion of the PRI has a higher probability of being accessed by viewers. PRI hypermedia can be divided into a set of nodes called a *time-lined segment*. Time-lined hypermedia is PRI hypermedia composed of time-lined segments.

3.3.1 Time-Lined Segment with Lifetime

The following steps show how to create time-lined segments.

Step 1. Define program parts in a broadcast program: *Tps* and *Tpe* are the start time and the end time of a broadcast program, respectively. The contents of the program is divided into several parts according to the semantics of the contents. Each program part Pn $(1 \leq n \leq N)$ has its own time period $Tn\{tns \leq t \leq tne\}$, where $Tps \leq tns \leq tne \leq Tpe$ and N is the number of program parts in the broadcast program. In the tour information program, there are three program parts.

Step 2. Create a Time-lined segment: Pick up nodes closely related to the content of each *Pn*. This set of nodes is called a segment *Sn*. Fig. 2 shows typical association level[2] of each segment *Sn* with the program part *Pn*. For example, in the period of *T2*, nodes in *S2* are considered to have higher possibility of being accessed by viewers, because these nodes have the most closely related information to the program part *P2*. We add a restriction to nodes in each segment that presentation of the nodes in *Sn* at the viewer's site can be done only during the period of *Tn*. *Tn* is called the lifetime of *Sn* which means *Sn* is valid only during *Tn*.

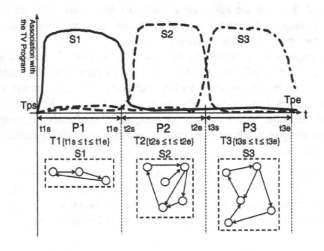

Fig. 2. Time-lined segments and their association with the program

In the case of the tour information program, we have time-lined hypermedia H' with three time-lined segments as described in Fig. 3.

---

[2] Deciding the association level depends on the content and its creator, and its definition is not in the scope of this paper.

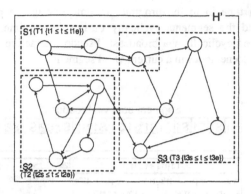

Fig. 3. Time-lined Hypermedia

3.3.2 Version Update of the Time-Lined Segment

Even if the number of time-lined segments is one, it is often the case that nodes in the segment have changing information while the link relationship with other nodes remains the same. Consider a financial news program with only one segment for PRI which has real-time updated stock data. Only the stock data is changing while the structure of the segment does not change. To support this kind of finer grain changes of the content, a time-lined segment has an attribute *version*.

Version-update of a segment Sn means that information in the nodes in Sn can be changed according to the sub-period $Tn'\{tn's \leq t \leq tn'e\}$ $(tns \leq tn's \leq tn'e \leq tne)$. An example of version-update is illustrated in Fig. 4. The structure of $S1$ remains the same during $T1$ but the information contained in each node is updated at $t1''s$ and $t1'''s$.

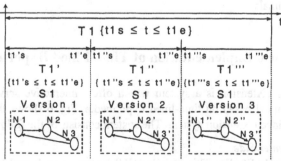

Fig. 4. Version-update of a segment S1

3.4 Pseudo Interactive Broadcasting

Broadcasting time-lined hypermedia is done by sending each time-lined segment to viewers according to its lifetime period. As broadcasting is mono-directional media and viewers can tune in to the program anytime during the program time, it is necessary to send the segments cyclically during the lifetime. Fig. 5 shows the segments delivered cyclically via broadcast. Pseudo interactive means viewers can navigate through scene nodes in a current valid segment.

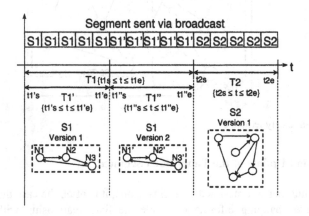

Fig. 5. Cyclic segment broadcasting

A Viewer's TV terminal should always keep watching whether either 1) the lifetime of a current valid segment has expired or 2) a newer version of the current segment has been received. In both cases, the TV terminal refreshes its presentation according to the new version of PRI. This mechanism enables an efficient way of delivering PRI to viewers in terms of broadcast bandwidth and decoding cost at the viewers' terminals.

4　A Practical Implementation of Time-Lined Hypermedia

Digital Video eXtension is a system which offers interactive services on a digital broadcast channel based on time-lined hypermedia. Fig. 6 shows the architecture of the Digital Video eXtension system. The system consists of the following subsystems:

Broadcast Center Side
Interactive Data Studio: This includes GUI authoring tools and format converters. PRI is prepared from the databases or outside networks, and time-lined segments are generated with user interface information for each scene node by the authoring tools. Then the PRI hypermedia is converted into interactive data.
Interactive Data Caster: This receives the interactive data from the Interactive Data Studio, and synchronously broadcasts them along with video and audio streams.

Viewer's Side

Interactive Data Player: This is a software implemented in the STB[3]. It decodes the interactive data and activates it synchronously with video and audio according to the time-line attributes of the PRI hypermedia represented in the interactive data.

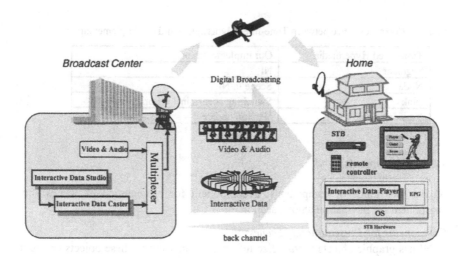

Fig. 6. Digital Video eXtension system

4.1 Contents Model

An ordinary TV program includes video and audio streams. On the other hand, an interactive TV program includes interactive data in addition to the TV program. The data is called Navigation Information (NI), which represents several scenes including user interaction information, presentable objects, and procedural codes. The NI corresponds to a segment, and the scene included in the NI corresponds to a node in the time-lined hypermedia model. The scene in the NI is overlaid on the video of the program on the TV screen.

The maximum size of the NI is determined in terms of memory size in the STB and broadcast bandwidth. If a single segment is too large, it is mapped to several pieces of NI, each of which has own identifier. A link between two nodes corresponds to either a link between two scenes within the single NI or a link between the scene in one NI and the scene in another NI. Viewers can navigate through the scenes according to the links among the scenes. In the following section, we describe the NI format, which is the implementation of time-lined hypermedia.

---

[3] STB is an acronym of Set Top Box. An STB is a decoder terminal which is connected to a TV.

4.2 Implementation of Time-Lined Hypermedia

Table 1 shows the correspondences among Time-lined Hypermedia and our implementation. NI corresponds to a segment in time-lined hypermedia, and consists of a set of scenes which correspond to nodes in the segment.

Table 1. Correspondence between Time-lined Hypermedia and our implementation

Time-lined Hypermedia	Our implementation
Segment	NI
Node	Scene in NI
Link	Switching scenes in 1 NI or across 2 NI
Version update	Version update

4.2.1 Components in NI

NI has two kinds of components; a component for presentation of scenes and a component for navigation among scenes.

Components for Presentation of Scenes
   NI has graphic objects to visualize information in scenes. These objects are button object, text object, and entry object. They can have four states; 1) normal, 2) focused, 3) selected, and 4) focused-and-selected. In each state, the graphic objects can have a different appearance such as different bitmaps, cell animations (a series of bitmaps to represent simple animation). A state can be changed by key events using a remote controller or bytecode programs described below. Furthermore, graphic objects have states of visibility; a) visible, b) invisible, and their positions when displayed on TV screen. NI is composed of several scenes, each of which is represented by making the set of the graphic objects visible and arranging their positions.

Components for Navigation among Scenes
   NI has the following components so that viewers can navigate among scenes:

Focus Control Table(FCT): This is used to enable viewers to navigate among graphic objects on the TV screen with the remote controller. The FCT specifies the procedure of navigation among the graphic objects to be selected. If graphic objects are presented on the OSD plane, viewers can focus on one of them using the remote controller. When the viewers use the arrow-keys of the remote controller, the current focused object is unfocused and a new one is set as focused according to the description in the FCT.

Bytecode Handler: Procedural behavior of each scene in NI is described by bytecodes which are independent of hardware. The bytecodes can change the attributes of graphic objects, switch the scenes, and other system control tasks. These bytecode handlers are driven by such events as the remote control and changes of attribute of the graphic objects.

Hyperlink Information Table(HIT): This is used to realize the links between two scenes in two separated NI in the case that a segment is mapped into several pieces of NI. HIT specifies link information about the other NI.

4.2.2 Lifetime of NI

Based on the time-lined hypermedia approach, NI has an attribute: *lifetime*. Lifetime indicates an period between start_time and end_time, and NI can be presented only during this lifetime. The NI may be associated with an arbitrary interval or frame of video stream. This means that each piece of NI becomes active or inactive synchronously as the deployment of the A/V stream proceeds.

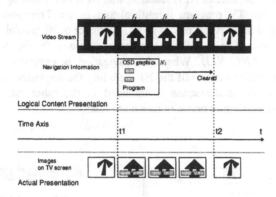

Fig. 7. Lifetime of NI and synchronization with video

Fig. 7 shows a typical example of the lifetime of NI. For the sake of simplicity, the content in the example consists of video stream containing 5 frames and single NI. The logical semantics which the interactive TV programs provider wants to give is:
1. A certain interactivity and presentation of *N1* is activated just when the time *t1* is reached.
2. The above interactivity and presentation of *N1* is de-activated just when the time *t2* is reached.

The data and program to provide interactivity for the period from *t1* to *t2* is stored in *N1*, and when the time *t2* comes, *N1* is discarded from memory. The period between *t1* and *t2* in the example is the lifetime of *N1*. The lifetime can be specified either in a relative time offset from the start time of the interactive TV program or an absolute time.

4.2.3 Version of NI

NI has an attribute as well as lifetime: *version*. Suppose a content provider intends to offer a new segment, in which the relationship among the nodes remains the same and only the contents of the nodes are changed. In this case, the content provider can create a new segment as a different version of the current NI. When the new version of NI arrives at the receiver, the current version of NI becomes inactive. This version-update mechanism is directly mapped onto the section data version update in MPEG2-TS.

4.2.4 Carousel Delivery Mechanism

In interactive TV programs, video and audio are multiplexed in MPEG2-TS in the same way as those of ordinary TV programs. On the other hand, NI is multiplexed in MPEG2-TS as MPEG2-TS private section. NI is delivered repeatedly during their lifetime. After the end time of NI is reached, that NI is never sent again. Fig. 8 shows how the interactive TV program is multiplexed in the Transport Stream. If one segment is mapped into several pieces of NI, all pieces of NI are delivered repeatedly during their lifetime. The set of NI is called a *carousel*. In Fig. 8, the carousel is composed of *{N11,N12,N13}*. When the navigation is requested by viewers, the Interactive Data Player waits until the NI including the requested scene is delivered and retrieves the NI if the scene is included in the other NI. This navigation corresponds to the change of nodes within a segment.

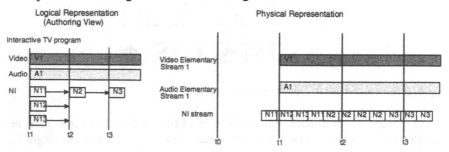

Fig. 8. Correspondence between logical interactive TV program and physical representation

5 Discussions

We have proposed our approach, time-lined hypermedia and developed a practical implementation based on our approach. In this chapter, we discuss the evaluation of our approach and lessons learned from real service examples.

5.1 Real Service Examples

We have developed the following types of interactive TV programs. All of them are working examples and some of them are real commercial ones.

- Gambling Information (horse races, boat races) : runner information and real-time odds
- Sports Information (Soccer, Baseball) : player information and real-time game scores
- Weather Forecast : interactively select a region's weather forecasts in Japan
- Tour Guide Information : sight seeing information, hotel reservation and shopping
- Music CD Promotion : trial listening of CDs by selecting audio streams
- Movie Promotion : interactively select movie information

In the service examples above, the gambling program example has the most nodes, 120 nodes with version-update at 60 second intervals. Fig. 9 shows some screen shots of the service examples.

MPEG Video   NI(OSD)

Fig. 9. Service examples (Restaurant Guide and Weather Forecast)

We introduce the tour guide service example below. This program is the tour guide for a tourist spa resort in Japan. It has 3 program parts; an introduction to the Japanese style hotel with a spa, restaurants in the resort, and a souvenir store. This example is 10 minutes long, and it is composed of a part of the real original program, which is 45 minutes long. In this example, PRI Hypermedia H is composed of 18 nodes, and H has 3 segments. The NI represents each segment. Fig. 10 shows the segments in PRI Hypermedia H and their structures and nodes in each program part. Fig. 11 shows the time line and TV screens of this example.

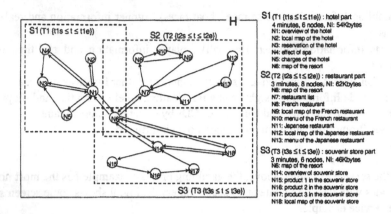

S1 (T1 {t1s ≤ t ≤ t1e}) : hotel part
4 minutes, 6 nodes, NI: 54Kbytes
N1: overview of the hotel
N2: local map of the hotel
N3: reservation of the hotel
N4: effect of spa
N5: charges of the hotel
N6: map of the resort

S2 (T2 {t2s ≤ t ≤ t2e}) : restaurant part
3 minutes, 8 nodes, NI: 62Kbytes
N6: map of the resort
N7: restaurant list
N8: French restaurant
N9: local map of the French restaurant
N10: menu of the French restaurant
N11: Japanese restaurant
N12: local map of the Japanese restaurant
N13: menu of the Japanese restaurant

S3 (T3 {t3s ≤ t ≤ t3e}) : souvenir store part
3 minutes, 6 nodes, NI: 46Kbytes
N6: map of the resort
N14: overview of souvenir store
N15: product 1 in the souvenir store
N16: product 2 in the souvenir store
N17: product 3 in the souvenir store
N18: local map of the souvenir store

Fig. 10. PRI Hypermedia and the segments and nodes in it

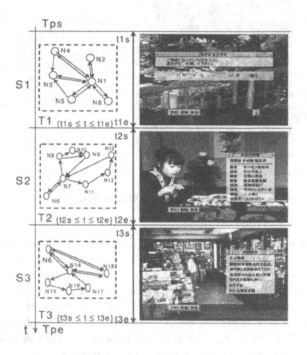

Fig. 11. Time-lined segments and TV screens of the tour guide service example

5.2 Issues Learned from the Real Service Examples

To verify our implementation with service examples, at first we assumed an experimental condition: size of NI, bandwidth for delivering NI, memory of STB, processing power of STB, and response time. The response time means interval between the time a pieces of NI is requested and the time the NI is represented on the TV screen. According to the conventional TV environment, the response time should be short enough to satisfy the viewer's operability.

The response time Tr is the sum of Tw and Tp as shown in equation (1). The waiting time Tw indicates the time interval taken to retrieve NI as shown in equation (2). In equation (2), D is the size of NI, and B is the bandwidth for delivering NI. The processing time Tp indicates the time interval taken to decode and represent NI. Tp depends on the size of NI and the processing power of the target STB. In order to estimate the response time, we assumed each parameter.

$$Tr = Tw + Tp \tag{1}$$

$$Tw = D / B \tag{2}$$

The bandwidth of NI took 0.5Mbps, and total bandwidth to transmit a service took 6Mbps. The maximum bandwidth of each program is 6Mbps in current digital satellite broadcasting systems, and from five to six programs are multiplexed into one MPEG2-TS with statistical multiplexing. Because of cost factors, the bandwidth of NI must be less than 0.5Mbps according to current bandwidth usage.

In the implementation, the maximum size of NI is assumed as 64Kbytes for the following reasons: 1) All the segments of the service examples could be created within 64Kbytes; 2) The response time could be estimated to be short enough for practical use; 3) The size of memory of the target STB for storing the NI must be less than 64Kbytes RAM because of cost factors.

Under these conditions, we measured response times on the service examples with the target STB. Table 2 shows the conditions and the experimental values. We got 1.6 seconds minimum, 2.2 seconds maximum, and 1.8 seconds on average as the results, respectively. Therefore, we have concluded that our implementation is practical and usable enough to be incorporated in the current digital broadcasting system under the limitations of the broadcast bandwidth and STB hardware specification.

Table 2. Conditions and experimental values of the service examples

Broadcast environment	Digital satellite broadcasting system
Hardware specification of target STB	Memory : 2MB ROM and 1MB RAM CPU : 32bit RISC
Broadcast bandwidth (NI / service total): B	0.5 Mbps (64 Kbytes/sec.) / 6 Mbps
Size of NI: D	64 Kbytes maximum
The result of response time: Tr	1.8 seconds in average

According to the characteristics of TV, the following issues should be considered:
- Under the limitation of the resolution of TV screen, one node may be mapped onto several scenes in NI, because the size of the node is too large to display on one screen.
- Because of navigation with a remote controller, we had better reduce the number of choices for links in each scene.

Still, there are some issues to be pursued to realize various services for Time-lined Hypermedia.

- In some cases, there may be some nodes which are always necessary while the program is being broadcast. In the case of the tour guide example, node N6 appears throughout the program. It should be considered to add a direct mechanism to represent such a *time independent node* to our approach.
- In our approach, nodes are valid only while the segment including the nodes are delivered, and they expire when the next segment is delivered. We should consider an *expiration mechanism* for time-lined hypermedia to allow the node to have an end time exceeding the current segment.

5.3 Evaluation

We have been starting the following evaluation:

Human factors: To prove our assumptions in our approach and verify the usability and practicality of our implementation, we have prepared an examination of the viewers' tendencies for navigation on the hypermedia space with our real service examples. For example, we are going to examine which nodes viewers will select and in which program part the selections will be made when the viewers see the service example in which all nodes can be represented throughout the program if they so request.

Response time intervals: We have prepared measurements of how long it takes to retrieve the NI in both cases of delivering the NI generated by applying Time-Lined Hypermedia approach and delivering the NI which consists of all the nodes.

Guideline of grouping the PRI: We have to create segments manually from the entire hypermedia created by interactive program providers. We have started to discuss guidelines on how to group the PRI into segments. For example, when one node is picked up, some nodes with a similar lifetime to the node or other nodes which are linked with that node may be added into the same segment.

6 Conclusions

This paper presented a broadcast based time-lined hypermedia approach to provide interactive services in digital broadcasting, where viewers can choose information from a pre-selected set of information related to audio-visual programs, called Program Related Information (PRI). In order to make effective use of broadcast bandwidth and minimize decoding cost, PRI hypermedia is divided into segments associated with their lifetime and each segment is transmitted during its lifetime. Furthermore, we proposed a version update mechanism to support changes in PRI belonging to the same segment.

This paper also presented a practical implementation of the time-lined hypermedia approach. Its cost conscious solution and usability are recognized by DIRECTV Japan[9], who started digital broadcasting using a communication satellite in

December 1997 in Japan, and they choose Digital Video eXtention system as the solution for their interactive services. Using the interactive service infrastructure with a hundred thousand STBs, some of the interactive services, which were described as service examples in this paper, were commercially started in October 1998. All the STBs for DIRECTV Japan are already equipped with the Interactive Data Player, and they are ready to offer the interactive services. We believe that our ideas on time-lined hypermedia could be applied to digital interactive broadcasting services, especially services based on MHEG5 type content models.

References

1.   DAVIC 1.2 Specification. Digital Audio Video Council, 1996. http://www.davic.org
2.   ISC/IEC 13522. Information technology - coding of multimedia and hypermedia information: Part 5 - support for base level interactive applications. Technical report, IS, 1996. (MHEG5)
3.   A. Antoniazzi and G. Schapeler. An open software architecture for multimedia consumer terminal. In proceedings of the Second European Conference on Multimedia Applications, Services and Techniques (ECMAST'97), pp. 621-634, May 1997.
4.   S. Acharya, R. Alonso, M. Franklib, and S. Zdonik. Broadcast disks: Data management for asymmetric communication environment. In proceedings of the ACM SIGMOD Annual International Conference on management of Data, pp. 199-210, May 1995.
5.   Kate Gerwig, "The push technology rage...so what's next?" ACM netWorker: The Craft of Network computing, Vol. 1, No. 2 (July/Aug. 1997), Pages 13-17.
6.   DVB-SI. Digital broadcasting systems for television, sound and data services: Specification for service information (SI) in digital video broadcasting (DVB) system. Technical report, ETSI ETS 300 468, TM1217, October 1995.
7.   ISO/IEC 13818-1. Information Technology - Generic Coding of Moving Pictures and Associated Audio Recommendation H.222.0 (systems), 1994.
8.   ISO/IEC 13818-6. Generic coding of moving pictures and associated audio information - part 6: Extension for digital storage media command and control (DSM-CC), 1997
9.   DIRECTV Japan. URL: http://www.directv.co.jp

Active Image Capturing and Dynamic Scene Visualization by Cooperative Distributed Vision

Takashi Matsuyama, Toshikazu Wada, and Shogo Tokai

Department of Intelligence Science and Technology
Kyoto University, Kyoto 606-8501, Japan
{tm, twada, tokai}@i.kyoto-u.ac.jp

Abstract. This paper addresses active image capturing and dynamic scene visualization by *Cooperative Distributed Vision* (CDV, in short). The concept of CDV was proposed by our five years project starting from 1996. From a practical point of view, the goal of CDV is summarized as follows: Embed in the real world a group of network-connected *Observation Stations* (real time video image processor with active camera(s)) and mobile robots with vision. And realize 1) wide-area dynamic scene understanding and 2) versatile scene visualization. Applications of CDV include real time wide-area surveillance, remote conference and lecturing systems, interactive 3D TV and intelligent TV studio, navigation of (non-intelligent) mobile robots and disabled people, cooperative mobile robots, and so on. In this paper, we first define the framework of CDV and give a brief retrospective view of the computer vision research to show the background of CDV. Then we present technical research results so far obtained: 1) fixed viewpoint pan-tilt-zoom camera for wide-area active imaging, 2) moving object detection and tracking for reactive image acquisition, 3) multi-viewpoints object imaging by cooperative observation stations, and 4) scenario-based cooperative camera-work planning for dynamic scene visualization. Prototype systems demonstrate the effectiveness and practical utilities of the proposed methods.

1   Introduction

This paper addresses active image capturing and dynamic scene visualization by *Cooperative Distributed Vision* (CDV, in short). The concept of CDV was proposed by our five years project starting from 1996.

From a practical point of view, the goal of CDV is summarized as follows (Fig. 1):

Embed in the real world a group of network-connected *Observation Stations* (real time video image processor with active camera(s)) and mobile robots with vision, and realize

1. wide-area dynamic scene understanding and
2. versatile scene visualization.

We may call it *Ubiquitous Vision.*

S. Nishio, F. Kishino (Eds.): AMCP'98, LNCS 1554, pp. 252–288, 1999.

Fig. 1. Cooperative distributed vision.

Applications of CDV include

- Real time wide-area surveillance and traffic monitoring systems
- Remote conference and lecturing systems
- Interactive 3D TV and intelligent TV studio
- High fidelity imaging of skilled body actions (arts, sports, medical operations)
- Navigation of (non-intelligent) mobile robots and disabled people
- Cooperative mobile robots.

The aim of the project is not to develop these specific application systems but to establish scientific and technological foundations to realize CDV systems enough capable to work persistently in the real world.

From a scientific point of view, we put our focus upon *dynamic integration of visual perception, action, and communication.* That is, the scientific goal of the project is to investigate how the *dynamics* of these three functions can be characterized and how they should be integrated *dynamically* to realize intelligent systems [1].

From a technological point of view, we design and implement hardwares and softwares to embody these three functions:

Visual Perception : versatile and high precision visual sensors, parallel and distributed real time vision systems.

Action : active camera heads, mobile robots with vision, and their dynamic control systems.

Communication : high speed wired and wireless network systems, communication protocols for cooperation, and cooperative distributed problem solving methods.

In this paper, we first define the framework of CDV and give a brief retrospective view of the computer vision research to show the background of CDV. Then we present technical research results so far obtained: 1) fixed viewpoint pan-tilt-zoom camera for wide-area active imaging, 2) moving object detection and tracking for reactive image acquisition, 3) multi-viewpoints object imaging by cooperative observation stations, and 4) scenario-based cooperative camerawork planning for dynamic scene visualization. Prototype systems demonstrate the effectiveness and practical utilities of the proposed methods.

2   Background and Basic Idea

Roughly speaking, the history of the computer vision research can be summarized as follows (Fig. 2):

–1970s: Image Processing: 2D Image $\mapsto$ 2D Image
A *given* input image is transformed into an output image to enhance its quality and to detect image features.
1980s: Computer Vision[1]: 2D Image $\mapsto$ 3D Scene
Recover 3D scene information from observed 2D image(s) based on geometric and photometric models of the imaging process.
1990s: The following two disciplines are being studied:
  1. **Active Vision**: Computer Vision $\times$ Physical Action $\mapsto$ Active Scene Understanding
     Integrate visual perception and physical action for active exploration of complex scenes [2], [3].
  2. **Image Media Processing**:
     Computer Vision $\times$ Computer Graphics $\mapsto$ Versatile Scene Visualization
     Integrate image analysis and synthesis methods to realize versatile scene visualization. Fig. 3 illustrates an example of the integration process: 3D Scene — *Imaging* $\to$ 2D Image(s) — *Computer Vision* $\to$ 3D Scene Description — *Edit* $\to$ Augmented 3D Scene Description — *Computer Graphics* $\to$ Image(s) of Virtualized/Augmented Scene.

The key idea of CDV is to *introduce network communication capabilities into active vision and image media processing*. That is, with the introduction of network communication capabilities, CDV systems are endowed with three

---

[1] Here we use "computer vision" in a narrow sense denoting computational and physics-based vision.

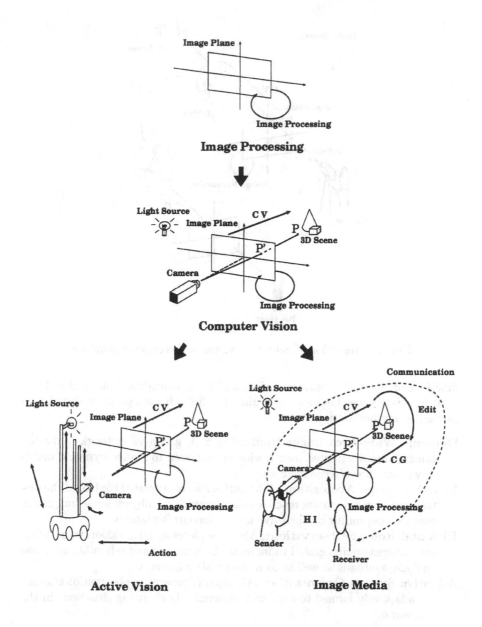

Fig. 2. History of the computer vision research.

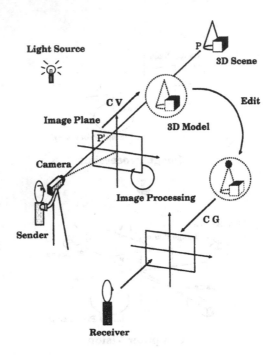

Fig. 3. Integration of computer vision and computer graphics.

functions of Visual Perception, Action, and Communication. The goal of CDV is to integrate these functions to realize the following cooperative distributed processing mechanisms:

Dynamic Wide-Area Image Capturing : A group of network-connected cameras are distributed over a wide spread area to realize dynamic multi-viewpoint object/scene imaging.

Reactive Image Acquisition : The active and coordinated control of the distributed cameras enables reactive image acquisition: object/scene images are captured depending on their dynamic behaviors/situations.

Rich and Robust Observation : Multiple pieces of information from different cameras are integrated to increase the accuracy and reliability of image analysis/synthesis as well as to measure 3D information.

Adaptive System Organization : Groups of cooperative observation stations are adaptively formed to cope with dynamically changing situations in the real world.

Using these mechanisms, both wide-area dynamic scene understanding and versatile scene visualization systems can be implemented.

We believe CDV offers a fundamental framework of visual information processing systems in the 21st century.

3 Fixed-Viewpoint Pan-Tilt-Zoom Camera for Wide-Area Active Imaging

3.1 Realization of Wide View Cameras

To develop wide-area video monitoring systems, we first of all should study methods of expanding the visual field of a video camera:

1. Omnidirectional cameras using fish-eye lenses and curved mirrors[4], [5], [6], or
2. Active cameras mounted on computer controlled camera heads[7].

In the former optical methods, while omnidirectional images can be acquired at video rate, their resolution is limited. In the latter mechanical methods, on the other hand, high resolution image acquisition is attained at the cost of limited instantaneous visual field.

In the CDV project, we took the active camera method;

- High resolution images are of the first importance for object recognition and scene visualization.
- Dynamic resolution control can be realized by active zooming, which increases adaptability and flexibility of the camera system.
- The limited instantaneous visual field problem can be solved by incorporating a group of distributed cameras.

Then, the next issue to be studied is how to design an active camera system. In this section, we first present an idea of a fixed viewpoint pan-tilt camera[8] and show the active camera head designed based on this idea. In the latter half of the section, we describe a sophisticated camera calibration method to make a commercial active video camera work as a fixed viewpoint pan-tilt-zoom camera. Experimental results demonstrate its practical utilities.

3.2 Fixed Viewpoint Pan-Tilt-Zoom Camera

Suppose we design a pan-tilt camera, where its optical axis is rotated around pan and tilt axes. This active camera system includes a pair of geometric singularities: 1) the projection center of the imaging system[2] and 2) the rotation axes. In ordinary active camera systems, no deliberate design about these singularities is incorporated, which introduces difficult problems in image analysis. That is, the discordance of the singularities causes photometric and geometric appearance variations during the camera rotation: varying highlights and motion parallax. In other words, 2D appearances of a scene change dynamically depending on the 3D scene geometry. To cope with such appearance variations, consequently, sophisticated image processing should be employed[7].

The following active camera design eliminates the appearance variations and hence greatly facilitates the image processing [8].

---

[2] We model the optical process of a camera by the perspective projection.

1. Make pan and tilt axes intersect with each other. The intersection should be at right to facilitate later geometric computations.
2. Place the projection center at the intersecting point. The optical axis of a camera should be perpendicular to the plane defined by the pan and tilt axes.

We call the above designed active camera the *Fixed Viewpoint Pan-Tilt Camera* (FV-PT camera, in short).

Usually, zooming can be modeled by the shift of the projection center along the optical axis[9]. Thus to realize the *Fixed Viewpoint Pan-Tilt-Zoom Camera* (FV-PTZ camera, in short), either of the following additional mechanisms should be employed:

- Design such a zoom lens system whose projection center is fixed irrespectively of zooming.
- Introduce a slide stage which adjusts the projection center fixed depending on zooming.

3.3   Image Representation for FV-PTZ Camera

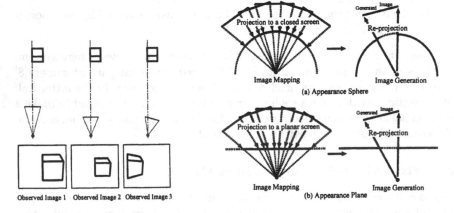

Fig. 4. Images observed by an FV-PTZ camera.

Fig. 5. Appearance sphere and plane.

While images observed by an FV-PTZ camera do not include any geometric and photometric variations depending on the 3D scene geometry, object shapes in the images vary with the camera rotation (Fig. 4). These variations are caused by the movement of the image plane, which can be rectified by projecting observed images onto a common virtual screen. On the virtual screen, the projected images form a seamless wide-area panoramic image.

For the rectification, we can use arbitrarily shaped virtual screens. The following are typical examples:

APS: When we can observe the 360° panoramic view, a spherical screen can be used (Fig. 5 (a)). We call the omnidirectional image on the spherical screen *APpearance Sphere* (APS in short).

APP: When the rotation angle of the camera is limited, we can use a planar screen (Fig. 5 (b)). The panoramic image on the planar screen is called *APpearance Plane* (APP in short).

As illustrated in the right side of Fig. 5, once an APS or an APP is obtained, images taken with arbitrary combinations of pan-tilt-zoom parameters can be generated by re-projecting the APS or APP onto the corresponding image planes. This enables the virtual look around of the scene.

The above mentioned omnidirectional image representation is equivalent to those proposed in [10] ∼ [12] in Computer Graphics and Virtual Reality. Our objective, however, is not to synthesize panoramic images natural to human viewers but to develop an active camera system that facilitates the image analysis for wide-area video monitoring. That is, in our case both the image acquisition and the projections onto/from virtual screens should be enough accurate to match well with physical camera motions. To attain such accuracy, we have to develop sophisticated camera calibration methods.

Fig. 6. Developed FV-PT camera head.

Fig. 7. High resolution APS representation of Kyoto University Clock Tower scene.

Fig. 8. FV-PTZ camera.

3.4 Camera Calibration

Fig. 6 shows the FV-PT camera head we developed, where a video camera is mounted on a group of adjustable slide and slant stages. We developed a high-precision camera calibration method using a laser beam to make the projection center coincide with the rotation center [8]. The wide rotation angles (i.e. -180° ≤ pan ≤ 180° and 0 ≤ tilt ≤ 45°) enables the APS representation of a scene

(Fig. 7). Note that using this camera head, any (compact) video camera with any lens system can be calibrated to realize an APS camera.

Fig. 8, on the other hand, illustrates an off-the-shelf active video camera, SONY EVI G20, which we found is a good approximation of an FV-PTZ camera ($-30° \leq$ pan $\leq 30°$, $-15° \leq$ tilt $\leq 15°$, and zoom: $15° \leq$ horizontal view angle $\leq 44°$) . We developed a sophisticated camera calibration method for this camera, with which we can use it as an FV-PTZ camera [1]. Note that this calibration method does not require any reference objects, and can be conducted automatically without any human support.

Fig. 9 illustrates a group of observed images with different (pan, tilt) parameters. Fig. 10 show the generated APP image. We verified that the physically accurate image mosaicing is realized on the APP image.

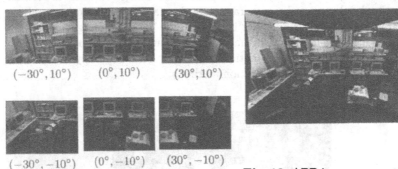

$(-30°, 10°)$ $(0°, 10°)$ $(30°, 10°)$

$(-30°, -10°)$ $(0°, -10°)$ $(30°, -10°)$

Fig. 9. Observed images.

Fig. 10. APP image generated from those in Fig. 9.

4   Moving Object Detection and Tracking for Reactive Image Acquisition

Since scenes in the real world are dynamically changing, image acquisition for computer vision and scene visualization should be done adaptively to dynamic situations. We call such adaptive image acquisition *Reactive Image Acquisition.*

This section first proposes a real time active vision system for object detection and tracking using the FV-PTZ camera. The tasks of the system are 1) detect an object which comes into the scene, 2) track it by controlling pan-tilt parameters, and 3) capture object images in as high resolution as possible by controlling the zoom. The system incorporates a sophisticated prediction-based dynamic control method 1) to cope with delays involved in image processing and physical camera motion and 2) to synchronize image acquisition and camera motion. Experimental results demonstrated that the proposed dynamic control method greatly improves the performance of the object tracking.

In the latter part of the section, we develop a dynamic scene visualization system using the above proposed active object tracking method. With this system, we can monitor detailed high-resolution object behaviors as well as its surrounding wide-area environments by a single FV-PTZ camera.

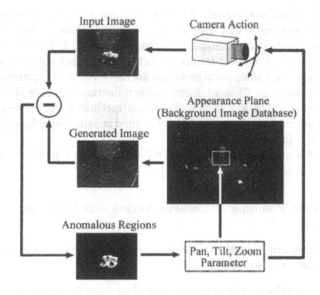

Fig. 11. Basic scheme of the object detection and tracking system.

4.1 Basic Scheme of Object Detection and Tracking

Fig. 11 illustrates the basic scheme of real-time moving object detection and tracking by the FV-PTZ camera:

1. Generate the APP image of the scene.
2. Extract a window image from the APP according to the current pan-tilt-zoom parameters and regard it as the background image.
3. Compute difference between the background image and an observed image.
4. If anomalous regions are detected in the difference image, select one and control the camera parameters to track the selected target.
5. Otherwise, move the camera along the predefined trajectory to search for an object.

This scheme is too naive and should be augmented in the following points:

Robust background subtraction : Although the background subtraction is a useful method to detect and track moving objects in video images, its effectiveness is limited; the stationary background scene assumption does not hold always in the real world.

System dynamics : The system dynamics realized by repeating the above steps sequentially is too simple to make the system adaptable to dynamically varying target object behaviors.

To augment the background subtraction for non-stationary scenes, [13], [14], and [15] employed probability distributions to model intensity variations at each pixel and used probabilistic anomaly computation methods for object detection. In [16], we proposed a novel robust background subtraction method for non-stationary scenes, where non-stationarities are modeled by 1) variations of overall lighting conditions and 2) local image pattern fluctuations caused by soughing leaves, flickering CRTs and so on. Since this method is time consuming, the current system employs the standard background subtraction followed by several auxiliary image processing operators.

In what follows, we concentrate ourselves on the design of the system dynamics.

4.2 Dynamic Planning of Camera Action and Image Acquisition Timing

The basic scheme requires that the image acquisition should be done taking the following points into account:

- **State of Action**: To prevent motion blurs from being included in an observed image[3], the image acquisition should be done when the camera stops or its speed is very slow. This means that the image acquisition cannot be done based on periodic clocks but should be triggered depending on the state of camera motion.
- **State of Target**: The image acquisition is to be done only when observed images are meaningful. That is, the images should include the target object in good appearance.

Thus, the determination of the image acquisition timing becomes a major concern in designing the system dynamics.

Fig. 12 shows the time chart of the perception-action cycle. Suppose the image acquisition is initiated at t_0. The right vertical bar in Fig. 12 illustrates the video cycle, which is not synchronized with the system; our FV-PTZ camera cannot accept the external trigger. Then, what the system has to determine are

1. $t_0 + \hat{t}_d$: the next image acquisition time and
2. such camera control command that satisfies
   1) A good target object image is taken at $t_0 + \hat{t}_d$.
   2) The camera motion is enough slow to apply the background subtraction at $t_0 + \hat{t}_d$.

---

[3] Motion blurs in an observed image incurs many false alarms in the background subtraction.

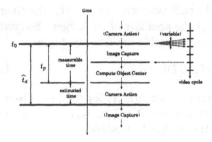 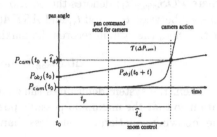

Fig. 12. Time chart.

Fig. 13. Estimation of the next view direction and image acquisition timing.

To solve theses problems, we first estimate the camera action dynamics. We conducted extensive experiments to model the dynamics of our FV-PTZ camera and obtained the following linear model:

$$t = T(\Delta P_{cam}, \Delta T_{cam}) = 0.007745 \times max\{\Delta P_{cam}, \Delta T_{cam}\} + 0.2986, \quad (1)$$

where $T(\Delta P_{cam}, \Delta T_{cam})$ denotes the time required to change pan and tilt angles by $(\Delta P_{cam}, \Delta T_{cam})$ and t is measured in second.

During tracking, the system measures t_p (see in Fig. 12) based on its internal clock and estimates the 2D target motion from the centroid displacement between object regions in a pair of consecutive video frames.

Then the system estimates both $(\Delta P_{cam}, \Delta T_{cam})$ to guide the camera toward the next view direction and $\hat{t}_d$, the next image acquisition timing in the following way. Suppose $\Delta T_{cam} < \Delta P_{cam}$. Fig. 13 graphically illustrates the dynamics of the target motion and camera action. That is, $\hat{t}_d$ and ΔP_{cam} are determined by the intersection point between the straight line representing the predicted target motion and the bent line representing the camera dynamics.

4.3 Dynamic Zoom Control

The dynamic zoom control should be implemented taking into account the following trade-off:

- To keep the target captured in observed images, wider view angles should be used; wider view fields can accommodate errors involved in the target motion and camera action estimations as well as image processing.
- To acquire detailed object images, larger zooming factors should be used.

To solve this trade-off, we employed the following dynamic zoom control method. During tracking, the system computes the instantaneous uncertainty degree at the ith observation time t_i, $\Delta UD(t_i)$:

$$\Delta UD(t_i) = \frac{POS_{error}(t_i)}{T(t_i) \times \sqrt{AREA(t_i)}}, \quad (2)$$

where $POS_{error}(t_i)$ denotes the positional prediction error at t_i, $T(t_i)$ the time interval between t_{i-1} and t_i, and $AREA(t_i)$ the area size of the target observed at t_i. Then, the system records the maximum possible uncertainty degree

$$\Delta UD_{max} = max\{\Delta UD(t_i)\}. \tag{3}$$

Then, the system determines the zooming factor $\alpha(t_{i+1})$ for the next observation so that the maximum possible position error, $POS_{error}^{max}(t_{i+1})$, defined by the following equation becomes less than the prefixed threshold.

$$POS_{error}^{max}(t_{i+1}) = \Delta UD_{max} \times (t_{i+1} - t_i)\sqrt{AREA(t_{i+1})} \tag{4}$$

$$AREA(t_{i+1}) = AREA(t_i) \times \alpha(t_{i+1}). \tag{5}$$

We conducted experiments to investigate the dynamics of the zoom control mechanism of our FV-PTZ camera and got the following observations:

- The zoom control can be done independently of the pan-tilt control.
- After the latency of about 0.05 sec, the zooming factor changes almost linearly.

Considering these observations and equation (1), which represents the dynamics of the pan-tilt control, the following zoom control method was implemented. 1) The pan-tilt control should have higher priority than the zoom control. 2) The former requires at least 0.2986 sec. Consequently, 3) the zoom can be changed in parallel with the pan-tilt control if the zoom control time is less than 0.2986 sec (see the bottom of Fig. 13). That is, after computing $\alpha(t_{i+1})$, the system modifies the zooming factor only by such an amount that satisfies this temporal constraint.

4.4 Performance Evaluation

To demonstrate the effectiveness of the proposed object tracking method, we conducted experiments to detect and track a radio controlled toy car. The car is manually controlled by a human; it moves around on the 4m × 4m flat floor avoiding several obstacles and sometimes stops and changes directions. The FV-PTZ camera is placed at about 2.5m above the floor corner looking downward obliquely. Fig. 14 shows a sequence of observed images and detected target silhouettes. Figs. 15 and 16 illustrate the histories of pan-tilt and pan-zoom controls during the tracking, respectively. The number i in the figures means the ith observation. The vertical axis of Fig. 16 denotes the horizontal view angle, which is inversely proportional to the zooming factor.

The entire tracking period is 13.77 seconds (i.e. about 2.1 image-acquisitions / second in average). Fig. 17 illustrates the dynamics of the image acquisition timing control. The solid line denotes the timing error, i.e. the difference between the predicted and practical image acquisition times. It almost stayed less than ±0.05 sec, the inevitable temporal fluctuation involved in the mechanical camera motion. The dotted line shows the time interval between a pair of consecutive

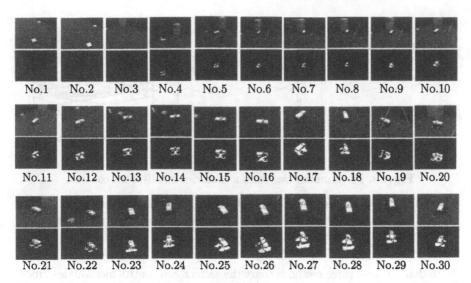

Fig. 14. Images observed during tracking(Upper:input images, lower:detected object silhouettes).

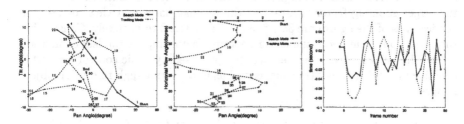

Fig. 15. History of pan-tilt control.   **Fig. 16.** History of pan-zoom control.   **Fig. 17.** Dynamics of the system (see text).

image acquisitions, where 0 denotes the average. These results verify that the adaptive system dynamics is realized depending on the target motion and the camera action.

To evaluate the effectiveness of the proposed dynamic control method, we conducted the following comparative study. The car is controlled to move continuously along almost the same circular track. Three FV-PTZ cameras, placed at almost the same position and with almost the same viewing direction, simultaneously track the car. The following three control methods are employed respectively.

Method 1 : Control the view direction to $(P_{obj}(t_0), T_{obj}(t_0))$, i.e. observed target location, without taking into account the target motion and the camera dynamics. The next image acquisition is done when the camera almost stops.

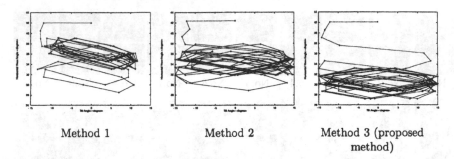

| Method 1 | Method 2 | Method 3 (proposed method) |

Fig. 18. Performance evaluation: histories of tilt-zoom controls.

Method 2 : Control the camera view direction by predicting the target motion while assuming the camera dynamics is constant. In the experiment, the camera motion is assumed to complete in 0.5 sec.

Method 3 : The proposed method.

Note that all these three methods share the same zoom control method described before.

Fig. 18 illustrates the histories of the tilt-zoom controls by these three methods. As is obvious from the figure, the more sophisticated control is employed, the larger zooming factor is attained; the average horizontal view angles (the vertical axis of the figure) are 35.7 °, 34.4 °, and 31.2 ° respectively. Considering the zoom control method, the larger zooming factor implies the less estimation error. This quantitatively verifies that the proposed dynamic control method is effective in moving object tracking as well as in capturing high-resolution object images.

4.5   Dynamic Scene Visualization by an FV-PTZ Camera

As is seen from the image sequence in Fig. 14, while the images taken by the tracking system nicely capture the target in very high resolutions, human viewers cannot understand the global target trajectory or the surrounding scene configuration. That is, foveated images are not enough for dynamic scene visualization.

[17] showed that the dynamic integration of foveated and peripheral views greatly facilitates tele-operations. They used a wide angle fixed camera for the global scene visualization and a pan-tilt-zoom camera for the local object visualization. 3D camera calibration establishes the correspondence between the images taken by these two cameras.

Using an FV-PTZ camera, on the other hand, we can easily realize seamless integration of foveated and peripheral views. That is, as is obvious from Fig. 11, foveated dynamic object images captured by the tracking system can be back-projected onto the APP image, which gives the peripheral view of the global scene. In other words, foveated and peripheral views are integrated on the APP. Fig. 19 shows an image sequence synthesized by this method, where a white quadrangle in each image illustrates the foveated person image captured by the tracking system.

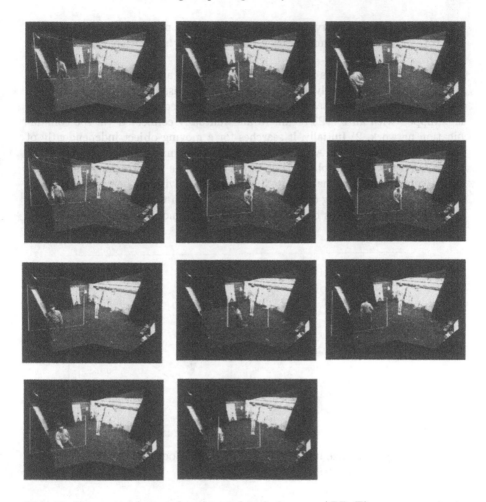

Fig. 19. Integrated foveated and peripheral views on APP. The sequence starts from the top-left and goes down followed by the next right column.

5   Multi-viewpoints Object Imaging by Cooperative Observation Stations

To realize versatile visualization of complex dynamic scenes, we have to employ a group of observation stations which cooperatively track objects and capture multi-viewpoints object images;

- Obstacles and other moving objects often interfere the view from a camera.
- Without specialized video cameras like [18], it is difficult to obtain 3D object information by a single camera.

Here we call an observation station with visual perception, camera action control, and network communication capabilities *Active Vision Agent* (AVA in short).

This section addresses a multi-AVA system (i.e. a group of communicating AVAs) which cooperatively detects and tracks a focused target object to obtain its 3D information. The task of the system is specified as follows: 1) Each AVA is equipped with an FV-PTZ camera and mutually connected via the communication network. 2) Initially, it searches for a moving object independently of the others. 3) When an AVA detects an object, it navigates the gazes of the other AVAs toward that object (Fig. 20). 4) All AVAs keep tracking the focused target cooperatively to measure its 3D information without being disturbed by obstacles or other moving objects (Fig. 21). 5) When the target goes out of the scene, the system returns back to the initial search mode.

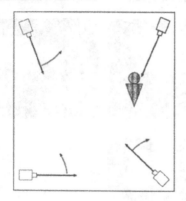
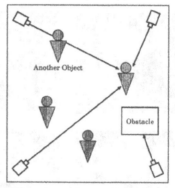

Fig. 20. Gaze navigation   **Fig. 21.** Cooperative gazing

The object detection and tracking by each AVA is realized by the same method as described in Section 4. We assume that while all FV-PTZ cameras are calibrated, 3D geometric configurations of the scene and obstacles are not known a priori. This is because the widely distributed camera arrangement makes it hard to employ stereo matching.

5.1   Integrating Visual Perception, Action, and Communication for Cooperative Object Tracking

In the cooperative object tracking, the following interactions among perception, action, and communication modules should be realized:

1. When no object appears in the scene, each AVA should search for an object autonomously by repeating its own perception-action cycle.

2. To realize the gaze navigation (Fig. 20), the camera actions of those AVAs which have not detected the target should be controlled by the information transmitted from the AVA that detected the target. This implies that the communication module in an AVA should be able to control its action module directly.
3. To realize the cooperative gazing (Fig. 21), the object identification should be established across multiple AVAs. Since all cameras are calibrated, if multiple AVAs capture object images simultaneously, the 3D location of the object can be computed, based on which the object can be identified. That is, for the object identification, the perception module of each AVA should be synchronized. Such synchronization is to be realized by communication among AVAs. Thus, the communication module in an AVA should be able to control its perception module directly.

Based on these considerations, we took the integration scheme where the communication module subordinates the perception and action modules.

5.2   Cooperative Object Tracking Protocol

In the above mentioned scheme, the design of the communication protocol becomes of the first importance in the system development. In designing the protocol, in turn, the ontology used for describing messages should be determined. Here we first propose a novel representation of the target object in the multi-AVA system, *Agency*, and then describe a cooperative object tracking protocol in terms of the agency.

Target Object Representation The most important ontological issue in the cooperative object tracking is how to represent the target object being tracked. In our multi-AVA system, "agent" means an AVA with visual perception, action, and communication capabilities. The target object is tracked by a group of such AVAs, whose perceptions and actions are tightly coupled (e.g. synchronized) by inter-AVAs communications.

Based on this consideration, we represent the target object by an *agency*, a group of those AVAs that are observing the target. With this object representation, specialized communication methods can be employed in the intra-agency communication: high-speed and low-latency communication methods to realize real-time synchronized behaviors of the member AVAs in the agency.

The above definition of the agency implies that the agency is not a static data structure but a dynamic entity with its own dynamics. We define its dynamics by the following two protocols:

Agency Formation Protocol: how and when the agency is formed.
Role Assignment Protocol: what roles the member AVAs in the agency take to cooperate.

Agency Formation Protocol Specifically speaking, the task of the prototype system is to track cooperatively by all AVAs such object that is first detected.

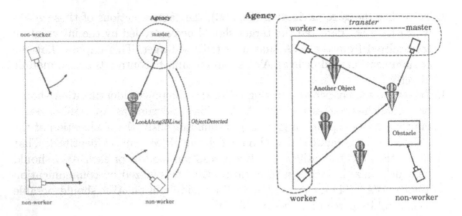

Fig. 22. Agency formation.   **Fig. 23.** Role assignment.

That is, while multiple moving objects can appear in the scene, the system tracks just one of them without paying any attention to the others. This task specification greatly simplifies the agency formation protocol.

1. Agency Generation Suppose no agency is generated yet. Note that as will be explained below, all AVAs know whether or not an agency is formed already. When AVA_i detects an object, it broadcasts the object detection message. If no other AVAs detect objects, then AVA_i generates an agency consisting of itself alone (Fig. 22). When multiple object detection messages are broadcast simultaneously, AVA_i can generate an agency only if it has the highest priority among those AVAs that have detected objects. That is, even if multiple AVAs detect objects simultaneously, which may or may not be the same, only one of them is allowed to generate an agency.

2. Joining into the Agency Once AVA_i has generated an agency, the other AVAs can know it by receiving the object detection message broadcast from AVA_i. Then they stop the autonomous object search and try to join into the agency.

Gaze Navigation :After generating an agency, AVA_i broadcasts the 3D line, L_i, defined by the projection center of its camera and the object centroid in the observed image. Then, the other AVAs search for the object along this 3D line by controlling their cameras respectively (Fig. 22).

Object Identification : Those AVAs which can successfully detect the same object as AVA_i are allowed to join into the agency. This object identification is done by the following method. Suppose AVA_j detects an object and let L_j denote the 3D view line directed toward that object from AVA_j. AVA_j reports L_j to AVA_i, which then examines the nearest 3D distance between L_i and L_j. If the distance is less than the threshold, a pair of detected objects by AVA$_i$ and AVA$_j$ are considered as the same object and AVA$_j$ is allowed to join the agency.

Object Tracking in 3D : Once multiple AVAs join the agency and their perception modules are synchronized, the 3D object location can be estimated by computing the intersection point among a group of 3D view lines emanating from the member AVAs. Then, the 3D object location is broadcast to the other AVAs which have not detected the object. The communication protocol among the member AVAs in the agency will be described later.

3. Exit from the Agency When the object goes behind an obstacle, some AVA in the agency may fail to track it. Then, such AVA exits from the agency and again searches for the object guided by the information broadcast from the agency. When all AVAs in the agency loose the object (e.g. when the object goes out of the scene), the agency dies out.

Role Assignment Protocol Once the agency is formed, its member AVAs works cooperatively to track the target object. To realize efficient cooperation among the member AVAs, we assign them different roles depending on situations. Here we address the role assignment protocol by which the role of each member AVA is specified. Note that since situations change dynamically, the roles of member AVAs are to be changed dynamically through mutual communications.

Since the agency represents the target object being tracked, it has to maintain the object motion history, which is used to guide the search of non-member AVAs. Such object history maintenance should be done exclusively by a single AVA in the agency to guarantee the consistency. We call the right of maintaining the object history the *master authority* and the AVA with this right the *master* AVA. The other member AVAs in the agency without the master authority are called *worker* AVAs and AVAs outside the agency *non-worker* AVAs (Fig. 23).

The transition between worker and non-worker is defined before in the agency formation protocol. So what we have to specify here is the protocol to transfer the master authority.

When an AVA first generates the agency, it immediately becomes the master. The master AVA conducts the object identification described before to allow other AVAs to join the agency, and maintains the object history. All these processings are done based on the object information observed by the master AVA. Thus, the reliability of the information observed by the master AVA is crucial to realize robust and stable object tracking. In the real world, however, no single AVA can keep tracking the object persistently due to occluding obstacles and interfering moving objects.

The above discussion leads us to introducing the dynamic master authority transfer protocol. That is, the master AVA always checks the reliability of the object information observed by each member, and transfers the master authority to such AVA that gives the most reliable object information (Fig. 23).

The reliability can be measured depending on observed object characteristics (size, speed), scene situations (occluding objects, local lightings), AVA's visual

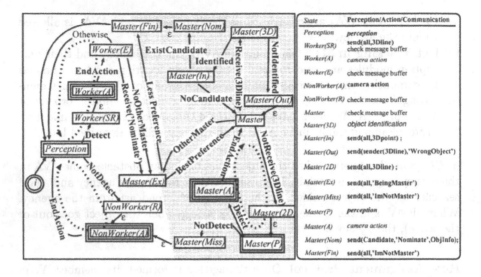

State	Perception/Action/Communication
Perception	*perception*
Worker(SR)	send(all,3Dline) check message buffer
Worker(A)	*camera action*
Worker(E)	check message buffer
NonWorker(A)	**camera action**
NonWorker(R)	check message buffer
Master	check message buffer
Master(3D)	*object identification*
Master(In)	send(all,3Dpoint) ;
Master(Out)	send(sender(3Dline),'WrongObject')
Master(2D)	send(all,3Dline) ;
Master(Ex)	send(all,'BeingMaster')
Master(Miss)	send(all,'ImNotMaster')
Master(P)	*perception*
Master(A)	*camera action*
Master(Nom)	send(Candidate,'Nominate',ObjInfo);
Master(Fin)	send(all,'ImNotMaster')

Fig. 24. State transition network for the cooperative object tracking.

perception capabilities (size of view field, view direction) and action characteristics (camera head speed), and so on. The prototype system employs a simple method: the master AVA transfers the authority to such member AVA whose object observation time is the latest in a predefined time period, since the latest object information may be the most reliable. Note that using this role assignment protocol, the master authority is continuously transfered around among member AVAs.

5.3   Implementation by a State Transition Network

Fig. 24 illustrates the state transition network designed to implement the above mentioned cooperative object tracking protocols. The network specifies event driven asynchronous interactions among perception, action, and communication modules as well as communication protocols with other AVAs, through which behaviors of an AVA emerge.

In Fig. 24, state i in the double circles denotes the initial state. Basically the states in rectangular boxes represent the roles of an AVA: master, worker, and non-worker. Since the master AVA conducts several different types of processing depending on situations, its state is subdivided into many substates. Those states in the shaded area show the states with the master authority. Each arrow connecting a pair of states is associated with the condition under which that state transition is incurred. ε means the unconditional state transition.

The right side of the figure shows what kind of processing, i.e. perception, action, receive, or send, is executed at each state. Those state in double rectangular boxes denote the states where perception is executed, while at those states in triple rectangular boxes, the camera action is executed. Thus, each state has its own dynamics and dynamic behaviors of an AVA are fabricated by state transitions.

Note that the prototype system assumes that the communication network is free from failures and delays. More robust and real time communication protocols should be developed for real world applications.

5.4   Experimental Results

While the prototype system is far from complete, we conducted experiments to verify its potential performance. Two persons walked around a large box located at the center of the room (5m × 6m). Four FV-PTZ cameras are placed at the four corners of the room respectively, looking downward obliquely from about 2.5m above the floor. The person who first entered in the scene was regarded as the target. He crawled around the box not to be detected by the cameras. The other person walked around the box to interfere the camera views toward the target person. Then, both went out from the scene and after a while, a new person came into the scene.

Fig. 25 illustrates partial image sequences observed by the four cameras, where the vertical axis represents the time when each image is captured. Each detected object is enclosed by a rectangle. Note that while some images include two objects and others nothing, the gaze of each camera is directed toward the crawling target person. Note also that the image acquisition timings of the four cameras are almost synchronized. This is because the master AVA broadcasts the 3D view line to or the 3D position of the target to the other AVAs, by which their perception processes are activated. This synchronized image acquisition by multiple cameras enables the computation of the 3D target motion trajectory (Fig 26).

Fig. 27 illustrates the dynamics of the system, the state transition histories of the four AVAs. We can see that the system exhibits well coordinated behaviors as designed. That is, the entire system works in the following three modes:

Mode 1: All AVAs are searching for an object.

Mode 2: The master AVA itself tracks the object since the others are still searching for the object.

Mode 3: All AVAs form the agency to track the object under the master's guidance.

The zigzag shape in the figure shows the continuous master authority transfer is conducted inside the agency.

Once a group of multi-viewpoints object images are obtained, we can generate a 3D object shape as well as measure its 3D location. We developed a sophisticated camera calibration method among widely distributed cameras and an efficient 3D shape reconstruction algorithm based on the 3D shadow volume intersection method[19]. Fig. 28 shows multi-viewpoints APP images of a mannequin and its reconstructed 3D shape. Currently we are developing a real time 3D shape reconstruction system using the multi-AVA system.

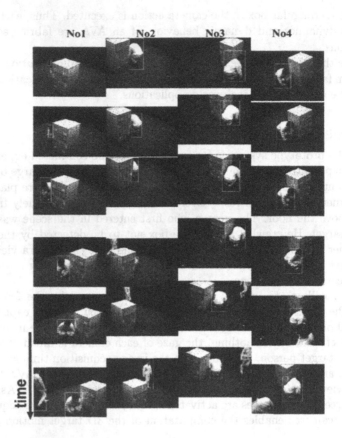

Fig. 25. Partial image sequences observed by four cameras. The vertical length of an image represents 0.5 sec.

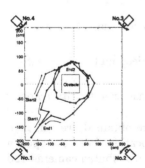

Fig. 26. 3D target motion trajectories.

Fig. 27. State transition histories of the four AVAs.

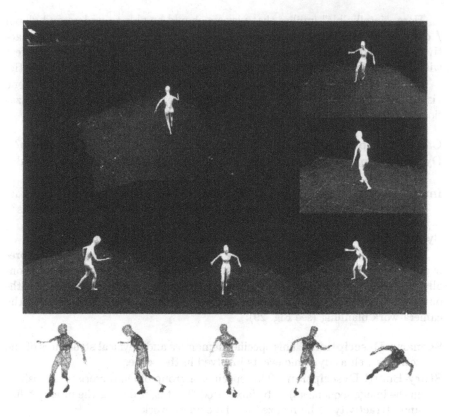

Fig. 28. Multi-viewpoints APP images and the reconstructed 3D shape of a mannequin. Top-left: entire APP image and windowed images extracted from the other five APP images.

6 Scenario-Based Cooperative Camera-Work Planning for Dynamic Scene Visualization

6.1 Camera-Works for Intelligible and Attractive Scene Visualization

Most of active vision systems developed so far including the ones described in Sections 4 and 5 capture images to control cameras and understand scene structures. This section, on the other hand, addresses active camera control methods for dynamic scene visualization. There exists a large difference between these two tasks; while the former throws away observed image data after processing, the latter puts its focus upon how we can fabricate image sequences intelligible and attractive for human viewers.

Here, "intelligible" implies that viewers should be able to understand global / dynamic scene contexts from limited sequences of captured images. "Attractive" means fabricated image sequences should keep attracting viewers' interest without being felt tired or boring. If possible, moreover, they should be artistic.

As discussed in Section 1, CDV offers a fundamental framework for scene visualization as well as scene understanding. To realize versatile scene visualization, we have to solve the following problems:

Camera Layout : How many and where should we put a group of cameras?

Dynamic Camera Control : How should we control camera parameters dynamically?

Image Sequence Fabrication : How should we fabricate intelligible and attractive image sequence(s) from raw image data observed by the cameras?

By *camera-work planning* we mean methods to solve these problems.

Since the real world includes a wide spectrum of dynamic scenes and moreover, the intelligibility and attractivity are too abstract to define computationally, it is almost impossible to attain the meaningful camera-work planning without knowledge. The following three types of knowledge can be used for the camera-work planning (see Fig. 29):

Scenario Description : This specifies semantics and physical structures of the scene as well as dynamic events involved in the scene.

Story-Board Description : This specifies a group of characteristic snapshots in the image sequence(s) to be fabricated. That is, it defines the intelligibility and attractivity to be realized by the camera-work.

Know-Hows about Camera-Works : Many effective camera-works have been developed in cinematography[20]. They include a variety of camera layout, image framing, and camera switching techniques. We can use such know-hows for the camera-work planning.

Note that the first one is described in terms of abstract semantic and/or 3D physical scene features, the second 2D image appearances taking into account psychological effects onto human viewers, and the third includes transformation rules between them.

Camera-work planning systems incorporate these three types of knowledge to solve the above mentioned three problems. In general, the planning should be done in the following two stages:

Off-Line Planning : Given a scenario description to be visualized, the system first makes a camera-work *plan* based on the knowledge.

On-Line Camera Control : Since the scenario is just a rough model of the real world scene, real world situations usually deviate from the scenario. Thus, on-line adaptive camera controls should be conducted during the scene visualization process. CDV systems such as those described in Sections 4 and 5 support such on-line adaptive camera controls.

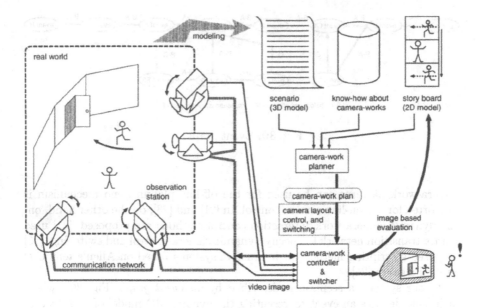

Fig. 29. Framework of the scenario-based cooperative camera-work planning for dynamic scene visualization.

Fig. 29 summarizes the framework for the dynamic scene visualization discussed above.

In this section, we describe a scenario-based dynamic scene visualization system being developed in the CDV project, where major emphasis is put upon dynamic cooperation between distributed cameras (i.e. observation stations). That is, we believe that to fabricate intelligible and attractive image sequence(s) from those observed by the cameras, flexible inter-camera coordinations are required as well as individual dynamic camera controls.

6.2   System Organization

Here we describe specifications of each component of the scenario-based scene visualization system in Fig. 29.

Knowledge Sources As discussed before, three types of knowledge is give to the system:

1. Scenario Description There have been proposed several scenario/camera-work description methods and camera-work planning systems [21], [22], [23]. In [22], Christianson et al proposed the Declarative Camera Control Language, with which various types of camera-work patterns can be described. While the

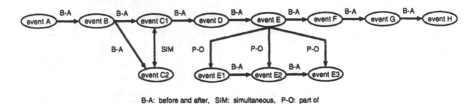

B-A: before and after, SIM: simultaneous, P-O: part of

Fig. 30. Event graph.

camera-work patterns can be used for the off-line planning, no mechanism is supported for the on-line camera control. In [21] and [23], on the other hand, on-line dynamic camera-work/interaction control methods are proposed. [21] used a state transition network to specify dynamic camera control and switching. [23] proposed a scripting method for interactive systems based on Allen's temporal interval algebra [24].

In our system, a scenario is described by an *event graph* (Fig. 30), where each node denotes an event representing the dynamic 3D model of a real world scene and an arc a temporal/geometric/semantic relation between events. The simplest but most popular event graph is a series of event nodes connected by a chain of directed arcs denoting the temporal order (i.e. B-A arcs in Fig. 30). Various types of semantic arcs, such as retrospection, hearsay, and illusion, may be used to enrich scenario contents.

An event node includes:

- Semantic Scene Features: type of the event and/or atmosphere of the scene, e.g. fighting, thrilling adventure, happy dining, solemn ceremony and so on.
- Background Scene Characteristics: overall geometric and illumination structures of the scene and their dynamic variations: e.g. soccer field, crowded downtown, conference room, and so on.
- Foreground Object Characteristics: attributes and dynamics of objects requiring focused imaging. Sometimes mental features and moods of objects may be associated with physical characteristics. For example, a tall man in a red shirt rushes out through the door crying loudly.

2. Story-Board Description This is described by a series of 2D sketches specifying how each *shot* in the finally fabricated image sequence looks like (Fig. 29). That is, it is the goal specification for the scene visualization. It contains the information about viewing angle, image framing, camera position, motion, and switching.

In addition, each sketch is associated with an event ID(s) in the event graph. Note that in general, associations between events and sketches are M : N. That is, it is very popular that an event is visualized by a series of shots taken from different cameras. We call a continuous video sequence taken by a camera *physical shot*. On the other hand, a single image frame sometimes includes multiple physical shots representing different events: e.g. a group of scientists in a conference room are discussing about the earth looking a video image transmitted from a satellite. We call such composite shot *logical shot*. That is, each sketch specifies characteristics of a logical shot, while an event corresponds to a physical scene.

3. Know-Hows about Camera-Works Since the story-board is just a rough and abstract goal specification, we need additional knowledge to attain the goal under the given scenario description. Know-hows about camera-works specify heuristic rules to take intelligible and attractive image sequences under various scenario situations. They include rules for camera layout, dynamic camera control, and switching. Note that camera control and switching rules specify not only actions of each individual camera but also coordination methods among distributed cameras.

Camera-Work Planning

Off-Line Planning Given three knowledge sources described above, the camera-work planner (Fig. 29) reasons about effective camera-works for the scene visualization.

1. First, for each event in a given scenario, the planner determines geometric camera layout, dynamic camera action, and temporal camera switching and coordination. Since there exist many different possible camera-works to visualize a given event, the planner uses sketches in a given story-board to select the most effective camera-work rule. Note that the camera-work plan generated at this stage specifies physical shots obtained by the cameras placed in the scene.
2. Then, the planner determines an image composition plan to fabricate logical shots specified in the story-board. Note that while most of logical shot compositions are realized by 2D image processing, virtual images may be synthesized based on the 3D scene information restored from multi-viewpoints image sequences. Note also that the planner should make camera coordination plans across multiple events to generate well synchronized/organized logical shots.

On-Line Camera Control After placing a group of cameras according to the designed camera layout, the camera-work and logical shot composition plans are loaded onto a group of observation stations and the camera-work controller & switcher respectively (Fig. 29). Then,

1. The cameras stand by and objects in the real world start the actions specified in the scenario.
2. Each observation station captures an image sequence by controlling camera parameters according to the camera-work plan. The acquired image sequence is delivered to the controller & planner. As noted before, each observation

station should adaptively control its camera since the scene usually deviates spatially and temporally from the plan. Moreover, multiple observation stations should cooperate with each other through communications to control their cameras. These adaptive and cooperative camera controls are realized using sketches in the story-board as goal specifications.

3. In the camera-work controller & switcher , a series of logical/physical shots are fabricated from a group of raw image sequences captured by observation stations. Note that the camera-work controller & switcher itself may generate virtual image sequences based on the 3D scene information restored from multi-viewpoints image sequences. Thus, it should dynamically communicate with observation stations to realize smoothly connected and/or well synchronized logical shots. The smoothness and synchronization are evaluated at the 2D image level referring the story-board (Fig. 29). In this sense, we may call it a director and/or a composer.

6.3   Prototype System

Currently we are developing a prototype system based on the framework proposed above. Here, we show two simulation results of the camera-work planning: (1) camera layout for 2D static scenes including multiple objects and obstacles and (2) scenario-based camera control and switching in 3D dynamic scenes. Simulation results demonstrate that our approach is very promising.

Planning Layout of Multiple Cameras in 2D Static Scenes Where to place a group of cameras is one of major problems to be solved in the off-line camera-work planning. We developed an optimization method for the camera layout.

First we assume the followings:

- The scene is two dimensional and static.
- The background scene is defined as a rectangular area, in which foreground objects, obstacles, and cameras are placed (Fig. 31(a)).
- An object is represented by a circle with a specific "front face" direction (Fig. 31(a)). Each point on the circular object surface is associated with a weight representing the importance for the visualization. In the current simulation, we used the following function to model the weight distribution over the surface:

$$Importance(\alpha) = \frac{1 + \cos(\alpha)}{2}, \qquad (6)$$

where α denotes the angular distance from the front face direction.

- An camera is modeled by a projection center with a fixed viewing angle (i.e. fixed zoom, Fig. 31(b)(c)). Note that this angle specifies the size of the image frame (i.e. area covered by an image). In addition, each camera is associated with a list of foreground objects to be imaged.

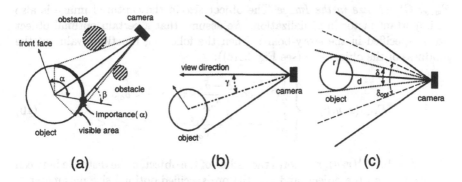

Fig. 31. Evaluation functions for the camera layout.

First, the size and shape of the background scene, locations and characteristics of objects and obstacles, and the number and viewing angles of cameras are given to the camera layout planner. Then, the position and viewing direction of each camera, (x, y, θ), is determined by optimizing the following evaluation function.

$$E_{total}(x, y, \theta) = \sum_{i \in Object-List} \{E^i_{visibility}(x, y, \theta) \times E^i_{position}(x, y, \theta) \times E^i_{size}(x, y, \theta)\},$$
(7)

where $Object - List$ denotes a list of the objects to be imaged by the camera.

Each component evaluation function is defined for an object-camera pair as follows:

$E_{visibility}$: *Object Surface Visibility* This evaluates how well the object looks in the captured image. We used the following function (see Fig. 31(a)):

$$E_{visivility}(x, y, \theta) = \int_{visible} Importance(\alpha) \, \cos\beta \, \cos\alpha \, d\alpha,$$
(8)

where α denotes the angular distance of a surface point from the front face direction and β the angle between the surface normal at that point and the view direction from the camera. The integral covers only those surface points that can be seen from the camera without being interfered by obstacles.

$E_{position}$: *Object Position in the Image* We assume the object is to be captured at the center of the image. Then, the following function evaluates the goodness of the object position (see Fig. 31(b)):

$$E_{position}(x, y, \theta) = \frac{1}{2}(1 + \cos\gamma),$$
(9)

where γ denotes the angle between the object center direction from the camera and the view direction of the camera.

E_{size}: *Object Size in the Image* The object size in the captured image is also an important factor in visualization. We assume that a certain optimal object size is specified in the story-board. Then the following function evaluates the goodness of the object size (see Fig. 31(c)):

$$E_{size}(x, y, \theta) = \begin{cases} \frac{1}{2}(1 + \cos \frac{\delta - \delta_{opt}}{\delta_{opt}} \pi) & \text{if } \delta \leq \delta_{opt} \\ \frac{1}{2}(1 + \cos \frac{\delta - \delta_{opt}}{\pi - \delta_{opt}} \pi) & \text{if } \delta_{opt} < \delta \leq \pi \end{cases} \tag{10}$$

where $\delta = 2\sin^{-1}(r/d)$, r denotes the radius of the object, d the distance between the camera and the object, and δ_{opt} the pre-specified optimal size parameter.

We conducted several simulations to examine the effectiveness of the above mentioned camera layout method. Fig. 32(a) illustrates the geometric configuration of a pair of objects to be visualized. Fig. 32(b) shows (1) the spatial distribution of $E_{total}(x, y, \theta)$ and (2) the optimal camera position and its viewing direction when both objects are required to be imaged simultaneously by a single camera. To depict (1), the optimal view direction, θ^*, is first computed at each position and $E_{total}(x, y, \theta^*)$ is encoded by the gray level: the brighter the gray level is, the higher value the evaluation function takes. (2) is depicted by a group of three line segments: their intersection point denotes the camera position, the central segment the view direction, and the pair of marginal ones the viewing angle. Figs. 32(c) and (d) show the optimal camera layouts when each object is required to be imaged by a single camera, respectively.

Fig. 33 illustrates the optimal layout of a pair of cameras when camera-A and camera-B are used for imaging {object-0, object-1} and {object-2, object-3}, respectively, where {·} denotes the list of objects to be imaged by a camera.

While these simulation results are simple and include many points to be improved, we believe they showed practical utilities of our framework. Currently we are developing a novel camera layout method which utilizes the story-board as the evaluation function.

Dynamic Camera Coordination for Smooth Camera Switching Here we demonstrate the importance of the on-line coordinated camera control and switching in visualizing dynamic scenes.

Suppose a scenario description specifies that "A man is running along the long straight path at the constant speed." and the story-board requires that his zoomed-up face should be captured continuouly since changes of his facial expressions are the crucial factor for visualizing the scene.

Based on these knowledge sources, the camera-work plan illustrated in Fig. 34 is generated at the off-line planning stage. The plan specifies (1) a pair of cameras are placed at the same side along the path, (2) each camera tracks the face by dynamically rotating the view direction[4], and (3) the image sequence taken by camera-1 should be switched to that taken by camera-2 when both image sequences can be smoothly connected. Here we assume the smoothness is

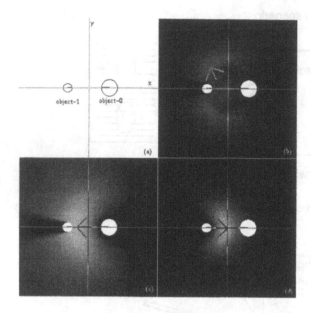

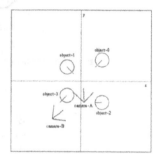

Fig. 32. Optimal camera layouts: (a) a pair of objects to be visualized, (b) optimal camera layout for simultaneous imaging of the objects, (c),(d) optimal camera layouts for object-0 and object-1, respectively.

Fig. 33. Optimal layout of a pair of cameras for four objects.

evaluated by the apparent face motion against the backgound scene in captured image sequences.

This camera-work plan is loaded onto a pair of observation stations and the camera-work controller & switcher in Fig. 29. When the action in the scene is started, the object detection and tracking process such as described in Section 4 is executed at each observation station. Then, the camera-work controller & switcher monitors a pair of image sequences captured by the observation stations and determines the optimal camera switch timing.

As noted before, the actual scene usually deviates spatially and temporally from the scenario. Fig. 35 illustrates the geometric configuration of the scene, the camera layout, and the object motion path described in the scenario. Here we assume that the actual object motion path deviates from the plan as shown in the figure. In what follows, we will demonstrate the importance of the on-line camera control in determining the optimal camera switch timing.

In the current simulation, the switch timing is evaluated by the difference in the camera rotation speed. The reason for this is as follows. Firstly, since both cameras are tracking the object, the object image stays fixed at the center

---

[4] For simplicity, we assume only the 2D panning is allowed for each camera.

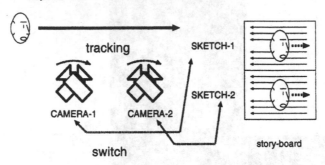

Fig. 34. Camera-work plan for dynamic scene visualization.

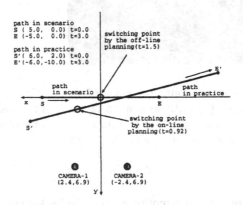

Fig. 35. Deviation of object motion.

of the image frame. Thus, human viewers perceive the object motion speed based on the optical flow of the background scene. Assuming the distance of the background scene from the cameras is constant, the camera rotation speed uniquely determines the strength of the optical flow. In other words, by switching the cameras when their rotation speeds coincide with each other, human viewers perceive the object as moving at the constant speed even if the camera is switched from one to the other. Note that to realize more smooth camera switching, we should control the zoom so as to make the object sizes in the pair of captured image sequences coincide.

Fig. 36 illustrates temporal variations of the rotation speeds of camera-1 and camera-2 when they are tracking the object along the path specified in the scenario. The optimal camera switch timing is determined as $t = 1.5$ sec and the object location at that time is shown in Fig. 35. Fig. 38(a) shows the image sequence fabricated from the pair of image sequences taken by camera-1

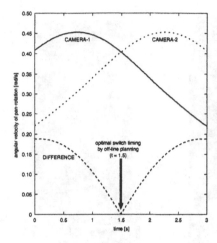
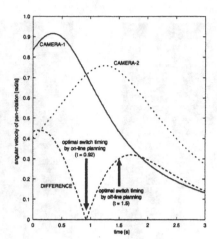

Fig. 36. Optimal camera switch timing determined by the off-line camera-work planning.

Fig. 37. Optimal camera switch timing determined by the on-line camera control.

and camera-2, assuming the object moves as specified in the scenario and the camera is switched at $t = 1.5$ sec.

If we directly applied this planned camera-work to the actual scene, we would obtain such a meaningless image sequence as shown in Fig. 38(b), which demonstrates the necessity of the on-line adaptive camera control.

Fig. 37 illustrates temporal variations of the rotation speeds of camera-1 and camera-2 when they are adaptively tracking the actual object motion shown in Fig. 35. The camera-work controller & switcher dynamically monitors these camera motion speeds and switches the cameras at $t = 0.92$ sec (see Fig. 37). Fig. 38(c) illustrates the image sequence fabricated by this on-line adaptive camera control and switching method, where the soomthly connected image sequence is fabricated.

6.4 Discussions

In this section we proposed a framework of senario-based cooperative camera-work planning for dynamic scene visualization. Its novel features are

- Introduction of three types of knowledge sources: scenario, know-hows about camera-works, and story-borad.
- Off-line camera-work planning followed by on-line dynamic camera control and switching.
- Cooperation among distributed active cameras (i.e. observation stations) to adaptively capture intelligible and attractive image sequences.
- Logical and virtual image shots fabrication from multi-viewpoint image sequences.

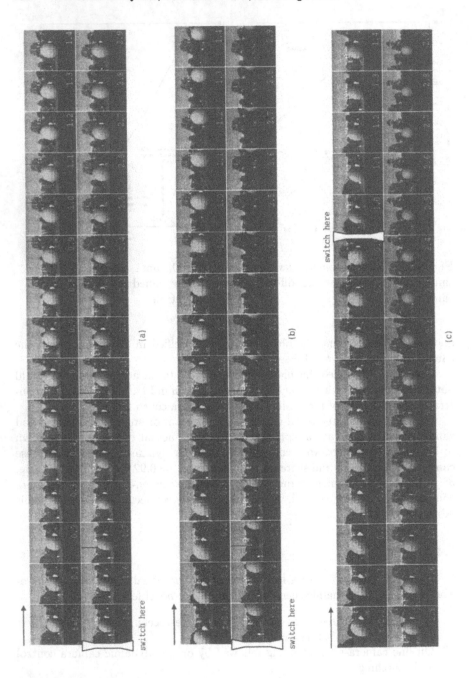

Fig. 38. Fabricated image sequences. Each sequence starts at the left of the upper row followed by the lower row including a mark denoting the point of the camera switching.

While we have shown practical utilities of our approach with several simulations, the following technical developments are required to implement a scene visualization system that can work in real world scenes.

- Description languages for the knowledge sources and the camera-work plan
- Knowledge-based camera layout and dynamic camera-work planning for 3D dynamic scenes
- Plan-guided dynamic camera control for scene/object visualization
- Dynamic cooperation protocols for well organized/synchronized multi-viewpoint visualization
- Image sequence switching and virtual image generation for intelligible and attractive image sequence fabrication
- Computational method of evaluating intelligibility and attractivity.

7   Concluding Remarks

This paper describes the idea and goal of our five years project on cooperative distributed vision and shows technical research results so far obtained on active image capturing and dynamic scene visualization: 1) fixed viewpoint pan-tilt-zoom camera for wide-area active imaging, 2) moving object detection and tracking for reactive image acquisition, 3) multi-viewpoints object imaging by cooperative observation stations, and 4) senario-based cooperative camera-work planning for dynamic scene visualization. Prototype systems demonstrate the effectiveness and practical utilities of the proposed methods.

The project holds annual international workshops, where research results are presented with working demo systems. All research results and activities of the project are shown in the homepage (URL: http://vision.kuee.kyoto-u.ac.jp/CDVPRJ).

This work was supported by the Research for the Future Program of the Japan Society for the Promotion of Science (JSPS-RFTF96P00501). Research efforts by all members of our laboratory and the assistance of Ms. H. Taguchi in preparing figures are gratefully acknowledged.

References

1. Matsuyama, T.: Cooperative Distributed Vision - Dynamic Integration of Visual Perception, Action, and Communication -, Proc. of Image Understanding Workshop, Monterey CA, 1998.11
2. Aloimonos, Y. (ed.): Special Issue on Purposive, Qualitative, Active Vision, CVGIP: Image Understanding, Vol.56, No.1, 1992.
3. Aloimonos, Y. (ed.): Active Perception, Lawrence Erlbaum Associates Publisher, 1993
4. Yagi Y. and Yachida M.: Real-Time Generation of Environmental Map and Obstacle Avoidance Using Omnidirectional Image Sensor with Conic Mirror, Prof. of CVPR, pp. 160-165, 1991.

5. Yamazawa K., Yagi Y. and Yachida M.: Obstacle Detection with Omnidirectional Image Sensor HyperOmni Vision, Proc. of ICRA, pp.1062 - 1067, 1995.
6. Peri V. N. and Nayar S. K.: Generation of Perspective and Panoramic Video from Omnidirectional Video, Proc. of IUW, pp.243 - 245, 1997.
7. Murray,D. and Basu,A.: Motion Tracking with an Active Camera, IEEE Trans. of PAMI, Vol. 16, No. 5, pp. 449-459, 1994.
8. Wada T. and Matsuyama T.: Appearance Sphere: Background Model for Pan-Tilt-Zoom Camera, Proc. of ICPR, Vol. A, pp. 718-722, 1996.
9. Lavest, J.M., Delherm, C., Peuchot, B, and Daucher, N.: Implicit Reconstruction by Zooming, Computer Vision and Image Understanding, Vol.66, No.3, pp.301-315, 1997.
10. Hall R.: Hybrid Techniques for Rapid Image Synthesis, in Image Rendering Tricks (Whitted T. and Cook R. eds.), Course Notes 16 for SIGGRAPH'86, 1986.
11. Greene N.: Environment Mapping and Other Applications of World Projections, CGA, 6 (11), pp. 21-29, 1986.
12. Chen S.E.: QuickTime VR – An Image-Based Approach to Virtual Environment Navigation, Proc. of SIGGRAPH'95, pp. 29-38, 1995.
13. Nakai, H.: Robust Object Detection Using A-Posteriori Probability, Tech. Rep. of IPSJ, SIG-CV90-1, 1994 (in Japanese).
14. Grimson, E.: A Forest of Sensors, Proc. of VSAM Workshop, 1997.
15. Davis, L.: Visual Surveillance and Monitoring, Proc. of VSAM Workshop, 1997.
16. Habe, H., Ohya, T., and Matsuyama, T.: A Robust Background Subtraction Method for Non-Stationary Scenes, Proc. of MIRU'98, Vol.1, pp.467-472, 1998 (in Japanese).
17. Yamaashi, K., Cooperstock, J.R., Narine, T., and Buxton, W.: Beating the Limitations of Camera-Monitor Mediated Telepresence with Extra Eyes, Proc. of CHI, pp.50-57, 1996.
18. Hiura, S. and Matsuyama, T.: Depth Measurement by the Multi-Focus Camera, Proc. of CVPR, pp.953-959, 1998
19. Mikoshi, Y.: 3D Image Measurement Based on Planes, Master Thesis, Kyoto University, 1998 (in Japanese)
20. Arijon, D.: Grammar of the Film Language, Focal Press Ltd., London, 1976
21. He,L., Cohen,M.F., and Salesin, D.H.: The Virtual Cinematographer: A Paradigm for Automatic Real-Time Camera Control and Directing, SIGGRAPH'96, pp.217-224, 1996.
22. Christianson,D.B., Anderson,S.E., He,L., Weld,D.S., Cohen,M.F., and Salesin, D.H.: Declarative Camera Control for Automatic Cinematography, Proceedings of AAAI '96, pp.148-155, 1996.
23. Mase,K., Pinhanez,C.S., and Bobick, A.F.: Scripting Method Based on Temporal Intervals for Designing Interactive Systems, Trans. of IPSJ, Vol.39, No.5, pp.1403-1413, 1998 (in Japanese).
24. Allen, J.F.: Towards a General Theory of Action and Time, Artificial Intelligence, Vol.23, pp.123-154, 1984.

Videoplex: A New System Framework for Constructing Video-Based Three-Dimensional Space

Go Nishimura, Tamio Kihara, and Ryoji Kataoka

NTT Information and Communication Systems Laboratories
1-1 Hikari-no-oka, Yokosuka-shi, Kanagawa, 239-0847 Japan
{go, tamio, kataoka}@dq.isl.ntt.co.jp

Abstract. By placing "motion objects" in a three-dimensional space, interactive videos can be easily produced. By managing both the walk-through images and the motion objects inside the three-dimensional space, and also by synthesizing the displayed video according to the user's location and actions, the system enables a more natural synthesized video experience to be created. In this paper, we propose a new video system framework called *Videoplex* for a video-based application that realizes both free movement of the user's viewpoint and object motion independent of viewpoint movement.

1. Introduction

This paper presents *Videoplex,* our new video system framework for an interactive video-based application. Its goals are to allow users to interact with individual objects in a 3D space and to provide functions that will make a stroll-type information guidance service using natural motion video more lifelike and attractive.

Video information is suitable for presenting detailed movement and expressions of objects in the real world. Therefore, an information guidance system based on natural motion video can provide lifelike information that cannot be equaled by computer graphics. Traditional video-based information systems can, however, only play video as it was recorded. Their interactivity is so low that users are compelled to passively watch the video in accordance with its author's scenario.

In an attempt to resolve this problem, much research has focused on allowing users to get information actively from a video-based information system by enhancing its interactivity. General approaches for this can be classified into two types. One involves the preparation of multiple video streams, each of which contains a different view of the same object; the Zapping-Video and the Multiple-Stream Video are examples of this type. Users can interactively change their viewpoint by selecting a playback video stream containing images of their favorite angle, and enjoy the video in accordance with their personal scenario. Suppose that a baseball game is captured at the same time from various angles, such as a close-up of each player, a view of the whole stadium and a shot of the audience. This method allows users to see their scene of choice, like the pitcher's expression or joyful countenance of the audience after a home run, since they can freely change their viewpoint by changing the playback video stream interactively. Increasing the number of viewpoints, however, requires

S. Nishio, F. Kishino (Eds.): AMCP'98, LNCS 1554, pp. 289-300, 1999.
© Springer-Verlag Berlin Heidelberg 1999

huge storage capacity. Consequently, the practical number of viewpoints in this method is limited to several at most.

In the other type of interactivity enhancement, video operations like forward play, reverse play, and stop, are entrusted to users. A typical system using this method is the video walk-through system Movie-Maps[1]. It stores video streams corresponding to motion images of scenery obtained while walking or driving along a particular street in a town. It correlates the time axis of the video streams with the space axis of the captured street. It thus enables users to interactively move through the virtual town with matching video displays. This method, however, is not suitable for presenting the behavior of objects in the real world since it cannot manage simultaneously both the spatial and temporal relations among video streams.

Videoplex incorporates the advantages of both of the above methods. It offers interactive movement in a virtual video space like the video walk-through system as well as more sophisticated object interaction like the Multiple-Stream Video. According to the users location and activity, it extracts and combines images from multiple video streams. Video streams for the walk-through component are stored separately from those of the objects. *Videoplex* provides a function for placing the images of objects within the walk-through image. In addition, the behaviour of each object in the three-dimensional space, described in a scenario, can be very sophisticated.

This paper presents the design and implementation of *Videoplex* as well as a prototype application. The rest of the paper is organized as follows: The main idea is described in Section 2. Section 3 describes the key components of our *Videoplex* system architecture and the implementation. *Videoplex* is compared with existing multimedia information systems in Section 4.

2. Main Idea

2.1. Concept

Videoplex aims at a new video system framework for a video-based application that realizes interactive movement in virtual three-dimensional space; it manages the behavior of objects independently of user movement.

Its basic concept is, based on the user's location and action, extracting the relevant object images and placing them appropriately within the current walk-through image. As one example, even if the user stops moving, one or more objects may continue their activity.

2.1.1. Walk-Through Images

Videoplex makes the position of the camera taking the the walk-through images and user's viewpoint in three dimensional space correspond by logically placing each frame of a walk-through video within the three dimensional space. The camera track is defined by arranging lanes, which show the road, and terminals, which show the termination point of each lane in the three-dimensional space. Figure 1 shows the concept of locating a walk through video within the three-dimensional space. The walk-through images are allocated in the lane of the camera track set in the three-dimensional space. Each frame of a walk through video is assigned a coordinate in the three-dimensional space by interpolating between the terminals of a lane. Accordingly, as the user views the walk-through images, his position in three-dimensional spaces is known.

2.1.2. Object Images

The movement routes of objects, as well as that of the user, is defined in the three-dimensional space. Object speed and orientation may be separately set before playing, and object location is updated during play. When the user moves in the three-dimensional space, background and object view images must be changed to reflect the movement. The view of an object can be made to vary according to both the user viewpoint and object's own movement. If multiple object videos are available for an object, each captured from a different angle, *Videoplex* can, in realtime, select the most appropriate video and extract the object images desired. This realizes virtual 3D video objects.

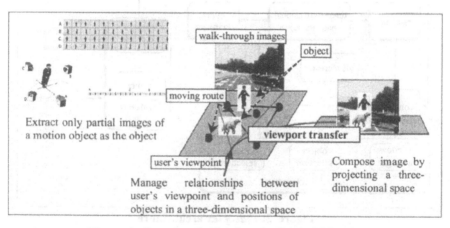

Figure 1: Concept of video integration in *Videoplex*

2.1.3. Control of Event Occurrance

A special feature of a stroll-type information system is that it enables users to experience chance encounters and discoveries. We realize this by preparing a scenario that describes the events that occur or can occur in various situations. For this purpose, *Videoplex* provides a function that allows the author to define the behavior of objects in the three-dimensional space. Since this function makes it possible to change the status of objects by passing messages among them, story variation is increased. Consequently, *Videoplex* can provide users with a high level of interactivity.

2.2. Architecture

In this subsection we outline the *Videoplex* architecture shown in Fig. 2. The main features of this architecture are as follows:

1. A three-dimensional space manager that manages the location of objects and user viewpoint. The resolver, view creator, and composer form the video according to the user viewpoint.
2. An event manager that manages event rules of objects.
3. A database subsystem that stores the raw video, stream data, the derived data generated by the author, and other meta-data input by the author with the help of an authoring system. It supports content-based query operations by software agents.
4. A graphical user interface that supports navigation and object selection .

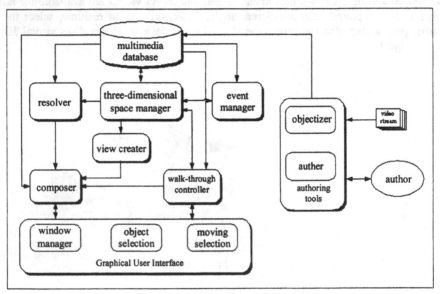

Figure 2: *Videoplex* **architecture**

This paper focuses on the first two features. Concepts are illustrated using the prototype that we have implemented.

All video data streams used by *Videoplex* are captured beforehand. The object module extracts objects from multiple stream video of motion object captured at the same time from various angles with information about the capture angle. In addition, the object module provides input to the database, which maintains both the annotated raw video data stream as well as information about video, such as the angle of view, frame rate, and frame size. The walk-through videos are also filmed beforehand, and input to the database with some meta-data. Information in the database can be queried by the three-dimensional space manager, the event manager, the resolver, and the composer.

The three-dimensional space manager module manages the location of each object and user view in the three-dimensional space, and updates them using information provided by the database, thewalk-through controller module, and the event handler module. The walk-through controller module is the core unit managing user movement. It orders the composer module to update playback frame positions and the three-dimensional space manager to update the user viewpoint. The event handler module manages event handling of the objects according to the user's location and/or actions.

The view creator makes the user view based on the placement information provided by the three-dimensional space manager. The resolver module selects the "best angle" movie for viewing each object from the user's viewpoint, and relays the video ID to the composer. The composer refers to the view created by the view creator and the video ID calculated by the resolver, and composes background images and objects retrieved from the database.

Finally, the user interface module displays the final image sequence to the user, and receives the selection and movement instructions from users through a graphical user interface.

Videoplex (a) extracts only the motion object area in multiple streams for objects, (b) stores objects and walk-thorough images, (c) relates the objects and the walk-through images to place on the same virtual three-dimensional space with their track, (d) describes the relationship between each object and its moving route in the three-dimensional space, triggered event rules, and the other meta-information, as a scenario, (e) manages the three-dimensional space with the scenario, (f) forms user views based on the virtual three-dimensional space, and (g) shows the movie created by the user view.

The following sections describe how the above-mentioned three key ideas were implemented as a prototype.

3.   Implementation of *Videoplex*

3.1.   Video-Based Virtual Environment

3.1.1.   Derivation of Walk-Through Images

The walk-through videos are assumed to be captured under constant velocity linear motion. Furthermore, the frames in a walk-through video are equally spaced based on the distance of sequential nodes. For instance, when the walk-through video from location A to location B consists of 100 frames, and where location A is set as (0, 0), and location B as (100, 0) in the 3D space, the 50th frame is associated with point (50,0). Walk-through images are stored O2 JPEG format. The images used by this prototype were captured at 10 frame/s, so the frame changes every 50cm in the real world.

3.1.2. Derivation of Object Images

To create a virtual 3D video object, the object is captured at the same time from various angles. The captured videos are stored together with information such as video identifier (video ID), filming angle (pitch, head, bank), and a start frame. They are related to meta-data, such as angle of view, frame rate, and frame size, in the database. The angle of an object to the user is determined by system. When a motion object is visible to the user, the object images whose capture angle is nearest to the object angle to the user is selected for display. They perceives the virtual 3D video object to be "coherent" since the system smoothly fuses the next object video into the current one .

For simplicity of explanation, we assume that the user and an object are on the same X-Y plane. When the coordinates (Xo, Yo) of observer's viewpoint O, the coordinates (Xa, Ya) of the object A, and angle Pa between the front direction of A and X-axis are determined on CG space, the angle θ_a between the front direction of A and the line OA is

$$\theta_a = Tan^{-1} \phi - P_a$$

$$where \quad \phi = \frac{Y_0 - Y_a}{X_0 - X_a}$$

The most-appropriate-angle video is determined from object An by selecting one whose camera angle is the nearest to θ_a.

3.1.3. Stories Created by Scenarios

Each object can be assigned its own route based and its own complex behavior. The document specifying all this control information is called the scenario.

Each scenario contains the following information:
1. Identifier of the route information.
2. ID of object placed on the route.
3. Velocity of the object.
4. Motion style on the route (loop, once, pingpong).
5. Status of the object.
6. Current position of the object in the three-dimensional space.
7. Rules for event processing.

The three-dimensional space manager manages the status of all objects and properly updates their position and status in the three-dimensional space.

To realize interactivity, event rules, such as playback conditions of an object, are described. Trigger events for playbacks are as follows:

State of video display. (e.g. If frame X is displayed)

Position of object. (e.g. If object A pass node X)

Position of object and point of view. (e.g. If user approaches object A, if object A catches up to object B)

Information from other scenario objects. (e.g. If object A sends instructions to begin moving)

Information from the outside. (e.g. If any key is pushed, if object A is selected)

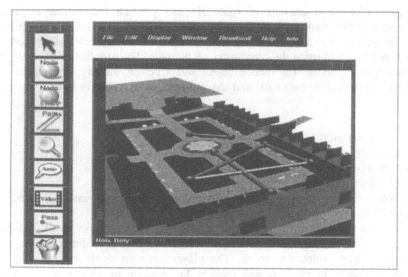

Figure 3: Screen shot of the authoring tools

3.2. Prototype

The effectiveness of the above-mentioned methods was evaluated by implementing a prototype.

We implemented it on an SGI *O2/R10000*. The prototype was coded in C language with OpenGL and digitalmedia libraries. The walk-through images, captured in O2 JPEG format, and the motion object's movie, captured in RGBA SGI format, were stored in local hard disks.

3.2.1. Authoring Tool
The prototype consists of authoring tools and the player.

Figure 3 shows a screenshot of the authoring tools. The author places video in a three-dimensional space, creates scenarios, and sets up the map to arrange the video. The route is set by using a GUI and arranging a 3Dmodel of the nodes and paths on the map. The coordinate value of each node is stored in a file in ASCII form with route ID. In this prototype, object behavior was separately prepared as a script, and manually related to the route.

A track for the walk-through video was defined in the following three steps:
1. Placing the nodes on a plate to define terminator of walk-through images.
2. Connecting two nodes by a path.
3. Associating the path with the walk-through images.
   The route of each object was defined using three steps:
1. Placing the nodes on a plate to define the ends of the route.
2. Connecting two nodes by a path.
3. Associating the nodes with the object.

Route information consists of route ID and position information of three-dimensional-CG models of multiple nodes. Node position information consists of a series of space coordinates ordered from the start point to the end point of the route. A path is not managed using the coordinates on its route. The coordinates of a path are derived from its nodes. The motion object was a person walking. The object was captured from eight orientations, and the captured videos were linked to the same object.

3.2.2. Player tool

In this prototype, the player has the following functions:
1. Scenario parsing.
2. Object resolving.
3. Video composing based on position information in the three-dimensional space.
4. Interactive event provision through scenarios.

Each scenario is stored in a C language structure. Each scenario specifies, for an object, its route, moving speed, moving pattern, location in three dimensional space, pointers for event rules, and so on. The player has own clock and refers to the scenario structure every case, and updates the position of objects in the three-dimensional space.

The three-dimensional space manager determines the current position of the object in three-dimensional space. After the three dimensional space is converted to the user view, the player resolves the movie of an object which becomes a drawing object, composes a two dimensional image by projecting the object images into the three dimensional space. Figure 4 shows a system flow overview of the resolver.

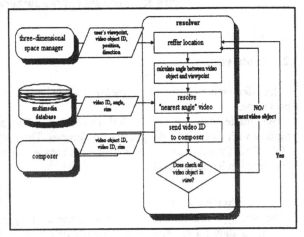

Figure 4: System flow overview of resolver

The prototype can handle two types of events: the user selection of an object using the mouse and the variation of object position. It also handles event processing for changing moving speed and replacing the selected object with another one as defined in the scenario. Figure 5 shows a screenshot of the player. In this figure, an object of a walking girl is composed into walk-through images.

Figure 5: Screen shot of player

3.3. Evaluation

(1) Object Resolving.

Because the number of objects and captured videos were small in the prototype, the resolving function worked smoothly. However, as the number of videos to be resolved increases, a more efficient resolution method will be needed.

(2) Video composing based on place information on three-dimensional space.

Composing video without incongruity is accomplished by placing the object and walk-through images appropriately in three-dimensional space. However, the placement must be done strictly for appropriate composing. The current prototype assumes that the movement of the user's viewpoint is linear and at a constant velocity. Therefore, if the camera capturing the walk-through video is moved at variable velocity, the final video will be incongruous.

(3) Interactive event provision with scenarios.

It was verified that scenarios make it possible to describe the behavior of objects with high flexibility. First, the position of the object is updated according to the system time. Therefore, the motion objects move regardless of user's movement. Second, defining the speed, orientation, and route of an object in a scenario allows changes to be made flexibly. The scenario-providing function is still being developed. In this prototype, the condition type, which the event controller can interpret, was simple. It will be necessary to test the performance of scenarios with the inclusion of the interpretation for practical use in the future.

4.   Comparison with Ordinary Systems

In this section, *Videoplex* is compared with other similar multimedia information systems in order to clarify the importance of this study.

4.1.   Comparison with Video Database Systems

Many video database systems supporting content-based retrieval of motion images stored in a database have been studied recently [2, 3, 4]. They are similar to *Videoplex* in the sense that all can retrieve and play motion images that satisfy the user requests. *Videoplex*, however, is very different in that the structure and usage of the video database differ. Video database systems generally divide video streams into logically meaningful units, such as scenes and cuts, and then organize them according to their temporal relation and the similarity of their contents. However, *Videoplex* organizes video streams according to their spatiotemporal relation, and provides a logical video space equivalent to a real world space.

4.2.   Comparison with Interactive Video Systems

Thus far, several interactive video systems that can compose video space corresponding to a real space from multiple video streams like *Videoplex* have been developed[1, 5, 6]. An example is the Movie-Maps system. It stores into optical videodiscs video streams corresponding to motion images of scenery obtained while driving along a particular street in a town, and allows users to drive in a virtual video town by interactively changing the video stream played according to the user's requests, such for left and right turns. In Movie-Maps, the driving speed is equivalent to the playback speed of the video streams. This means that the video space provided by Movie-Maps has no time axis. Unlike *Videoplex*, Movie-Maps is not designed to manage simultaneously both the spatial and temporal relationship among video frames.

Another example is the QuickTimeVR system. Like *Videoplex*, it allows users to interactively look around a video space as if they were standing at the center of the corresponding real space. It, however, composes the video space from video frames sampled by panning and tilting and moving a single video camera in various directions from a fixed position. Therefore, like Movie-Maps, QuickTimeVR cannot simultaneously manage both spatial and temporal relations among video frames since its video space has no time axis.

4.3. Comparison with Virtual Reality Systems

Videoplex is similar to virtual reality systems in the sense that both can provide a virtual space for users. The main difference is that virtual reality systems, in general, dynamically generate space images using computer graphics, while *Videoplex* composes the space from natural motion images stored in a database. Using natural motion images has several disadvantages: only the virtual spaces corresponding to the real spaces being captured using a video camera can be provided and a large number of video streams must be stored in order to compose a single virtual space. It, however, has some great advantages: the virtual space corresponding to a real space can be easily composed by just capturing the spatial images, and real movement, such as that of person, can be easily described in detail. Recently, creating models directly from photographs has considerable attention in computer graphics studies [7, 8, 9]. Since natural images are used as input in this method, it has the advantage of producing photorealistic renderings as output. However, it is still a difficult task to simulate real movement of the motion object perfectly using computer graphics.

5. Concluding Summary

This paper has presented the design and implementation of *Videoplex*. *Videoplex* aims at a new video system framework for a video-based application that realizes both free movement of user's viewpoint and object motion independent of viewpoint movement. The main concept is to extract only partial images of a moving object from all walk-through images and to manage them as an object. Then, it composes a series of images from the walk-through ones selected on the basis of user's action and from object ones selected independently of user's action. Thus, motion objects continue their action even if walk-through images are stopped. We have presented and implemented the following three main ideas:

- Management of relationships between user's viewpoint and positions of objects in a virtual three-dimensional space.
- Derivation of object images of a object according to user viewpoint.
- Expression of various stories using a scenario defined to each object.

Furthermore, their utility was verified using a prototype. In the current prototype, the number of the scenarios and objects is limited, and various restrictions exist. Additional problems for the future consideration include the treatment of large amounts of data, the preparation of an authoring environment, the handling of complex event rules, and the management of decentralized scenarios.

References

[1] Lippman, A.,"Movie-Maps: An Application of the Optical Videodisc to Computer Graphics", Proc. SIGGRAPH'80, pp.32-42, 1980.

[2] Gibbs, S., Breiteneder, C., and Tsichritzis, D., "Audio/Video Databases: An Object-Oriented Approach", Proc. 9th Data Engineering Conf., pp. 381-390, 1993.

[3] Oomoto, E., and Tanaka, K., "OVID:Design and Implementation of a Video-Object Database System", IEEE Trans. Knowledge and Data Engineering, Vol.5, No. 4, pp.629-643, 1993.

[4] Ardizzo, E., Gesu, D. V.,Casc, L. M., Valenti, C., "Content-Based Indexing of Image and Video Databases by Global and Shape Features", Proc. ICPR '96, pp. 140-144, 196.

[5] Teodosio, L. A., and Mills, M., "Panoramic Overview For Navigating Real-World Scenes", Proc. Multimedia'93, pp.359-364, 1993.

[6] Chen, E. S., "QuickTimeVR - An Image-Based Approach to Virtual Environment Navigation", Proc. SIGGRAPH'95, pp.29-38, 1995.

[7] Szeliski, R., "Image mosaicing for tele-reality applications", IEEE Computer Graphics and Applications, 1996.

[8] McMillan, L., Bishop, G., "Plenoptic Modeling: An Image-Based Rendering System", Proc. SIGGRAPH '95, p39-46, 1995.

[9] Debevec, E. P., Taylor, J. C., Malik, J., "Modeling and Rendering Architecture from Photographs: A hybrid geometry- and image-based approach", Proc. SIGGRAPH '96, pp11-20, 1996

[10] V.M.Bove,Jr, "Multimedia based on object models: Some whys and hows", IBM Systems Journal, Vol35, NOS 3&4, 1996.

[11] Kataoka, R., Satoh, T., and Inoue,U., "Video Reality: A Multimedia Information System Based on a Visual Conducting Model", Proc. ADTI '94, pp.207-214, 1994.

[12] Seitz, M. S.,Eyer, R. C., "View Morphing", Proc. SIGGRAPH '96, pp21-30, 1996.

[13] Horry,Y., Anjyo,K., Arai, K., "Tour Into the Picture:Using a Spidery Mesh Interface to Make Animation from a Single Image", Proc. SIGGRAPH 97, pp225-232, 1997.

[14] Tanemo, F., Yoshida, T., Kataoka, R.,"Managing Complex Object Information for Interactive Movie Systems", IEICE Trans. Inf. & Syst., Vol.E79-D, No.6, 1996.

[15] Kelly, H. P., Katkere, A., Kuramura, Y. D., Moezzi, S., Chatterjee, S., Jain, R., "An Architecture for Multiple Perspective Interactive Video", Proc. ACM Multimedia '95, pp201-212, 1995.

Construction of Virtual Environment from Video Data with Forward Motion

Xiaohua Zhang, Hiroki Takahashi, and Masayuki Nakajima

Graduate School of Information Science & Engineering
Tokyo Institute of Technology
2-12-1 O-okayama, Meguro-ku, Tokyo, 152-8552 Japan
{zhxh,rocky,nakajima}@cs.titech.ac.jp

Abstract. The construction of photo-realistic 3D scenes from video data is an active and competitive area of research in the fields of computer vision, image processing and computer graphics. In this paper we address our recent work in this area. Unlike most methods of 3D scene construction, we consider the generation of virtual environments from video sequences with a video-cam's forward motion. Each frame is decomposed into sub-images, which are registered correspondingly using the Levenberg-Marquardt iterative algorithm to estimate motion parameters. The registered sub-images are correspondingly pasted together to form a pseudo-3D space. By controlling the position and direction, the virtual camera can walk through this virtual space to create novel 2D views to acquire an immersive impression. Even if the virtual camera goes deep into this virtual environment, it can still obtain a novel view while maintaining high resolution.

Keywords: Virtual environment, Pseudo-3D structure, Image composition, Pyramid data structure, Camera motion, Image-based rendering

1 Introduction

The construction of virtual environments has attracted much attention in recent years, and has become an active area of research in the fields of virtual reality, computer vision, image processing and computer graphics. These areas are related in areas ranging from games and entertainment to business, education and telemedical surgery. A direct application would involve real-time interactive walk-through or fly-through within a virtual environment. The 2D views are created from 3D scenes of a virtual environment, which remains the same, by controlling view position and direction of virtual camera.

The traditional approach of obtaining such effect is by 3D model-based rendering with explicit 3D structure[1]. With this method, views are generated by directly rendering 3D geometric models of objects or scenes at specified viewpoints. In contrast with 3D model-based rendering, more another recent method for creating new images is called image-based rendering, which produces views directly from images, relying on interpolation using the original set of input images or re-projection from source images onto the target image. Image-based

S. Nishio, F. Kishino (Eds.): AMCP'98, LNCS 1554, pp. 301–312, 1999.
© Springer-Verlag Berlin Heidelberg 1999

rendering can relatively easily to produce photo-realistic images from images of real scenes. A survey of image-based rendering techniques is given in [1].

Techniques for creating novel views of objects are reported in many papers such as [2,3,4], and most well-known techniques for generating new 2D views of 3D scenes can be found in [5,6,7,8]. In particular, [8] only use one picture or photograph or painting to obtain new views with 3D effects. Research in [5,6,7] creates mosaic from images taken by translation and panning of 3D scenes, and also can generate new views of that scene with the viewpoint fixed. The method in [8] allows the animation of both background and foreground. Unfortunately, if the virtual camera goes forward too far from the original assumed viewpoint, the image will lose its resolution. The reason is that the information at a *far* place from the original viewpoint can not be recovered accurately.

Both how 3D scenes are projected on image plane and what the 3D scenes' structure are not known in advance. Whether to recover the motion of the camera, or reconstruct the 3D environment, or both, depends on the applications. It is known that in some cases motion estimation and 3D structure reconstruction may be difficult[9]. Over many years, researchers observed that such a full and precise reconstruction and motion recovery may not be necessary for many video-based tasks. In this paper we do not focus on structure from motion or motion recovery. Instead, we exploit the 3D pseudo-structure of 3D scenes to reconstruct the information of the 3D world, and endow a virtual camera with the capability to navigate in this constructed virtual environment to obtain an immersive impression.

As computer-based video becomes more ubiquitous with the development expansion of transmission, storage, and manipulation capabilities, video image sequence will offer abundant sources of information for many applications. In this paper, considering the fact that the nearer the camera is to the 3D scenes, the higher the resolution is, we proposed a method for constructing virtual environment from image sequences with forward motion based on image registration and composing techniques. Despite the fact that we limited ourselves to forward motion images, small rotation around the center of projection is also allowed. The virtual camera can walk-through or fly-through in this virtual space to acquire an immersive impression by changing the position and orientation of a virtual camera. Even when the virtual camera goes deeper into this virtual environment, it also can obtain a novel view while maintaining high resolution in the range between the first original viewpoint and the last original viewpoint.

After a review of related work in Section 2 and a brief explanation of the camera model in Section 3, we describe the 3D modeling from images and explain how to concatenate images with forward motion to form a virtual space in Section 4. Section 5 discusses rendering to create novel views and Section 6 shows experimental result. Finally Section 7 gives our conclusions and ideas for future work.

2   Related Work

As for the construction of virtual environments, it has not been very widely and deeply investigated. In R. Szeliski's work[10] related to creating large panoramic images of arbitrary shape and detail from video frames by panning a camera over a scene, forward camera motion was not mentioned. In [6], a virtual environment is represented by composed several overlapping images around a center point. An animation of this environment can be made with the viewpoint fixed. The method in [5] proposed animations, with many closely spaced images, and its theoretical background heavily depends on computer vision techniques. Forward motion was also not dealt with. With projective groups, [7] constructed high-stills for the two cases: multiple frames taken of a flat objects, and multiple frames taken from a fixed point. With a stereo-viewing system, Takeo Kanade[11] computed the 3D structure of events by using a stereo technique for capturing the scenes of events. In order to create a new image which also has 3D effects obtained from 2D images, morphing techniques including [12] can be employed, but they largely rely on a user to specify feature points or line segments in images. Recently, a system called TIP[8](Tour Into the Picture) appeared which allows animation to be made from only *one* image. TIP uses an attractive approach, but when camera goes too far from the original viewpoint into the image, the resolution becomes very low. In this paper we employ video data to generate a virtual environment which can maintain high resolution at a *far* place within the field from the first original viewpoint to the last original viewpoint when the virtual camera goes deeper into this environment constructed from an image sequence. Our approach can be regarded as an extension of [8]. However it is not just an extension, as the concatenations between images with forward motion are also addressed.

3   Preliminaries

In this section, we briefly review the camera model we used, and describe the relationship between world coordinates and view coordinates.

The camera model we used is the classical *planar pinhole model*[15]. The object space is considered to be the three dimensional Euclidean space $\mathcal{R}^3$, and the image space is a two dimensional Euclidean space $\mathcal{R}^2$ which is a subspace and located somewhere in $\mathcal{R}^3$. In our case, we set the image plane to be parallel to the XOY plane in 3D world coordinate as illustrated in figure 1. Let $P(X, Y, Z)$ denote the Cartesian coordinates of a scene point in 3D space and let $p(x, y)$ denote the corresponding coordinate in the image plane. The image plane is located at focal length $Z = f$, and without loss of generality, the center of projection, i.e., the viewpoint $(v_x, v_y, 0)$ of the camera is located on the XOY plane. It is obvious that the optical axis is parallel to the Z axis and perpendicular to the XOY plane.

With the arrangement shown in figure 1, the perspective projection of a scene point $P(X, Y, Z)$ onto the image plane at a point $p(x, y)$ can be expressed by following formulae:

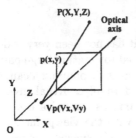

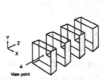

Fig. 1. Camera model and coordinate system. 3D point $P(X,Y,Z)$ is projected onto the image plane.

Fig. 2. Image concatenation from image sequence with forward motion.

$$x = \frac{f(X - x_v)}{f - Z} + x_v \qquad y = \frac{f(Y - y_v)}{f - Z} + y_v \qquad (1)$$

These formulae plays a very important role when modeling and navigating inside the virtual environment. The details for navigation are described later.

4   Concatenation of Images with Forward Motion

In this section we explain how to model 3D scenes from images with forward motion to acquire a 3D pseudo-model, in order to concatenate adjacent textures into a unified form. To do such work, we employed the Levenberg-Marquardt iterative algorithm to estimate the relationship between two sub-images of planar surface. A coarse-to-fine scheme is used to guess the initial translational parameters for the Levenberg-Marquardt algorithm.

4.1   3D Modeling from Images

When taking picture, an assumption is made that camera parameters have no sharp change, and that camera moves forward. Experiments show that we can allow a small rotation about the the center of projection. For each perspective picture, after roughly specifying the vanishing point, visual 3D geometry of the scene's background can be defined as a simple polygonal model. Assuming that 3D polygons are perpendicular to each other and giving the parameters of the camera, all of the 3D coordinates of the polygons can be found by inverse perspective transformation. The textures of 3D polygons are inherited from their corresponding 2D polygons on the modeled image. These 3D polygons construct a pseudo-3D structure. The discussion for details can be found in [8].

Note that this kind of structure is not precise, which is why we call it a pseudo-3D structure. Generated pseudo-3D structures are used for decomposition of each image into five or less sub-images $\{S_{ij} | 1 \le i \le N; j \le M\}$. Here N is the number of images in the video sequence, and M is the largest number of sub-images in one original image. These sub-images are reused for correspondingly composing more accurate texture of 3D space. That means sub-images S_{ij} with fixed j are pasted together to form one jth side of target 3D environment. Varying j from 1 to 5, the texture of each side of the 3D environment can be reconstructed. The rear wall of the 3D space remains the same as the rear wall of the last frame. This procedure is illustrated in figure 2.

4.2 Prewarping

As mentioned above we allow the video sequence has small rotation around the center of projection. In this case, image stabilization is needed to improve the results of correspondingly composing each piece $S_{ij}, (1 \leq i \leq N)$ with a fixed side j. Since we consider only small rotation, stabilization can be accomplished by employing the morphing technique proposed by Beier and Neely[12]. For good stabilization, all the images except the first can be warped to a position along the optical axis of the first image to cancel the small 3D rotation. Such prewarping results in a new sequence, looking as if it has been taken from a stabilized platform.

4.3 Registration and Concatenation

In the case that 3D scenes are not planar surfaces, it is difficult to directly paste these video sequence with forward motion into a large mosaic. In our case images have been predecomposed into sub-images regarded as planar surfaces $\{S_{ij} | 1 \leq i \leq N; j \leq M\}$. In the modeling procedure we know that these sub-images can be regarded to be planar or nearly planar surface. The registration between two planar scenes is relatively simple. For an ideal pinhole camera, there has a basic result of projective geometry[10] between 2D images of a 3D planar scene, which can be described as follows:

$$x'_k = \frac{m_0 x_k + m_1 y_k + m_2}{m_6 x_k + m_7 y_k + 1} \qquad y'_k = \frac{m_3 x_k + m_4 y_k + m_5}{m_6 x_k + m_7 y_k + 1} \qquad (2)$$

Here we use featureless method to estimate the motion parameters $m_l (l = 0, 1, \ldots, 7)$ based on the intensities of corresponding sub-images S_{ij} and $S_{(i+1)j}$ of adjacent images. We compute the transformation parameters such that the sum of squared pixel intensity differences between the two images is minimized:

$$E = \sum_k [I'(x'_k, y'_k) - I(x_k, y_k)]^2 = \sum_k e_k^2 \qquad (3)$$

To minimize E, we use the Levenberg-Marquardt iterative nonlinear minimization algorithm[13]. This algorithm requires computation of the partial derivatives of e_k with respect to the unknown motion parameters $\{m_0, m_1, \ldots, m_7\}$:

$$\frac{\partial e_k}{\partial m_l} = \frac{\partial I'}{\partial x'} \frac{\partial x'}{\partial m_l} + \frac{\partial I'}{\partial y'} \frac{\partial y'}{\partial m_l} \qquad (4)$$

The Levenberg-Marquardt algorithm computes an approximate symmetric matrix $\mathbf{A} = [a_{pq}]$ and the gradient vector $\mathbf{b} = \{b_p\}$ at each stage:

$$a_{pq} = \sum_k \frac{\partial e_k}{\partial m_p} \frac{\partial e_k}{\partial m_q} \qquad b_p = -2 \sum_k e_k \frac{\partial e_k}{\partial m_p} \qquad (5)$$

The equations solved at each stage belong to the linear system $(\mathbf{A} + \lambda I) \Delta m = \mathbf{b}$ where λ is a tuning parameter that is adjusted according to the change in the sum of squared differences at each stage. The initial small value of λ can be selected to be $\lambda = 0.0001$[13]. The advantage of using Levenberg-Marquardt over straightforward gradient descent is that it converges in fewer iterations.

The motion parameters $\{m_0, m_1, \ldots, m_7\}$ in formulae (2) describe the relationship between two sub-images S_{ij} and $S_{(i+1)j}$. Once $\{m_0, m_1, \ldots, m_7\}$ are found, it is easy to seamlessly concatenate two sub-images together to form a large image with high resolution.

Figure 3 shows an example of composition of images. (a) and (b) are two original images which are taken inside the building Venture Business Laboratory of our university. (c) is the result of composition. The geometric relationship between these two 3D planar surfaces is estimated by employing Levenberg-Marquardt iterative algorithm.

4.4   Coarse-to-Fine Scheme

Unfortunately, the Levenberg-Marquardt minimization algorithm can only guarantee a local minimum, i.e., the desired result can be attained only by giving a *good* initial guess. In the general case, finding a *good* initial guess also is problematic. In our case the corresponding sub-images have a dominant motion of translation, despite allowing only a small rotation. This is suitable for roughly estimating the translation first as an initial guess for the Levenberg-Marquardt algorithm. We use a coarse-to-fine scheme to do such work by employing a Gaussian pyramid method[14].

The pyramid data structure approach for finding initial translation is efficient, as the search range in each level is small, and the upper levels of the pyramids are ideal for getting an overview of the image scene. The coarse-to-fine scheme works as follow:

1. Create pyramids with n levels, and a resize factor of r (generally using $r = 2$) in both width and height. The upper or coarse resolution levels are obtained by averaging groups of adjacent pixels.
2. The initial translations at the top level can be guessed by image correlation. This can be done fast, as both of top coarse images have small size.
3. Propagate estimated translations to the next finer level, and check each option to modify translations until reaching finest level.

The first step in recursion stops when the size of two images is reduced below a given threshold. At this coarsest level all possible translations between two images are checked, and the results are used as initial values of the pyramid data structure. The resulted translations are taken to be the initial guess for the Levenberg-Marquardt algorithm.

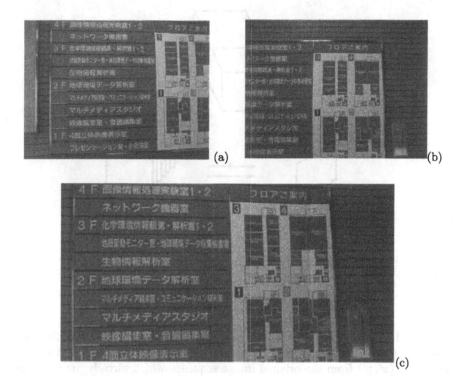

(a)   (b)

(c)

Fig. 3. Example: two planar surfaces are pasted together.

5   Image Rendering

At this stage, the construction of the virtual environment is completed. The 3D space has a pseudo-3D structure. The virtual camera can be controlled to navigate inside this pseudo-3D environment. Typically, parameters of the virtual camera are position, direction and focal length. When these parameters are provided, it is convenient to obtain new images by using formulae (1) to acquire immersive impressions.

The rendering method we used for obtaining new views is in fact a texture mapping. In order to avoid holes and empty curved strips in new images, backward mapping is employed, which is illustrated in figure 4. Here, a single arrow means the flow of finding a pixel and a double arrow indicates a pixel value transfer. By first scanning the new 2D image for each point (x, y) and using

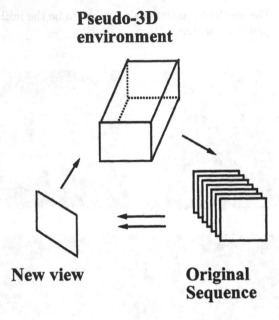

Fig. 4. Image rendering

formula (1), we can decide which side and which 3D point of 3D space that the position of this point comes from. Then the pixel value of the 3D point can be settled at (x', y') from the original image sequence which already are concatenated together; Finally, as an option to reduce the jagged phenomenon in new image, the average of the pixel values are used. These pixels are inside a small window with (x', y') as its center.

6   Experimental Results

For verifying the method proposed above, both of image sequence of real indoor scene and outdoor scene are used. Since small rotations around the center of projection are allowed, first each frame is prewarped using the first frame as reference. Then each image is partitioned into five sub-images, and these sub-images are seamlessly composited into four images except that the rear of 3D space remains the same as the rear of the last frame. The correspondingly composited images can be regarded as textures of five sides of a box-like pseudo-3D space. Changing the parameters of the virtual camera, new images are obtained (see figure 5, 6). User can designate an arbitrary view point position and an orientation. Designated view point should be within the range from the original viewpoint specified to the rear wall of the constructed virtual environment.

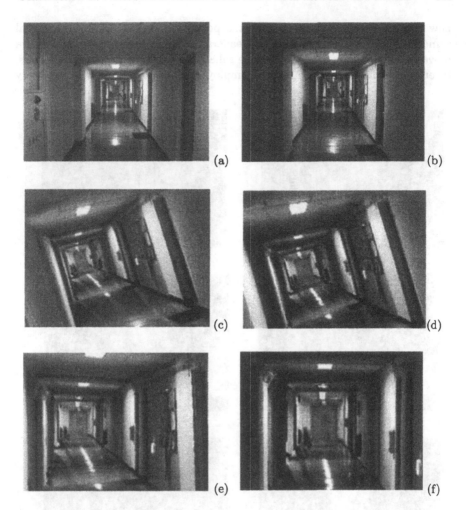

Fig. 5. Indoor scene: (a) and (b) are original image frames with forward motion, small rotations around center of projection exists.(c)-(f) Generated new images when walking through the 3D virtual space.

Figure 5 shows experimental results of indoor scene. (a) and (b) are two of original frames with forward motion along a corridor. (c)-(f) are new views generated when changing the position and direction of a virtual camera. For (c), the virtual camera moves ahead, shifts to left and rotates around optical axis; (d) has the same orientation, with camera going deeper into virtual environment; For (e), the virtual camera shifts to left sharply and goes deeper; And for (f),the camera goes more deeper into pseudo-space. Figure 6 shows other experimental results which are similar to figure 5 but with images of outdoor scenes. (g)-(h) again are two of original frames with forward motion a long a road. (i)-(l) are new views generated when changing the position and orientation of a virtual camera. From the images generated, readers can know how the virtual camera moves. It is worth to note that for the sake of saving time and avoiding holes, inverse projective transformation is employed. Since coordinates of arbitrary

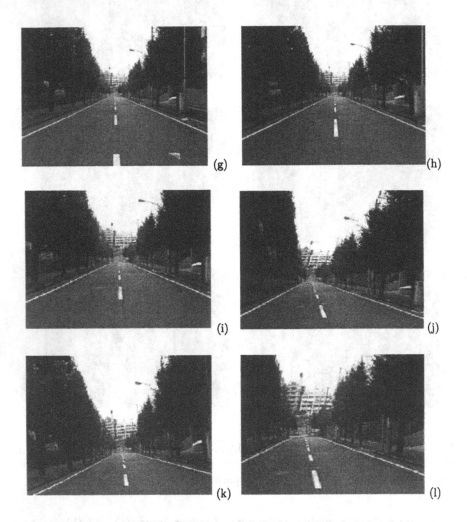

Fig. 6. Outdoor scene: (g) and (h) are two original image frames with forward motion, small rotations around center of projection exists. (i)-(l) are new images generated when walking through the 3D virtual space.

points in the 2D image are known, testing from which 3D point it comes, on which 3D wall it is located, new pixel values in the 2D image can be obtained from merged textures. Even when camera goes deeper into the pseudo-3D space, the resolution is maintained well within the limited range from the first original viewpoint to the last original viewpoint. The color compensation technique is used to decrease the differences between each pair of images as much as possible.

7 Conclusions and Future Work

We have presented a method to construct a virtual environment from reference image sequences with forward motion. These images are concatenated to form a pseudo-3D environment. The navigation of a virtual camera inside the constructed virtual environment is allowed to generated novel views to acquire an immersive impression. Even when the virtual camera goes deeper with a range from the first original viewpoint to the last original viewpoint into this virtual environment, it can obtain a novel view without losing resolution.

For the limited, restricted situation of moving down a hallway or down a road, we present hope that a user can take a sequence of real images and later generate a different sequence of moving down pretty close, but not exactly the same path and with different direction of viewpoint.

In practice, most applications cannot be restricted to only forward-moving paths, curved and branched camera paths should be considered. With a nonlinear camera path, a mechanism of controlling the virtual camera's passing from one subspace to another subspace is necessary. Only static environments are considered at the current stage, when there exists moving objects in the scene, a tracker should be applied to separate foreground from background based on image sequence. The pyramid data structure is sensitive to the intensities, phase correlation may be better than this one. The color differences are dealt with at the stage of registration by employing color compensation technique based on the intensities of pixels. The reflecting properties of illumination should be considered in practical applications, which will also be under investigation in our future work.

Acknowledgments The authors owe the anonymous reviewers many thanks for their invaluable comments.

References

1. S.B.Kang: A Survey of Image-based Rendering Techniques. Technical Report CRL 97/4, Cambridge Research Lab., Digital Equipment Corp., August 1997.
2. Chen,S.E., Willians,L.: View Interpolation for Image Synthesis. Proc. SIGGRAPH '93 (Anaheim, California, August 1-6,1993). In Computer Graphics Proceedings, Annual Conference Series, 1993. ACM SIGGRAPH, pp.279-288.
3. Seitz, S.M., Dyer, C.R.: View Morphing. Proc. SIGGRAPH '96 (New Orleans, Louisiana, August 4-9,1996). In Computer Graphics Proceedings, Annual Conference Series, 1996. ACM SIGGRAPH, pp.21-30.
4. S.Avidan, T.Evgeniou, A.Shashua, T.Poggio.:Image-based View Synthesis by Combining Trilinear Tensor and Learning Techniques. ACM VRST '97 conference(Also http://www.cs.huji.ac.il/~shashua/)

5. McMillan,L., Bishop,G.: Plenoptic Modeling: An Image-based Rendering System. Proc. SIGGRAPH '95 (Los Angels, California, August 6-11,1995). In Computer Graphics Proceedings, Annual Conference Series, 1995. ACM SIGGRAPH, pp.39-46.
6. Chen, S.E., Willians, L.: Quicktime VR - An Image-based Approach to Virtual Environment Navigation. Proc. SIGGRAPH '95 (Los Angels, California, August 6-11,1995). In Computer Graphics Proceedings, Annual Conference Series, 1995. ACM SIGGRAPH, pp.29-38.
7. S.Mann, R.W.Picard.: Virtual Bellows: Constructing High-quality Images from Video: In *first IEEE International Conference on Image Processing(ICIP-94)*, Volume I, pp. 363-367, Austin, Texas, November 1994.
8. Y.Horry, K. Anjyo, K.Arai: Tour Into the Picture: Using a Spidery Mesh Interface to Make Animation from a Single Image. Proc. SIGGRAPH '97 (Los Angels, California, August 3-8,1997). In Computer Graphics Proceedings, Annual Conference Series, 1997. ACM SIGGRAPH, pp.225-232.
9. W.Eric, L.Grimson.: Why Stereo Vision is Not Always About 3D Reconstruction. Technical Report, AILab.,MIT, A.I.Memo No.1435, July, 1993.
10. R.Szeliski: Video Mosaics for Virtual Environment. IEEE Computer Graphics and Applications, March 1996. pp.22-30.
11. T.Kanade, P.J.Narayanan, P.Rander: Virtual*ized* Reality: Concept and Early Results. IEEE Workshop on the *Representation of Visual Scenes*, Boston, June 24, 1995. (Also in URL http://www.cs.cmu.edu/Groups/VirtualizedR/index.html)
12. T.Beier, S.Neely: Feature-Based Image Metamorphosis. Proc. SIGGRAPH '92 (Chicago, Illinois, July 26-31,1992). In Computer Graphics Proceedings, Annual Conference Series, 1992. ACM SIGGRAPH, pp.35-42.
13. W.H.Press et al.: Numerical Recipes in C: The art of Scientific Computing. 2nd edition, Cambridge Unive. Press, Cambridge, England, 1992. pp.683-687.
14. E.H.Adelson, C.H.Anderson, J.R.Bergen, P.J.Burt, J.M.Ogden: Pyramid Methods in Image Processing. RCA Engineer, 29-6, Nov/Dec 1984, pp.33-41.
15. L.McMillan Jr.: An Image-Based Approach to Three-Dimensional Computer Graphics. PhD.Thesis, 1997, pp.31-40.

Spatial Browsing for Video Databases

Masatoshi Arikawa[1] and Tetsu Kamiyama[2]

[1] Faculty of Information Sciences, Hiroshima City University, Japan
arikawa@its.hiroshima-cu.ac.jp
http://www.db.its.hiroshima-cu.ac.jp/~arikawa
[2] Hochi Newspaper, Tokyo, Japan

Abstract. Time-series description data concerning a camera are expected to be automatically generated by various sensors in the near future, such as a camera's time-series positions, directions and zoom ratios so as to provide a rich environment for retrieving and browsing video data spatially. We also have used real-time 3D CG (three dimensional computer graphics) for interfaces to browse videos in a virtual space corresponding to an existent space in the real world. Cameras' movement or video sequences are represented as 3D icons in a virtual space. If we click one of the 3D icons, the corresponding video sequence will be replayed in a virtual space. This paper presents a basic principle of the 3D spatial hypermedia for video data browsing and reports some demonstrations of our prototype system.

1 Introduction

It is important to retrieve our intended scenes easily and naturally from a large amount of video data using their *description information*. Many research efforts have been made for creating description information for video data by hands and for making use of the hand-made description information in order to retrieve our intended video sequences from large scale video data [3]. For example, keywords are attached to frames of videos and users can use the keywords for their queries to retrieve their intended video sequences. Narrations of video data can also be used as description information. The keywords and narrations have been created by persons for news videos on television, but it is difficult to create them for all videos, including home made videos etc., by human's hands.

Data of a camera's condition may be useful for the description information of video data. The data of the camera's condition such as position, direction and zoom ratio are expected to be automatically generated in the near future by sensors. The position will be measured by some sensors such as high precision GPS (Global Positioning System). The precise direction can be available using optical fiber gyros. The zoom ratio can be taken from a digital camera itself. This paper discusses how to make use of these automatic generated data of the camera's condition to browse and retrieve video data spatially.

S. Nishio, F. Kishino (Eds.): AMCP'98, LNCS 1554, pp. 313–327, 1999.
© Springer-Verlag Berlin Heidelberg 1999

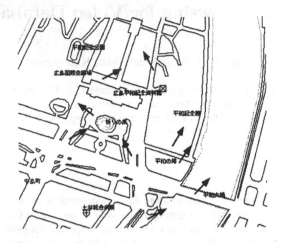

Fig. 1. 2D arrow icons representing pieces of the position and direction of a camera on a 2D map

2 An Overview of Spatial Browsing for Video Data

This chapter overviews the use of spatial description information of video data for browsing and retrieving video sequences of users' interests from the viewpoint of spatial queries. A video sequence is considered a collection of time-series images in this paper. First, we discuss the use of some spatial description information of *photo pictures*, that is, images, then extend it to the use for videos later. In the remainder of this paper, a picture means a photo picture which is one image. One picture corresponds to a camera at a certain moment. The camera at a certain moment is represented by some spatial attributes, such as its position, direction, zoom ratio and so on. 2D (dimensional) map data are also used for making their relations to some spatial attribute data of a camera. If the time-series positions of a camera for some duration are visualized on a 2D map, the visualization can show the distribution of the camera's movement which corresponds to a collection of time-series pictures. The position data can also be used to create clickable icons representing time-series pictures. If we click one of the position icons, the picture corresponding to the clicked icon will be shown on a screen. Also, we use the direction of the camera as well as its position to represent cameras or pictures (Fig. 1). In the case, a camera may be visualized as an arrow which can provide the information of the direction in addition to the position. The arrow on a 2D map enables a user to understand what direction's scene can be viewed in pictures. If information about the position, direction, height and zoom ratio of the camera are available, we can tell what region in the real world can be taken or viewed in a picture at a certain moment(Fig. 2). The region can also be used as a clickable icon for users' interactions. If we click the clickable region icon, the corresponding picture will be displayed on a screen.

The spatial data, such as position and direction of a camera, can be used for spatial queries. The spatial data are also used as clickable icons for hypermedia

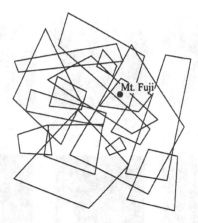

Fig. 2. A set of 2D regions viewed in a picture or a video frame by a camera at a certain moment

2D maps as we mentioned before. For example, if we want some pictures of Mt. Fuji, we can realize it by making a spatial query to find regions containing a point (Fig. 2). The regions represent pictures and the point represents Mt. Fuji in Fig. 2. Thus, we can indirectly retrieve pictures by queries on contents in pictures using spatial data corresponding to pictures. We can extend this idea to applications of video data. A video data can be considered a collection of images or pictures. The unit images comprising a video sequence are called *frames* of the video sequence. Each frame of a video sequence can correspond to a camera at a certain moment and it has its spatial description information (Fig. 3). Fig. 3 shows the correspondence between an image and its momentary camera's condition as well as some temporal relations of time-series images. We also make some spatial queries for video sequences. For example, we can make a query to select some video sequences which show Mt. Fuji. For cameras taking videos, their time-series positions and directions are expected to be automatically recorded by using sensors in the near future. The movement of the camera for the duration of playing a video sequence can be simulated. We can see the camera's movement and replay the corresponding video at the same time. If we visualize all spatial description data for all frames of a video sequence as arrow or region icons on a 2D map, the number of arrow or region icons becomes too large to browse and click. We must simplify such many icons for pieces of video sequences. For example, many arrow icons can be represented by a small number of arrow icons. A representative arrow icon should be considered to represent video sequences or a camera's movement for a certain duration. A video sequence should be replayed when the corresponding icon is clicked by a user.

There is a problem how to divide a video sequence into multiple video sub-sequences which should be represented by some icons. For example, each arrow icon may represent a segment of every 5 minutes video sequence. It is useful to divide a video sequence into more meaningful sub-sequences, but it will be much more difficult than dividing them into pieces of constant duration sub-sequences.

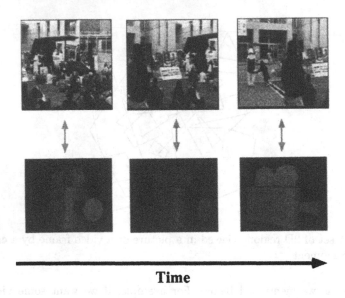

Time

Fig. 3. A sequence of time-series video frames and moving cameras in the real world

A simple method of generating a representative arrow icon for a video sequence is to use the average values of the positions and directions of a set of time-series momentary cameras. The method often fails because it is generally difficult that an average value represents all values for any case. Selecting methods of generating representative arrow icons should be adaptable for various cases.

We can extend the idea of spatial browsing of videos on a 2D map to the one for a 3D CG space. All 3D CG spaces discussed in this paper can correspond to existent spaces in the real world. Such 3D CG spaces corresponding to parts of the real world are called *3D virtual spaces* in this paper. We can walk through a 3D virtual space, and can browse or retrieve our intended video sequences which view similar to the current views in the current 3D virtual space. This user's requests of interactions can be interpreted as spatial queries which find arrows or regions closer to user's intention. The view in a 3D virtual space means the user's intended region and can serve as the selective condition of a spatial query. For instance, while walking through a virtual space, a user can click a 3D arrow icon representing a video sequence so as to replay it in another window on a screen (Fig. 4). Furthermore, it is also possible to incorporate replaying videos into a 3D virtual space as components of the virtual space (Fig. 5). Users appreciate past real-world videos in a 3D virtual space. This kind of application is called *Augmented Virtuality*[2]. It can provide users with more spatial experience. Another application of using spatial queries for 3D virtual spaces is to retrieve and replay videos which show a user's clicked objects in a 3D virtual space. This application uses a spatial query to select regions, which correspond to video sequences, including or intersecting the user's clicked objects.

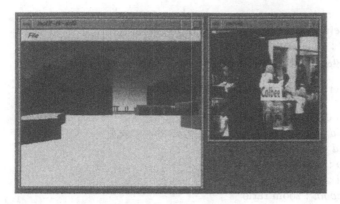

Fig. 4. A moving camera in a 3D virtual space and the corresponding video replayed in another window

Fig. 5. Replaying a video projected on a rectangle plane in a virtual space with the corresponding moving camera icon

3   A Model of Spatial Browsing for Video Data

3.1   Pictures and Cameras

A picture can be represented by condition of a camera. The condition includes position, direction, zoom ratio, characteristics of the lens, time and so on. The term *image* is used for a picture in order to construct definitions in this paper. An image can be identified by an image identifier, id_{image}. An image of id_{image} is denoted by $image(id_{image})$ and has two attributes concerning time, t and camera identifier, id_{cam}.

[Definition of time-series *image* taken by a camera]

$image(id_{image}) : (t, id_{cam})$

  t : time or duration

  id_{cam} : camera identifier

A camera at a certain time or duration is defined as follows.

[Definition of time-series *camera*]

$cam(id_{cam}, t) : (id_{image}, id_{space}, p, d, zoom, reg)$

  id_{space} : space identifier

  p : position

  d : direction

  $zoom$: zoom ratio

  reg : *region* which covers the area viewed in the image of id_{image}

A camera at a certain time has attributes of id_{image} which corresponds to one image, and id_{space} which addresses a space such as the real world, a 2D map and a 3D virtual space. The attributes p, d, $zoom$ and reg respectively denote the camera's position, its direction, its zoom ratio, and the region which is viewed through the camera.

3.2   Spatial Objects and Spaces

A spatial object (SO) at a certain time or duration is defined as follows.

[Definition of SO]

$SO(id_{SO}, t) : (id_{space}, geo, p, d, s, graf)$

  id_{SO} : spatial object identifier

  t : time or duration

  id_{space} : space identifier

  geo : geometry

  p : position

  d : direction

  s : scale factor

  $graf$: graphic properties

id_{SO} is a unique identifier for a spatial object as a component of a space such as a real space and a virtual space. geo represents the geometry for a spatial object and is used in a spatial query. p, d and s are used to define the geometry in a relative coordinate. $graf$ means a set of graphic properties such as its material color and texture. A space is defined as a set of spatial objects.

[Definition of *Space*]

$Space(id_{space}, t)$: a set of SO's (spatial objects)

For example, a 2D map of Hiroshima City in 1998 is a visualization of a space composed of spatial objects such as Hiroshima Baseball Stadium and Peace Park in Hiroshima in 1998. The time attribute values for the spatial objects are all the same, and can be considered that their time attribute values are inherited from the time attribute of the space.

It is useful to visualize cameras on a 2D map to access to pictures of our interests. For example, if we visualize a camera's position where a picture was taken from, it helps us realize the place where the picture was taken. Many positions are visualized on 2D maps as the places of pictures, and clicking one of the visualized positions of cameras can pop up the corresponding picture on a screen. If the direction of a camera is visualized as an arrow icon on a 2D map as well as its position, we can get more information from the visualization. Pictures can be selected from the viewpoint of the direction in addition to the position. It is useful to put some annotations beside the arrows. The examples of the annotations are the time when pictures were taken and names of persons who took the pictures. Furthermore, the pictures correspond to regions which are viewed in the pictures (Fig. 2). The region can also be a representative symbol of the picture. If we want to see some pictures of the regions of our interests, we just click one region as an icon so that the corresponding picture will be displayed on a screen. Thus, a camera can be visualized as a spatial object on a 2D map. The following is an example of visualizing a camera as an arrow.

$$SO(id_{SO_i}, t_k) : (\quad id_{space} = SPACE_j, geo = \text{ARROW\_GEOMETRY},$$
$$p = cam(id_{cam}, t_k).p, d = cam(id_{cam}, t_k).d,$$
$$s = \text{DEFAULT}, graf = \text{DEFAULT})$$

3.3 Spatial Query for Spatial Data of Pictures and Real World Objects

First, we discuss the use of only position data for spatial retrieval. For example, we are looking for some pictures which were taken from the bridge "A." The bridge "A" may be interpreted as a region and the query is made for finding a set of points representing camera's positions which are covered by the region. The query is realized in the following query formulation.

$$\{ c | \quad c \in CAM \text{ AND } c \in Space(id_{Hiroshima}, 1998)$$
$$\text{AND } SO(id_{Bridge"A"}, 1998).reg \textbf{ CONTAIN } c \}$$

"CAM" means a universal set of cameras. "$Space(id_{Hiroshima}, 1998)$" means a set of spatial objects in Hiroshima in 1998. "$SO(id_{Bridge"A"}, 1998).reg$" defines the region of the bridge "A" in 1998. "**CONTAIN**" is a spatial comparative operator which returns TRUE if its left side geometry contains the right one, and FALSE if not.

If direction data of a camera become available in addition to its position data, we can make more complex spatial query. For example, we can retrieve some pictures by a spatial query which formulates that the pictures were taken from the bridge "A" and show the north direction from the bridge. Furthermore, we can bind pictures with their corresponding *regions* which cover scenes of the pictures. We can make much more practical spatial query by using the region data for pictures. For example, we search for some pictures which show the building "B." To realize this query, we have to use spatial data of the building "B," that is, the location of it or the region of it. The position of the building can be addressed by specifying the name of the building "B." The spatial query will find a set of regions containing the position of the building "B." The selected set of regions are converted into a set of pictures which view the building "B." Then, the set of pictures viewing the building "B" can be displayed on a screen as a result of a user's request.

3.4   Video Data Extensions

This section discusses video data with a 2D map. Video data are considered as a sequence of pictures which are called *frames* of videos. Each frame has its time attribute. For example, a video is composed of 30 frames of pictures per second. The number of the corresponding spatial data for all frames of a video sequence may become large. If we put the corresponding representative symbols for all frames on a 2D map, the number of representative symbols for them becomes too large to browse and click. If the movement of the camera is very fast or we use only a short-duration video sequence, it will be better to use all representative symbols for all frames. In the case of a long duration video sequences such as 10 or 30 minutes videos, the number of the symbols becomes too large and the visual result becomes too dense or unreadable. For example, even if the camera is still or does not move, a large number of the same symbols for all frames of the video sequence are generated. It may be useless. In the case that the camera is still, it is reasonable to use only one representative symbol for crowded symbols corresponding to a certain duration video sequence.

It is difficult to decide what part of a video sequence should be represented by only one symbol. It is a naive approach to use only one symbol for a certain constant duration video sequence. For example, we can use a point icon for a set of frames of every 5 seconds video sequence in a virtual space. We can appreciate the path of the camera's movement through the time-series point icons. The movement of a camera can be represented by animating point symbols in real-time on a 2D map. The point symbol can be extended to an *arrow* symbol which represents direction of the camera as well as its position. If users can click one of the arrow symbols on a 2D map, the corresponding video sequence can be played back in another window on a screen. Region symbols are also useful and can correspond to all frames of a video sequence. It is good to use some of all region symbols for all frames. For example, representative region symbols can stand for segments of every 5 seconds video sequence in order to decrease the number of symbols on a 2D map. The regions can be animated in real time on a 2D map for representing the movement of the camera.

A video sequence should be divided into multiple meaningful segments of video sub-sequences. There are many ways to make representative symbols for the segments of video sub-sequences. For example, a line symbol can be used as a collection of point symbols, each of which corresponds to each frame of a video sub-sequence. If the camera pans frequently while taking a video, it is difficult to use only one still arrow to represent all directions of the camera for a certain duration. In that case, many arrows may represent the panning camera. It is also useful to make special symbols or animated symbols for a certain duration movement of a moving camera.

The previous paragraph has discussed the ways of visualizing the movement of a camera as components of a hypermedia 2D map to access to video sequences of our interest. Here, we discuss some use of spatial query for spatial data for video sequences. The basic principle of retrieving video sequences of our interests is almost the same as the one for retrieving pictures. For example, if we search for some video sequences which were taken from the bridge "A" and show the building "B," we should make some queries to find some frames which were taken from the place near the bridge "A" and view the building "B." The videos which include the frames selected by the spatial query should be replayed on a screen as a result of video data retrieval.

3.5   3D Virtual Space Extensions

If a 3D virtual space data corresponding to the real world and 3D data of movement of a camera are available, the movement of the camera should be visualized using 3D CG techniques for browsing and retrieving video data. Representative 3D icons can be used for video sequences. The basic principles of browsing and retrieving video sequences in a 3D virtual space are almost the same as the case of a 2D map. We walk through a virtual space and can execute some spatial queries to select video sequences which could show the same views as the current views in the virtual space. Videos can be incorporated into a 3D virtual space as components of it. It means videos can be replayed in a 3D virtual space in real time. We can experience the direction and position of the camera which took a video, and appreciate the video more spatially. It allows users to immerse themselves in a 3D virtual space with realistic video browsing. 2D maps can be considered a view from sky. If videos took geographic objects on a ground from the sky, the videos can be mapped and replayed on 2D maps. Most of videos, however, took geographic objects on a ground from the ground. It may be meaningless to map and replay these kinds of video frames on 2D maps. 3D virtual spaces are useful to map and replay these kinds of video frames.

4   An Experimental System

For our experiment, we used two sensors, a gyro and a GPS (Global Positioning System) to generate time-series spatial data for video sequences which were taken by a digital video camera. We collected some spatial data for videos taking

Fig. 6. A crew taking videos of scenes of campus festival of Hiroshima City University with a digital video camera and sensors

Fig. 7. A scene of campus festival of Hiroshima City University in October 1997, which was taken by a digital video camera

some scenes of campus festival of Hiroshima City University held in October 18th and 19th, 1997 (Fig. 6, 7). The gyro was precise enough to record the direction of the camera, but the GPS was not precise to measure the position of the camera. We compensated the position data of the camera by plotting every one second positions of it on a 2D campus map by hands later. The zoom ratio can be obtained automatically from the current style of digital camera. To simulate the continuous movement of the camera as real-time 3D computer graphics animation, we use the method of linear interpolation for the discrete time-series spatial data such as position, direction and zoom ratio.

We have implemented two typical applications for spatial browsing video data. One is to simulate the movement of the camera as real-time 3D CG animated camera icons in a virtual space. Users could appreciate the movement of

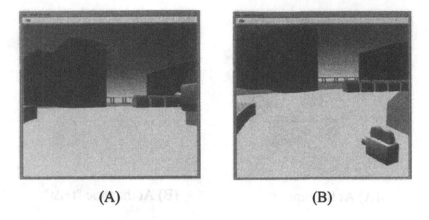

Fig. 8. Simulation of the movement of a real-world camera from two different user's views

3D CG animated camera icons in the virtual space from arbitrary views (Fig. 8). If we click a 3D CG animated camera icon at a moment, the video corresponding to the current 3D CG animated camera icon can be replayed on a CG rectangle plane appearing in front of the 3D CG animated camera icon in a virtual space (Fig. 9). The zoom ratio of the camera is used for the distance between the camera and the CG rectangle plane for replaying a video. Thus, we can appreciate spatially both the movement of the camera and replaying videos in a 3D virtual space.

The other application is to use 3D CG still arrows, each of which represents each of segments of video sequences. In the experiment, a video sequence has been divided into meaningful segments of video sub-sequences by hands. 3D CG still arrow icons representing the segments of the video sequences were automatically generated by visualizing the average values of the position, direction and zoom ratio of the camera. The length of the arrow represents the average value of the zoom ratio of the camera (Fig. 10). We can walk through the 3D CG still arrow icons which address video sequences and indicate the average values of the position, direction and zoom ratio of the movement of the camera (Fig. 11). If we click one of the arrow icons, we can appreciate videos being replayed in a virtual space (Fig. 12). In this application, the camera to view the virtual space is restricted to be set in the same position, direction and zoom ratio of the real-world camera. It guarantees that we can see the video in the middle of the scene in a virtual space. Compared with the previous application, we always appreciate videos from right angles, but our views cannot be controlled. Furthermore, we could have a wider view compared to only a video being played in another window on a screen. The video can be augmented as a wider view and be imposed of a virtual space so as to enable users to experience the video more spatially.

(A) At the time "t"   (B) At the time "t+Δt"

Fig. 9. Replaying a video on a rectangle plane with a moving camera icon in a 3D virtual space

5   Concluding Remarks

Sensors are expected to be affordable and become precise enough to generate spatial description data for objects in the real world in the near future. The spatial description data of a video camera's condition measured by the spatial sensors are also useful for browsing and retrieving video data spatially. The spatial data of the video camera enable richer visual index environment. We can retrieve both pictures and videos in 3D virtual spaces as well as on 2D maps. Also, real-time 3D computer graphics hardware enables us to walk through a 3D virtual space and to appreciate replaying videos in the 3D virtual space. This kind of environment can provide users with spatially browsing videos in a virtual space. We can experience virtual spaces in more realistic ways because real world videos can be replayed in a virtual space. While we walk though 3D virtual spaces which have their corresponding existent real world spaces, we can be aware of the existence of videos in the virtual spaces with watching 3D icons for video sequences, and appreciate replaying videos more spatially and naturally in virtual spaces. This application may promise to manage video data for some application domains, such as sight-seeing tours of the real world, simulations of the real world, and virtual disaster management systems.

ART+COM [1] has researched on spatialization of video sequences as a 3D volume objects in a 3D virtual space. The 3D volume object is composed of a collection of video frames which are located and directed in the corresponding momentary camera's location. They call the 3D volume objects of video sequences 'invisible shape. If a user clicks a part of some invisible shape, the end plane of the invisible shape is getting short because the end plane is used for replaying the selected video sequence. Historical films were used for invisible shapes in a virtual space, Virtual Berlin. They place multiple historical films

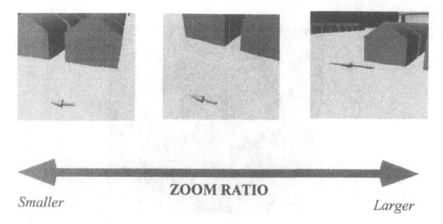

Fig. 10. Zoom ratio of a camera is represented as the length of a 3D arrow icon

Fig. 11. 3D CG arrow icons representing video sequences in a 3D virtual space

as invisible shapes in a virtual space. The virtual space can be considered a composite virtual space of multiple time-intervals. Their system provides a 3D time scale measure to change current time-interval of the virtual space. Only the invisible objects in the user's specified time-intervals are visualized with their original colors, others are visualized as transparent objects. Their approach is elegant and useful, but it is not perfect from the viewpoint of covering generic applications. For instance, their approach cannot guarantee cases of stopping cameras and many video sequences.

We explain some examples of more advanced applications based on our presented model. If a user clicks one of objects in a 3D virtual space, the clicking of one object can be interpreted as a spatial query to retrieve and replay video sequences which show the clicked object in the video. At the same time, the viewpoint of the user in the virtual space can be automatically moved to the

Fig. 12. A replaying video rectangle plane positioned in the center of a window because of synchronization of a virtual camera and a real camera

good position to appreciate the videos being replayed in a virtual space. Furthermore, if a user click part of videos replayed in a 3D virtual space, some objects corresponding the clicked part of the videos are selected for next operations. These applications are considered to have aspects of both augmented virtuality or augmented reality. Though there are many problems in managing a large amount of videos, we believe our model could give some ideas to solve some of the problems managing a large amount of video data. We plan to evolve our prototype system to more practical one from the viewpoint of functionality, easiness of use, and salability for a large amount of real video data.

Acknowledgements

We would like to thank Mr. Shigeru Kakumoto, Central Research Laboratory of Hitachi Ltd., for his ideas which motivated us to do this research. He told the ideas to one of the authors when we joined the IEICE Workshop of *Functional Graphics* in Lake Tazawa in November 1993. Mr. Steffen Meschkat, ART+COM, and we had a good discussion on managing time-series camera's various data in a 3D CG space. The projects of *Invisible Shape of Things Past* and *Terra_Vision* inspired us to generalize the concept of this research. We appreciate every effort and supports by staff at Database Systems Laboratory, Department of Intelligent Systems, Faculty of Information Sciences, Hiroshima City University. This work was supported in part by the Grant-in-Aid for Research for the Future Program of Japan Society for the Promotion of Science under the Project "*Researches on Advanced Multimedia Contents Processing* (JSPS-RFTF97P00501)" , Scientific Research on Priority Areas "*Advanced Databases*" of the Ministry of Education, Science, Sports and Culture of Japan, and the grant "*Specified Research*" at Hiroshima City University. Our 3D virtual campus data was created by Miss Hiromi Michiyori, a student of Faculty of Art at Hiroshima City University, using SOFTIMAGE, Microsoft Inc.

References

1. ART+COM, http://www.artcom.de/
2. Milgram, P., Kishino, F.: A Taxonomy of Mixed Reality Visual Displays, IEICE Transactions of Information Systems, Vol. E77-D, No. 12 (1994) pp. 1321 – 1329
3. Uehara, K. , Oe, M., Maehara, K.: Knowledge Representation, Concept Acquisition and Retrieval of Video Data, Proceedings of the International Symposium on Cooperative Database Systems for Advanced Application, Kyoto, Japan (1996) pp. 218 – 225

AI-STRATA: A User-Centered Model for Content-Based Description and Retrieval of Audiovisual Sequences*

Yannick Prié[1], Alain Mille[2], and Jean-Marie Pinon[1]

[1] LISI 502, INSA Lyon, F-69621 Villeurbanne Cedex, France
Yannick.Prie@insa-lyon.fr
pinon@if.insa-lyon.fr
[2] LISA, CPE-LYON, F-69616 Villeurbanne Cedex, France
am@cpe.fr

Abstract. We first insist on the need for conceptual and knowledge-based audiovisual (AV) models in AV and multimedia information retrieval systems. We then propose several criteria for characterizing audiovisual representation approaches, and present a new approach for modeling and structuring AV documents with Annotations Interconnected Strata (AI-STRATA). This consists in analyzing AV documents through analysis dimensions allowing the detection of objects of interest of any type (structural, conceptual,...). Annotations are structured by annotation elements (AE) representing both objects of interest and relationships. A knowledge base is used in order to monitor the annotation process. We show how to use annotations to link different strata on the base of explicit or implicit contexts and how AI-Strata can be used to build contextual views of a stratum, using both annotation and knowledge levels. We finally show how the model can efficiently support different description tasks such as indexing, searching and browsing audiovisual material.

1 Introduction

As computer and network capabilities grow, so do the size and the number of digital data repositories, while the notion of document evolves to include new technical developments. Terminologically speaking, a multimedia document is likely to become the standard, while mono-media documents may be considered as restrictive. So, there is an urgent need for multimedia information retrieval systems capable of dealing with new digital media like images or videos. In this emerging field of multimedia information management, we will focus on AudioVisual (AV) information systems.

After a short survey of recent trends in audiovisual information retrieval systems, we will present SESAME[1], a project with a user-centered and description-

---

* This work is partially supported by France Télécom (through CNET/CCETT), research contract N° 96 ME 17.

[1] Multimedia and Audiovisual Sequences Exploration enriched by Experience System.

S. Nishio, F. Kishino (Eds.): AMCP'98, LNCS 1554, pp. 328–343, 1999.

unified approach. We will then concentrate on audiovisual modeling and detail AI-STRATA, the SESAME modeling approach, before studying what services could be provided considering the fundamental task of description.

2 Audiovisual Information Retrieval Systems

Research in audiovisual information retrieval systems has greatly increased in the last decade, mainly among the image processing and the database communities. Many efforts have been done to compute image features in order to build tools using them as retrieval medium [3], or to propose audiovisual extensions to databases [17] [2]. As multimedia data do not fit exactly into classical database schemes, researchers are getting aware that tools for managing and organizing visual information could take advantage of using concepts and algorithms issued from other domains including information retrieval and artificial intelligence [13] [11]. These techniques should allow problems of similarity querying and visual browsing to be dealt with. Text-oriented query systems disappear behind environments allowing to visualize the "content" of multimedia documents, to browse visual objects, and to visually query or interpret results[2].

At the same time, guided both by this new awareness of the fundamental characteristics of visual data and by a current trend in information retrieval [4] (development of graphical interfaces and generalization of browsing to IR) novel approaches are focusing on *(re-)inserting the user in the heart of the system* for a real cooperation between human and machine. Relevance feedback [16] and machine learning are applied to similarity queries. The way people search a visual database is studied [12] and navigation through semantic ontologies of visual information are considered [5].

Studying current information retrieval systems leads to the observation than one has to *cooperate* with an audiovisual information retrieval system for performing the following tasks:

- *Indexing* is the task of describing a new document according to a model in order to organize its insertion in an index.
- *Querying* is the task of designing a query in order to find out what in the base could match it. A query can be very sharp or vague, or be an example.
- *Analyzing* consists in describing an AV document extensively in order to detect any regularities or known structural forms in AV documents concerning montage, stories, characters, camerawork, *etc.*
- *Browsing* can be seen as very precise querying (from the machine point of view), but also as wandering through the database (from the user point of view).
- *Editing* audiovisual documents is a task that can be performed to modify existing documents, to create new ones, but also to create visual summaries or clips [14] [6] guided by a set of selected descriptors.

---

[2] Though audiovisual systems are both concerned with visual and aural modalities, research has up to now mainly focused on the visual one. Audio information are nevertheless more and more being used.

We consider that all these tasks can be thought of as "describing a piece of video for...", which implies that the *description* of audiovisual documents should become a central task in a well-designed AV information retrieval system. Therefore AV representation (in the machine) and AV presentation (to the user) should be very close conceptually, so that user-centered visual interface could support real user-machine cooperation for description of AV documents.

Audiovisual representation should take into account every available feature provided by audio and video processing techniques. However this cannot be enough for sufficient and efficient description and representation of the audiovisual content. Higher level conceptual modeling should be used to organize this material, and a knowledge approach appears to be a necessity, either for itself [10] [22] or in order to reduce the search space before launching low-level features searches [5]. Current work on the future standard MPEG7 [18] focuses on this knowledge-based approach for conceptual modeling.

3 SESAME

SESAME is a project supported by the CNET[3] to propose and study a global approach to exploit the potentially huge repositories of audiovisual documents. The aim of the project is to take into account all facets of the problem and several research teams of different French laboratories[4] are involved in its development. Industrial partners such as INA[5] and FRANCE 3[6] are directly associated. First of all, a global architecture has been defined as sketched in figure 1. Audiovisual chunks stored in repositories are reached through High Speed Networks[7] (HSN). Audiovisual chunks are indexed or accessed on different types of clients: Indexing Clients (IC) and Accessing Client (AC), and are managed on servers by an Audiovisual Documents Base Management System (ADBMS) and a Parallel Access Engine (PAE).

The present paper focuses on the modeling approach proposed to describe audiovisual chunks in such an architecture.

The model aims to define basic elements of a content-oriented describing process. The level of granularity of an audiovisual chunk varies indeed from a whole document as it was produced originally to a small piece of an audiovisual stream focusing on a particular object. Moreover, as argued above, users accessing audiovisual servers have different needs and strategies depending on the task they want to achieve. A description model of audiovisual chunks has to consider such various levels of interest and has to be flexible enough to be used in a variety of tasks. Such a model must offer:

---

[3] CNET: France Télécom Center for Research and Development.
[4] LIP-ENS-Lyon, LISA-CPE-Lyon, LISI-INSA-Lyon and RFV-INSA-Lyon.
[5] INA: French TV and Radio Archives.
[6] FRANCE 3 is a french public T.V. channel.
[7] Local area networks and/or world area networks.

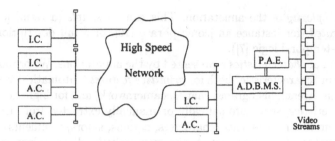

Fig. 1. *Sesame global architecture*

- a chunking process based on primary annotation and semantic annotations at any level of abstraction (from computed characteristics of images to complex concepts of the real world),
- canonical and end-user points of view to fit different goals and tasks,
- relationship links between annotations in order to build dynamic contexts as efficient user-centered filters.

The AI-STRATA model aims to offer such possibilities through task-oriented man machine interfaces, where several assistants could help the describing process. SESAME is concerned directly with image processing assistance to automatically or interactively annotate audiovisual chunks, whereas a case-based approach would help to design "winning" queries on the servers (primary indexed audiovisual chunks) from a local search description (secondary indexed audiovisual chunks).

4   Modeling Approaches of Audiovisual Documents

Audiovisual is both a visual and accoustic *sequential* medium, with a strong part of fixed *temporality* (unlike text) leading the viewer to the illusion of reality, and, by using montage, to the possibility of telling stories, with contextual relations between parts of documents.

4.1   Criteria for the Comparison of Modeling Approaches

The fundamental concept concerning a modeling approach for sequential media is *annotation*, which consists in attaching an annotation (a description) to a piece of the considered AV document. Each piece is bounded by two instants t_1 and t_2 separating it from the starting instant t_0 of the stream.

Main criteria to analyze the different approaches to describe contents of an audiovisual stream are:

- the *time granularity* of the representation model : from a time interval corresponding to a frame to a whole document (*e.g.* a two hours movie), everything is possible. *Cutting* can also be regular or not, and possibly multi-layered (*e.g.* the document is annotated as a whole plus its shots plus its sequences [7] [22]).

- the *complexity* of the annotation. This can be *simple* (a term, [6]) or *sophisticated* (for instance an iconic phrase [10], or a full description with a spatio-temporal logic [7]).
- the kind of *characteristics* represented by the annotations. We propose to use the term "characteristic" both for *primitive features* (automatically extracted from the stream: histograms, shapes, camerawork) and for higher conceptual characteristics, semi-automatically or manually extracted from the stream and from other documents (characters, actions, actors, comments).
- the *structure* of the document. As the granularity level goes thiner as one goes deeper in the document it becomes necessary to link up the different pieces that have been annotated. The resulting representation of the document can therefore be *implicitly* structured, as a result from time continuity (shots follow one another); or *explicitly* if a global structure is set up, which will be most of the time hierarchical (shot/sequence/document).

4.2   Two Main Structural Approaches for Annotation

Two main structural approaches arise from the litterature: *segmentation* and *stratification*. *Segmenting* an AV document consists in cutting it up into predefined pieces (*e.g.* shots), which will be annotated later. A structural organization can be set up to make explicit relations between pieces (as time granularity often corresponds to structural decomposition). This approach is most often used, and the present trend is to consider a three layers structure (shot/sequence /document, see figure 2), a structured annotation (usually records of attributes/ values) being associated with each piece of document (see Corridoni & al. and their "filmic grammar" [7] or [22]).

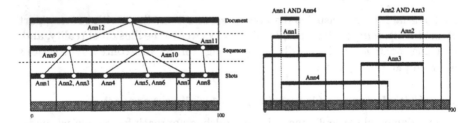

Fig. 2. *On the left, the segmentation approach: describing predefined AV pieces possibly organized in a structural hierarchy — On the left, the stratification approach: annotating (hence defining) pieces of AV documents, before a posteriori "useful" cutting up.*

The *stratification* approach [9] [10] differs from the segmentation one in several ways: in the segmentation approach, shots or sequences just represent themselves, the montage is forgotten, and there is a loss in the continuity of the AV media. Oppositely, a *stratum* can be any piece of an AV document to which an annotation is attached, mainly atomic, possibly expressing icon sentences. Queries result in "useful" cuttings of the document, through strata intersection (see figure 2). No structure-oriented annotation is considered.

The essential difference seems to lie in the definition of the temporally situated and annotated pieces of documents. In segmentation, the document is first tiled with generic tiles like shots and sequences, and so cutting exists before annotation. The pieces are annotated later (classical paradigm in document description). Oppositely, in stratification they are temporally situated annotations that create the strata, while other pieces (eventually structured) emerge *a posteriori* from queries[8]. There are in fact two different cuttings of the AV documents, *a priori* and *a posteriori*, reflecting two different status for the annotation.

We think that although the stratification approach is useful for taking into account AV temporality and does really fit it, it unfortunately does not allow to consider relations between strata, for instance well known structural data though considered in recent segmentation models.

As a conclusion, this study shows first that the *dynamic* aspect related to a temporal stream should be taken into account, which implies the use of *strata* and the *atomicity* of the annotations. Second, it seems necessary to structure the annotation in order to increase the expressive power of the description, at any level of complexity. Third, contextual relations in audiovisual documents have to be considered both explicitly and implicitly.

5   AI-STRATA

The general principle of our approach is simple: we consider everything that can be revealed or said about a video, features, high-level characteristics, structural notations as characteristics temporally situated in the stream, as terms in relationship with strata. Then, in order to increase the expressive power of the representation, we set up inter- and intra-strata relations between annotations, *in the same way as annotating*. Lastly we really consider any annotation (and its relations) as *a support for contextual relations* that allow and guide contextual annotation for supplementing primary annotation.

5.1   Definitions

We call *audiovisual objects of interest* the entities that can be spotted when watching/listening to an AV stream. They can be considered as the conceptual and cognitive facets of AV characteristics. An object of interest can refer to any AV feature, from a low-level abstraction (a color histogram) to a high-level one (an action). There are as many objects of interest as there are possible analyses, so we group into *analysis dimensions* the analyses that allow to spot the same kind of objects of interest. We can for example consider *analysis dimensions*

---

[8] Of any type: looking at two time lines and the associated annotations on Mediastream [10] is a kind of "visual" query.

linked with shots, faces, activities, people detection, or more specialized objects, like "President Clinton", or more general ones, like structural unit spotting. Analysis dimensions are just a way to group detection methods or types of characteristics in a relevant manner regarding the goal of the subjacent analysis.

An *audiovisual stream* is no more than a file with audio and video data, beginning at t_0. We define an *audiovisual unit* (AVU) as an abstract entity representing any stratum of the stream. An AVU is created whenever its existence becomes of interest, that is when it has been spotted as a stratum linked with an object of interest: "this is a part of an AV document called a shot", "this is a part of an AV document in which appears X", *etc.* Every AVU must, by definition, be annotated, associated with a characteristic from the spotting of which it has been established: this is the primitive annotation (see figure 3).

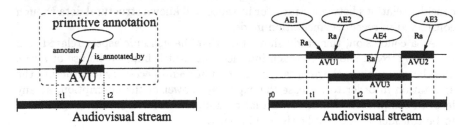

Fig. 3. *On the left: one audiovisual unit and the primitive annotation that creates it — On the right: audiovisual units and annotation elements*

We call an *annotation element* (AE) a *term* in *annotation relation* with the AV unit, as the symbolic expression of a characteristic. As analysis can spot many types of objects of interest, there are many annotation elements associated with color analysis, shapes, camera movements, audio features, objects, activities, types of documents, sensations, *etc.* For instance ⟨*Shot*⟩, ⟨*Clinton*⟩, ⟨*Round_object*⟩, ⟨*Zoom*⟩, ⟨*INCLUDED_IN*⟩, ⟨*Music*⟩ or ⟨*Sad*⟩. The fundamental principle lies here in the *atomicity* of the AE: it is possible to express any characteristic as an AE. AEs can have attributes for numerical expression of features (a color histogram will appear as an attribute of an AE indicating that this histogram was computed). Other AE attributes can also be added: texts, images, sets of features for similarity search, *etc.*

To complete the primitive annotation that defines an AVU, it is possible to add as many AEs as necessary to annotate it (*cf.* figure 3), in two differents ways. First, by *grouping* annotations of the same temporal range: we can add to the annotation element ⟨*Document*⟩ (annotating an AVU corresponding to a whole AV document) other AEs regarding for instance the author or the producer. Second, by *structuring* the annotation using annotation elements (see part 5.2). We call *direct annotation* from an AVU the set of AE in annotation relation R_a with an AVU.

5.2   Relations Between Annotations

We have seen earlier how it was important to structure the annotation to be able to express more complex information pieces of AV documents. To achieve this goal it is necessary to express relations between atomic annotations we have already set. If some relations can be considered as implicit (two AEs, corresponding to two people, annotating the same AVU probably shows that the two characters appear together in the video) while other ones have to be explicit. For instance "this shot is included in this sequence", "this character has that activity", "this object, the sun, is linked with that round, yellow form", "this shot is re-used in that document", and so on.

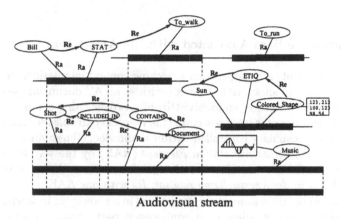

Fig. 4. *Structuring the annotation.*

This can be done by keeping the stratification principle, using the atomic annotations already set, *in the way* we have already annotated. We define one single relation between AEs (the *elementary relation R_e*), and choose to *name* relations that matter as annotation elements. For instance, to express the fact that "Bill is walking" starting from two AEs $\langle Bill \rangle$ and $\langle To\_walk \rangle$[9] annotating two different AVUs, we connect them by a "canonical" annotation element $\langle STAT \rangle$[10] which is used also to annotate the first AVU. We set up two elementary relations to express both the fact that "Bill's walking" (see figure 4). Considering a relation like "this shot is included in this sequence", we should use the canonical AEs $\langle INCLUDED\_IN \rangle$ and $\langle CONTAINS \rangle$.

Finally, annotating a stream consists in studying it through as many analysis dimensions as necessary, so that AEs and the associated AVUs emerge. Relations between AEs are then set up with canonical relation AEs and linked up by elementary relations, as illustrated by figure 5.

---

[9] Spotted along two analysis dimensions corresponding for instance to "characters" and "activities".

[10] Inspired by Sowa's Conceptual Graphs [20].

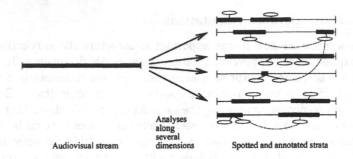

Fig. 5. *The annotation process: from raw AV material to annotated and linked strata*

5.3 Knowledge Base Associated with the Annotation

Annotation elements are supports for searching contextual relations, for infering from the network of annotations that constitutes an AV document (see figure 8) annotated by, for instance, an archivist from a broadcasting company. In order to facilitate and monitor later searches, it is necessary to consider AEs as terms issued from a vocabulary controlled within a knowledge base: an AE is instanciated from an *Abstract Annotation Element* (AAE) by the way *inscription in the stream*. An AAE is in *stream inscribing relation* R_{si} with one or several AE while an AE is in *decontextualizing relation* R_d with one AAE.

Abstract annotation elements are organized in a knowledge base with concept-relations (for instance specialization, equivalence, part-whole,...) between AAE-nodes, see figure 6. The knowledge base as an ontology appears like a thesaurus in classical information retrieval, but, as it is in fact an ontology, it should also be able to organize knowledge about possible AE attributes (value domains, by default or privileged relations between AE).

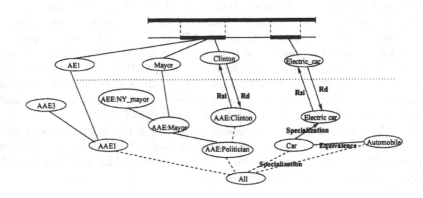

Fig. 6. *Abstract annotation elements and knowledge base*

The description of an AV document leads to a set of AVUs and AEs. This set is not *a priori* independant from the system, it is a part of it. It merges into the *Annotated Audiovisual Document Base* (AADB — set of all the AVUs and AEs of the system) taking into account the relations that some AEs have with other AEs from the base. The AV stream/file does not only appear *per se* in the base, but also as a set of annotated AVU in the AADB, linked with the knowledge base (see figure 7).

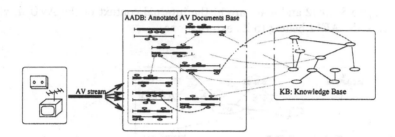

Fig. 7. *Stream annotation and system bases*

5.4 Context from an AVU and Construction of Views

We insisted above on the importance of context in the audiovisual domain, at the intra-document level: shot context ("this music goes together with this character"), montage context ("this shot is only understandable if preceeded by this") ; but also at the inter-document level ("this shot has been extracted from this document").

As it is possible to set up relationships between AEs annotating two different AVUs, we should be able to consider different AVUs in relation to each other, one being part of the context of the other. This context can either be a *temporal* one (one AVU has a temporal relation with all the AVU of the stream), a *structural* one (two AVU annotated by the AE ⟨*Sequence*⟩ and ⟨*Document*⟩) or — and this is the most general case — a *conceptual* one.

For example if many AEs corresponding to a given character are inscriptions (*i.e.* instances) in the stream of the same AAE, or if there is an explicit relation between the AEs ⟨*Zoom_in*⟩ and ⟨*Car*⟩ specifying that "the zoom-in of the camera focuses on the car".

We consider that every contextual relation we have evoked can be thought of as a relationship between AE annotating AVU spotted along well-choosen analysis dimensions. In other words, we think of *annotation elements* not only as simple annotations, but also as starting *elements* for more complex annotations in their net organization, and eventually, out of this organization, as *media* for contextual relations.

We define two contexts for an AVU:

– the context *internal to the annotated documents base*, composed of AVU related to each other with an *internal contextual relation $e_1 R_e \ldots R_e e_2$* between some of the AE annotating them.
– the context *external to the annotated documents base* concerns a set of abstract annotation elements (AAE) in *external contextual relation* with annotation elements from the AVU.

In figure 8, *avu2* and *avu4* are in the internal context of the AVU *u*, while AAE11 and AAE12 belong to its external one.

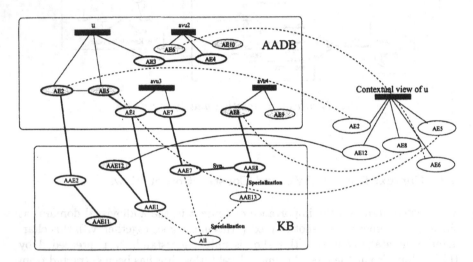

Fig. 8. *First step in view construction: building the context of an AVU ; second step: choosing the contextual AE*

The notion of context so defined, we now consider how we can build *contextual views* of AVUs, made up with *contextual annotation elements* coming from their context. These AEs can come from the internal context (AEs formerly annotating another AVU) or from the external one (inscriptions in the stream from AAEs).

A contextual view is constructed in two steps: the first one consists in constructing the context of the AVU and determining what AEs are candidates for contextual annotation: in grey in figure 8 we have AE2, AE5, AE3 (u is in its own context); AE4, AE6, AE10 (from *avu2*); AE8, AE9 (from *avu4*); AAE11 and AAE12. The second step consists in choosing the ones among the potential contextual AE which will actually belong to the view of the AVU: some issuing from the internal context will already exist (fig. 8: AE2, AE5, AE6, AE8), while others (coming from the external context) will be inscriptions in the stream of AAE (see AE12).

These two steps in the construction of a contextual view are controlled by two filters called the *contextual filter* and the *descriptive filter* together representing the *view filter*. We see that the simplest view filter would associate to an AVU the set of the AE annotating it. This is called the minimal view, as the filtered context would be minimal (the AVU itself) and the filtered description would be maximal (all the potential AE are selected).

View filters are defined according to what users want to do: for a large search in the context of an AVU, the scope of the context should be limited in depth by a maximum number of elementary relations between the elements of the context and the AVU. However, more specific search should narrow the context (for instance, only the context linked with camera movements and people) and the choice of particular potential AE (only politicians, or actions...). Using view filters allows any kind of describing task to fit with particular needs and goals. As the whole system (AVUs+AEs+AAE's) appears much like a graph, tools and algorithms that will be designed and used to manage it will be strongly related with graph theory and practice: graph manipulation and subgraph isomorphisms will indeed play an important role in AI-Strata management.

5.5   Discussion and (Closely) Related Work

We have presented our approach for AV modeling with Annotation Interconnected Strata, which consists in considering every possible "comment" about an AV document as an object of interest spotted along an analysis dimension. The object is represented by an annotation element defining an audiovisual unit. Annotations are structured thanks to relations between atomic annotations (hence between strata). This structuring is made by annotating. The vocabulary of the terms-AE is controlled with a knowledge base organized as a semantic net. The "deconstruction" of a document resulting from its annotation allows the explanation of every contextual relation inside the document (even between AVUs that do not temporally meet, which goes further than the simple juxtaposition of annotations), but also with other documents in the base. Then it becomes possible for any AVU to consider a context based on the relations its AEs have with other AEs or AAEs of the Knowledge Base. Annotations act as a medium for establishing conceptual contextual relations (among which temporal and structural relations take place), allowing propagation of annotations.

In [23] is presented a video description model called "time-stamped authoring graph", where textual annotations are attached to time instants. Annotations are then connected using three types of links: commonsense, generalization and normal links. Retrieval uses keyword match and time interval calculation with time-stamps. This approach is related to ours as it also leads to an annotation graph for AV documents, but it differs in at least two important features. First, no controlled vocabulary nor conceptual relations are available (for instance generalization only occurs in the context of a document, and is not a knowledge link). Second, the annotations are time-stamped, while we on the contrary introduced the concept of audiovisual unit, which acts as a mediation between the audiovisual stream and the annotation, with no intrinsic semantics[11]. Using this neutral and intermediate level between AV data and annotations could bring an elegant solution to the conflict between segmentation and stratification stressed ealier (see table 1).

	Segmentation	Stratification	AI-STRATA
Time-granularity	linked to hierarchical struct doc/seq./shots	not limited	not limited
Complexity	records Att:Val	single icon or icon sentence	not limited (context)
Characteristics	not limited	issued from conceptual categories	not limited
Structure	structural tree hierarchy	no structuration	not limited

Table 1. *Comparison between segmentation, stratification and AI-STRATA approaches*

Second remark: it is not mandatory to use temporal relations (co-occurrence [10], "interval-inclusion based inheritance" [17]) or structural relations (a shot can inherit annotations from a document it is a part of [21], a sequence is enriched by annotations of the shots that are part of it [6]) relations to propagate annotations between pieces of an AV document. On the contrary we extend all of these relations into *conceptual* ones, considering every propagation (whatever its type) as contextual annotation.

Third, the possibility to consider AEs' attributes like representative images, scripts, digital features, *etc.* could narrow the distance between AE and *semcons* [11], or evolution of MPEG4 objects. We insist nevertheless on the *term* annotating the document rather than on the object on the screen. The term can of course be represented by an icon [10], but as a symbol, not as screen representation of real-world object.

Fourthly, the strata approach, hence the annotation interconnected strata approach, is not limited to audiovisual medium, but can be applied to any *sequential* medium, be it audio or text. This is the ability to freeze a part of audiovisual media (for instance an image), and to build relations between annotations that allow the description of non-sequential material.

---

[11] The meaning of the AVU (hence of the AV piece of the AV document it represents) is provided by the AVU annotation (*i.e.* its set of annotation elements), but also by the relations this annotation has with other annotations (abstract ones in the knowledge base, or not). A shot can therefore be represented by an AVU u annotated with the AE $\langle Shot \rangle$: part of the meaning of u resides in this annotation, however other parts like structural shot-sequence relations, characters appearing in the shot, and others, are also in this annotation, in the sense that they are connected to it by some relation paths.

Fifthly, the whole set of AVUs of a specific repository is tightly coupled to the corresponding knowledge base. So meta-models can express specific indexing or accessing methods depending on explicitly wanted strategies. On the other hand, generic user-oriented meta-models could be defined in order to help and orientate users in their tasks depending on the knowledge available on indexing strategies.

6   An AI-STRATA Based Tool for Indexing, Searching, and Browsing Audiovisual Units

The aim of this part is to show how the AI-STRATA model is able to be an efficient base of conceptual modeling for different tasks and applications. As claimed in the introduction, any task exploiting an Annotated Audiovisual Document Base needs to some extend to describe audiovisual sequences. Our currently developped demonstrator includes the three main tasks to exploit an AADB:

The indexing task: this task involves the indexing of a sequence, where the person annotates with respect to some systematic procedure and anticipates later uses of the sequence. Systematic procedures are easy to describe (as meta-models) as predefined analysis dimensions, while anticipations of searches could be helped by "sounding" the AADB to retrieve how similar AVUs have been used (*i.e.* described for retrieval). We have also elaborated [19] the notion of *annotation assistants* as agents meant to help users to annotate sequences. The first and useful annotation assistant deals with automatic processing of video streams (cut and motion detection, similarity features extraction), leading to automatic ("visual" AE) or/and semi-automatic (eg. linking AE representating shapes/colors with AE denoting concepts) annotation of AV sequences. Other assistants could be added: sound annotation assistant, text-based annotation assistant, *etc.*

The search task: anyone searching for something is asked to describe what they are searching for. Ways to describe an AVU are infinite. We see the search process as an iterative process starting from a first description which depicts the first idea of the user, and is progressively enriched by results of the corresponding queries to the AADB. This process can be long and complex, but AI-STRATA should improve it in several ways by interconnecting first AEs with the available network of AEs (filters, contexts,*etc.*). Using local experience (*i.e.* previous searching episodes) or/and local examples (*i.e.* a local library of AVU) to elaborate the first description could greatly accelerate this process.

The browsing process: this is not exactly a task but a common way of searching documents by going from one point to another. Using AI-STRATA is straightforward for this activity. Starting from a first (weak) description the user can exploit existing AE relations from any retrieved AVU to slide from strata to strata. The ability to figure the starting typed links of each AVU as visual paths on a screen, and (in some way) to navigate at the knowledge level [15] gives more freedom to the person browsing and increases his/her interaction.

The editing process, that is the task of reusing and creating new audiovisual document taken from AV databases, can also rely on AI-STRATA model. Video abstracting, or storytelling can use annotations, and links between them to describe and find (as UAV) desired story units [8] and pieces of a document. Editing could then just be reorganizing AVUs and annotations through a story model.

All these tasks can be long, complex and tedious. Exploiting experience to establish "winning" descriptions should be an efficient assistance. The Case Based Reasoning paradigm [1] will be used to help the description process by reusing past similar descriptions as bases for new description.

7   Conclusion and Future Work

We have presented in this article AI-STRATA, a new approach for audiovisual document description. Main contributions of the paper are: the proposition of several criteria for characterizing audiovisual representation approaches (time granularity, complexity, kind of characteristics, structure) ; the Annotation interconnected strata as an original way to describe AV documents taking into account both the dynamic dimension of AV streams thanks to atomic annotation elements, and the necessary structuring of the representation to cope with the inherent complexity of audiovisual material ; a generalized notion of context, based on the existing annotation, which should allow temporal, structural and conceptual contexts to be considered in the same way and which will be useful for browsing and annotation propagation. We consider also that every task related to AV retrieval systems is in fact a description task, and we illustrate how AI-STRATA is a promising way to easy out these tasks.

Current work to integrate the AI-STRATA approach has several objectives: develop an annotation application with the first annotation assistants (JAVA/ Corba based) ; develop mechanisms of contextual inferences leading to contextual annotation ; design and test the first meta-models with I.N.A. and France3 specialists ; and map the AI-STRATA model with data model proposed by database searches (semi-structured databases look promising). These research themes are the subject of "transversal" research in collaboration with other teams of the Sesame project.

All these developments have now become possible thanks to the unifying nature of the proposed model. We actually claim that most exploitation of audiovisual material can be expressed through the AI-STRATA model.

References

1. A. AAmodt and E. Plaza. Case based reasoning: Foundational issues, methodological variations, and system approaches. *AICOM*, 7(1):39–59, Mars 1994.
2. S. Adali, K.S. Candan, S. Chen, K. Erol, and V.S. Subrahmanian. Advanced video information system: Data structures and query processing. *Multimedia Systems*, 4:172–186, 1996.

3. P. Aigrain, D. Petkovic, and H.J. Zhang. Content-based representation and retrieval of visual media : A state-of-the-art review. *Multimedia Tools and Applications special issue on Representation and Retrieval of Visual Media*, 1996.
4. N. Belkin. Braque : Design of interface to support user interaction in information retrieval. *Information Processing and Management*, 29(3):29–38, 1993.
5. S.F. Chang, J.R. Smith, M. Beigi, and A. Benitez. Visual information retrieval from large distributed online repositories. *Communications of the ACM*, 40(12):63–71, Dec. 1997.
6. T.-S. Chua and L.-Q. Ruan. A video retrieval and sequencing system. *ACM Transactions on Information Systems*, 13(4):372–407, October 1995.
7. J. M. Corridoni, A. Del Bimbo, D. Lucarella, and H. Wenxue. Multi-perspective navigation of movies. *Journal of Visual Languages and Computing*, 7:445–466, 1996.
8. G. Davenport. Indexes are out, models are in. *IEEE Multimedia*, pages 10–15, 1996.
9. G. Davenport, T. Aguierre Smith, and N. Pincever. Cinematic primitives for multimedia. *IEEE Computer Graphics and Applications*, pages 67–74, Jul. 1991.
10. M. Davis. Media streams: An iconic visual language for video annotation. In *Proceedings of the 1993 IEEE Symposium on Visual Languages*, pages 196–203, Bergen, Norway, August 1993. IEEE Computer Society Press.
11. W.I. Grosky. Managing multimedia information in database systems. *Communications of the ACM*, 40(12):73–80, Dec. 1997.
12. A. Gupta, S. Santini, and R. Jain. In search of information in visual media. *Communications of the ACM*, 40(12):35–42, Dec. 1997.
13. R. Jain. Visual information management. *Communications of the ACM*, 40(12):31–32, Dec. 1997.
14. R. Lienhart, S. Pfeiffer, and W. Effelsberg. Video abstracting. *Communications of the ACM*, 40(12):55–62, Dec. 1997.
15. J. Nanard and M. Nanard. Adding macroscopic semantics to anchors in knowledge-based hypertext. *Int. J. Human-Computer Studies*, 43:363–382, 1995.
16. C. Nastar, M. Mitschke, C. Meilhac, and N. Boujemaa. Surfimage: a flexible content-based image retrieval system. In *ACM Multimedia 98*, Bristol, Sept 1998.
17. E. Oomoto and K. Tanaka. Ovid: Design and implementation of a video-object database system. *IEEE Transactions on Knowledge and Data Engineering*, 5(4):629–643, Aug. 1993.
18. F. Pereira. Mpeg-7 : A standard for content-based audiovisual description. In *2nd Int. Conf. on Visual Information Systems*, pages 1–4, San Diego, Dec. 1997.
19. Y. Prié, J.-M. Jolion, and A. Mille. Sesame: audiovisual documents conceptual description model and content annotation assistants. In *CORESA'98, 4th. Workshop on COmpression and REpresentation of Audiovisual Signals*, Lannion — France, Juin 1998. CNET — France Télécom.
20. J.-F. Sowa. *Conceptual Structures — Information Processing in Mind and Machine*. Addison-Wesley, 1984.
21. R. Weiss, A. Duda, and D.K. Gifford. Composition and search with a video algebra. *IEEE Multimedia*, 2(1):12–25, 1995.
22. B.L. Yeo and M.M. Yeung. Retrieving and visualizing video. *Communications of the ACM*, 40(12):43–52, Dec. 1997.
23. K. Zettsu, K. Uehara, K. Tanaka, and N. Kimura. A time-stamped authoring graph for video databases. In *Databases and Expert Systems Applications, LNCS 1308*, pages 192–201. Springer-Verlag, 1997.

Use of Action History Views for Indexing Continuous Media Objects

Kaoru Katayama[1], Osami Kagawa[2], Yasuhiro Kamiya[3], Hideki Tsushima[4],
Takuya Yoshihiro[4], and Yahiko Kambayashi[4]

[1] Department of Information Science, Graduate School of Engineering,
Kyoto University, Yoshida-Honmachi Sakyo-ku Kyoto 606-8501, Japan
kaoru@isse.kuis.kyoto-u.ac.jp
[2] Department of Computer Science, Faculty of Engineering, Hiroshima Denki
Institute of Technology, 6-20-1 Aki-ku Nakano Hiroshima 739-0321, Japan
kagawa@isse.kuis.kyoto-u.ac.jp
[3] Toyoda Automatic Loom Works, Ltd., 2-1 Toyoda-cho Kariya Aichi 448, Japan
kamiya@isse.kuis.kyoto-u.ac.jp
[4] Department of Social Informatics, Graduate School of Informatics,
Kyoto University, Yoshida-Honmachi Sakyo-ku Kyoto 606-8501, Japan
{htsushim,tac,yahiko}@isse.kuis.kyoto-u.ac.jp

Abstract. As it is very difficult to find proper contents from continuous
media objects such as videos and audios, we need to develop appropriate
indexing methods. In this paper we will discuss methods utilizing "action history views" for retrieving proper contents from continuous media
objects. Histories of users' actions shown on a screen are recorded as
lists called "action history" in a symbolic form and each element of an
action history is marked by a time stamp. The action history view mechanism deals with such histories of actions observed on screens. When
continuous media objects are generated synchronizing with action histories, we can use them to search proper portions of continuous media
objects. Recorded sequences of actions satisfying specified conditions can
be selected and played back together with the corresponding continuous
media objects synchronously. In this paper we will discuss basic mechanisms and their applications to distance education.

1 Introduction

In this paper how to retrieve continuous media objects utilizing the corresponding "action histories" is discussed. "Action histories" are sequences of users'
operations of computers with their mouses, keyboards and so on. When continuous media objects are generated, they are recorded synchronizing with them in a
symbolic form and each element of an action history is marked by a time stamp.
A basic idea is illustrated in Figs.1 and 2. They show the case when our method
is applied to a discussion using a shared (hypermedia) document. Each user has
her/his own pointer, pen and keyboard so that s/he can work on the common
documents and modify them. We assume that display contents such as users'
images, editing documents, and users' actions such as motion of pointers and

S. Nishio, F. Kishino (Eds.): AMCP'98, LNCS 1554, pp. 344–355, 1999.
© Springer-Verlag Berlin Heidelberg 1999

Fig. 1. Discussion using a shared document

pens during the discussion are recorded on videos together with sounds. Important snapshots of the documents are stored like conventional database back-up systems. The portions that are different from the previous snapshot are shown by a special background color in Fig.2. Besides such videos with sounds and text information we assume that action histories (i.e. sequences of users' operations of pointers, pens and keyboards) are recorded. Intermediate contents can be recovered by using the corresponding action histories. And editing process of documents can be replayed for some extent, that is, redundant operations can be removed and we can speed-up replay speed when required.

Videos with sounds, documents and action histories are recorded synchronizing each other, so that we can identify appropriate locations of continuous media objects like video with sounds by specifying locations of action histories and/or documents as follows.

1. Specification by action histories.

   For example, we can find the interval when user A modifies section 2 and after that user B modifies section 2 using word 'databases'. Another example is to find video portions where there are many pointers.

   By recording status of each user's microphone (on or off), a user can specify the location of videos by a combination of their status (on/off). For example, a user may be able to find the locations where two specified users are discussing.

2. Specification by text strings.

   Users can specify a set of words appearing on documents and utilize conventional information retrieval technology to search the corresponding videos. Extension is required since some words may appear only in intermediate versions of the documents and not in the snapshot versions. From the action history we can generate such intermediate snapshots. We also need to handle intervals, such as "find the portions of videos when 'database security' appears in the documents".

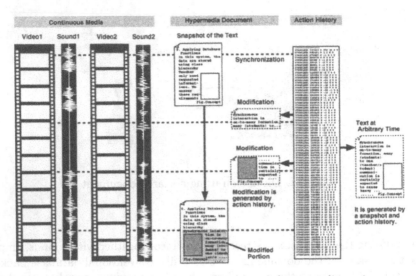

Fig. 2. Relationships among data generated during a discussion

In this paper we introduce an algorithm to retrieve continuous media objects utilizing the corresponding action histories and its application to an education system for distance learners which we are developing. We call the system VIEW Classroom. VIEW stands for Virtual Interactive Environment for Workgroups. In section 2 we describe concept of action histories, how to record them and play back sequences of actions on displays using them. In section 3 we introduce VIEW Classroom and its search facility of recorded lectures based on action histories. In section 4 we survey related works and conclude in section 5.

2   Action History Views as Index

2.1   Concept of Action Views and Action History Views

The idea of the action view and action history view mechanisms[3] is based on the analysis of popular scenarios of CSCW(Computer Supported Cooperative Work). Common hypermedia-based presentation systems are based on the WYSIWIS principle (What You See Is What I See): the users share the same displays. For many applications such a work mode is undesirable, e.g., for security/privacy reasons, or because of overloading of a work partner's screen by irrelevant information. Thus it is necessary to restrict data objects that can appear on remote displays. This can be done by several methods, for example, by explicit determining which objects have to be exported outside, or by subdividing the screen area into private and public parts. In many situations, however, these methods are cumbersome and unreliable. We propose the method based on *object views*, which restrict kinds of objects that may appear on remote screens.

Another problem concerns a situation when collaborative work has to be done by the partners at different time. In this case a remote work partner A should see not only objects that appeared previously on the screen of B, but should also have the possibility to *reply* all the actions performed by B. Again, it would be desirable to restrict the information shown in these replays, because of security/privacy and possible overloading of a screen.

The action views and action history views are methods of the information export restriction for CSCW systems. They are based on the well-known idea of *database views*. A view for a database (object base) can be defined as a mapping of each possible database state (a database instance) into another (virtual) state. For the purposes of CSCW it is necessary to represent not only database states, but also operations that have been performed to change the states. Thus we introduce the concept of *database actions* as a combination of objects' states and operations that have been applied to the objects. As in case of objects shown on screens, for user coordination it may be necessary to present database actions at other workstations. Again, due to security/privacy constraints it is undesirable to present on other workstations all such actions. For example, assume that users A and B work together for project α, but user A also works for project β while user B does not. Hence operations related to project β should not be seen by user B. Thus, selection of database actions is necessary. An *action history* is defined as a sequence of actions shown on a display and a *database action history* is defined as a sequence of database actions. An *action history view* is a function that maps such a database action history into its part. This concept corresponds to a view for a temporal database [8], defined as a part of data in the whole history of the database.

2.2   Record and Play Back of an Action History

Operations on the screen are recorded in the database as database structures that can be accessed and processed by standard data manipulation methods. If the operations are recorded with time stamps, we can replay previous work. This makes it possible various view functions to be realized; in particular, data at any moment can be selected. This method requires structurization and separation of objects shown on a display, thus practically cannot be applied for windows which display video pictures.

We view a history as a *sequence of snapshots*, where each snapshot reflects either objects' states, actions performed on objects, and time dependencies. There are several possibilities to determine time moments when snapshots have to be made and what a snapshot should contain. We considered that for our system the event-driven paradigm is the most suitable. Because any change to a database is triggered by a user operation, we associate a snapshot with such an event. The objects' states between events are not changed, hence snapshots made between events cannot bear a new information. For implementation we only store operation sequences and snapshots taken by some intervals. All required snapshots can be recovered utilizing a snapshot and following operation history.

An event is represented as a *message* sent to an object; a message triggers some *method* associated with an object's class. Following this concept and terminology, snapshots of our action histories have the following components:

R: an object identifier of an object, a receiver of a message.

M: a message having an object identifier as parameter.

A_{old}: it represents attribute values of objects before the method determined by a message is executed.

A_{new}: it represents attribute values of objects after the method is executed.

I: a user's input which triggered an operation.

T: time when execution of a method is finished.

S: a name of a user who triggered a operation.

Definition 1 (Action History) *An Action History H is defined as follows.*

$$H = [(R_1, M_1, A_{old_1}, A_{new_1}, I_1, T_1, S_1), \ldots, (R_n, M_n, A_{old_n}, A_{new_n}, I_n, T_n, S_n)]$$

To avoid redundancies in action histories, if some attribute values are not changed, instead of copying A_{old} or A_{new}, a link to previous values is generated.

A stored action history can be used to play back sequences of displays at different time. Let a history $h = <h_1, h_2, ..., h_k>$ be a sequence of snapshots h_i. The user can play back the whole history, or only such its snapshots that satisfy a given condition; in such cases the order of displayed snapshots is preserved. The user can also replay the history in the reverse order, or in arbitrary order, possibly with the duplication of some snapshots.

The algorithm of playing back action histories has peculiarities connected with the necessity to restore objects' states. Before each snapshot is replayed, a state of objects should be restored according to the A_{old} component of this snapshot. Then, the method determined by the message component can be executed with proper object's states. If the method depends on absolute time and/or other variable or undetermined factors of the system's environment, the result objects' state can be different from A_{new}. Thus, the A_{new} component makes it possible to restore the final objects' state which was at the original time, and then, to show it to the user.

Algorithm 1 (Simple Play Back) *An action history is played back in the following manner.* ← *denotes a substitution.*

1. $h_1(R_1, M_1, A_{old_1}, A_{new_1}, I_1, T_1, S_1)$ ← *the first record of the action history.*
2. *Restore attribute values A_{old_1} of the object R_1.*
3. *Send a message M_1 with I_1 to the object R_1.*
4. $h_i(R_i, M_i, A_{old_i}, A_{new_i}, I_i, T_i, S_i)$ ← *the next record of the action history.*
5. *Restore attribute values A_{old_i} of the object R_i.*
6. *Wait for $T_i - T_{i-1}$.*
7. *Send a message M_i with I_i to the object R_i.*
8. *Back to the step 4 until $i = n$*

2.3 Retrieve of Continuous Media Objects Using Action History

Action histories are generated simultaneously synchronizing with continuous media objects when they are recorded or edited as shown in Fig.2. Therefore we can retrieve portions of continuous media objects which a user want by searching action histories. For example, we can search scenes of a movie in which a person wrote the specified sentence in a document or which the pointer of the specified person are moved if s/he can use it.

We need methods to retrieve proper portions of an action history to search continuous media objects. We may use regular expressions as a way to realize it. For example, assume that in the history $<h_1, h_2, h_3, h_4, h_5, h_6>$ the snapshots are marked by $<A, B, C, A, C, D>$ correspondingly. A regular expression BA selects the sub-sequence of snapshots $<h_2, h_4>$; a regular expression $C*$ selects the sub-sequence $<h_3, h_4, h_5, h_6>$ (it must start from C, and then $*$ accepts every string of symbols), and the regular expression $(AC)^+$ selects the subsequence $<h_1, h_3, h_4, h_5>$ (because $(AC)^+$ produces all strings $AC, ABAC, ACACAC, ...$).

3 Application to an Education System for Distance Learners

Education systems for distance learners are realized by using personal computers or workstations connected with computer network such as Internet. As in such systems teachers and students have to use their computers to interact with each other, it is easy to record various kinds of information produced during lectures, for example, teachers' videos, motions of pointers and pens, questions and answers, etc. Recorded lectures can be regarded as continuous media objects. They are used in order that students review past lectures and teachers reuse them in their future lectures. Facilities to retrieve past lectures which are continuous media objects are needed for these purposes.

3.1 Outline of VIEW Classroom

Fig.3 illustrates the conceptual image of an education system for distance learners which we call VIEW Classroom. In VIEW Classroom teachers and students in distributed location are supposed to be in virtual classroom connected via network. Each of them has a personal computer or workstation with a video camera and a microphone. They can use pointers to specify a location of a teaching material and pens to write some sentences, underlines, figures and so on by hand on it. Students can not only attend from the beginning of a class but also participate from the middle of the class by tracing the process of the lecture until then. Therefore, as time goes by, numbers of attendants may be changed. Fig.4 shows an user interface of the prototype system we are developing.

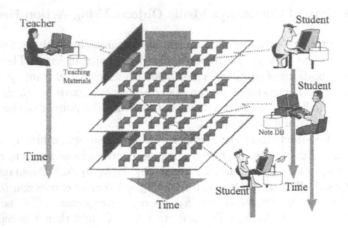

Fig. 3. Concept of VIEW Classroom

3.2   Recorded Lectures as Continuous Media Objects

Relationships between data generated in VIEW Classroom are illustrated in
Fig.5. A teacher prepares slides as teaching materials in advance and explains
them during a lecture. Users' actions such as motion of pointers and pens, change
to the next slide, change of status of users' microphones (on or off), selection of
buttons are recorded with time stamps as action histories. A teacher's video and
voice are recorded synchronizing with each slide and action history. Students'
images and voices are also recorded when they talk with a teacher to ask their
questions and so on.

Definition 2 *Recorded lectures are defined as follows.*
   LECTURES = {*Slides, Actions, Movies*},
   $sort(Slides) = \langle S_{id}, S_{img}, S_{txt}, S_{att_1}, \ldots, S_{att_i} \rangle$,
   $sort(Actions) = \langle L_{id}, S_{id}, Act_{name}, Act_{time}, Act_{type}, Act_{att_1}, \ldots, Act_{att_j} \rangle$,
   $sort(Movies) = \langle L_{id}, M_{start}, M_{end}, M_{mv}, M_{name} \rangle$,
   *In Slides data related to teaching materials are stored. S_{id} is an identifier of
a slide, S_{img} an image of a slide, S_{txt} texts included in a slide. $S_{att_1}, \ldots, S_{att_i}$
are attributes related to a slide. In Actions actions histories are stored. L_{id} is
an identifier of a lecture, Act_{name} a name of a person who operates a computer,
Act_{time} time when a event is occurred, Act_{type} a type of a event such as clicking
right button of a mouse. $Act_{att_1}, \ldots, Act_{att_j}$ are attributes of an event. In Movies
videos are stored. M_{name} is a name of a person who is taken in a video, M_{mv}
data of a video, M_{start} time when recording a video is started, M_{end} time when
recording a video is finished.*

Example 1 *The Table1 is an example of an action history in VIEW Classroom.
Kaoru explains the slide whose identifier is S1 using a mouse. The button1 of a*

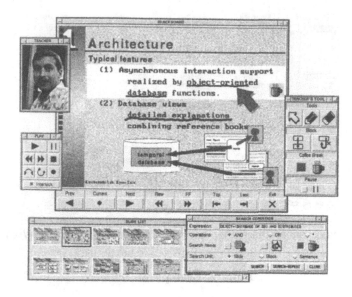

Fig. 4. User interface of the prototype system

mouse is selected during the explanation. Then he changes the slide to the one whose identifier is S2.

3.3 Search and Play Back Using an Action History

In distance education systems support of searching and playing back lectures is important. Students who review past lectures may need a part of lectures, for example, the part explaining about relational databases. In VIEW Classroom the following facilities are prepared to specify a part of lectures except a method based on action histories[5, 4].

1. partial match retrieval with complicated expressions for texts included slides.
2. selection from slide list in which thumbnails of all slides used during a lecture are displayed.
3. fast-forwarding and rewinding videos.
4. matching for non-character information such as 'slide with a figure', 'the slide where teacher talked on something not directly related to the contents'.

In this paper we focus a method of search based on action histories. Motion of pointers and pens, changes of slides and so on are recorded as action histories. When a lecture is played back by using the corresponding action history, pointers or pens can be displayed on slides (teaching materials) synchronizing with movies of teachers or students. We can use search conditions as follows to retrieve proper portions of lectures by using action histories.

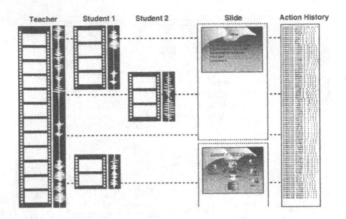

Fig. 5. Synchronous constraints between data generated in VIEW Classroom

Table 1. Example of *Actions*

L_{id}	S_{id}	Act_{name}	Act_{time}	Act_{type}	x	y	$slide\_id$
L1	S1	Kaoru	09:10:11	mouse_button1_down	123	456	
L1	S1	Kaoru	09:10:12	mouse_button1_move	200	456	
L1	S1	Kaoru	09:10:13	mouse_button1_release	300	456	
L1	S1	Kaoru	09:15:20	next_slide			S2
L1	S2	Kaoru	09:22:30	mouse_button2_down	456	100	
L1	S2	Kaoru	09:22:31	mouse_button2_move	456	150	

1. A teacher explained a location of a slide.
2. A teacher wrote something in slides with his/her pen.
3. A teacher used a slide for explanation more than once.
4. A teacher discussed with a student.

We explain how to realize the search using an action history according to an example in case where a user specifies a range of a slide related to her/his interest (the first condition of the above list). Fig.6 shows the process in searching the corresponding part of a video in this case. Let the identifier of the slide which a user needs be $S1$, the specified range of the slide $S1$ be from (x_1, y_1) to (x_2, y_2) $(x_1 < x_2, y_1 < y_2)$ and the name of a teacher be "kaoru". The concrete algorithm to search and play back parts of lectures is as follows.

1. Search the necessary data to play back.
   (a) Retrieve image data of slides.
       Retrieve S_{img} from *Slides* where $S_{id} = S1$.

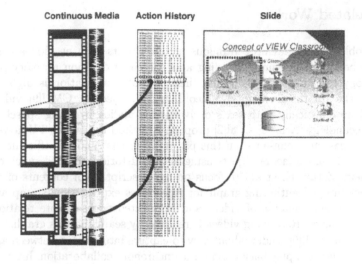

Fig. 6. Search of a lecture using an action history

(b) Retrieve action histories.

Retrieve Act_{time}, Act_{type}, Act_{att_1}, ..., Act_{att_j} from *Actions* where $S_{id} =$ $S1$ and $Act_{name} = kaoru$ and $x_1 < x < x_2$ and $y_1 < y < y_2$.

If retrieved action histories consist of several groups of continuous actions, the start and end time of each groups must be computed. Now we suppose that a retrieved action history consists of only group in which three actions Act_1, Act_2, Act_3 are included and their Act_{time} are t_1, t_2 and t_3 ($t_1 < t_2 < t_3$) respectively.

(c) Retrieve videos.

Retrieve M_{mv} from *Movies* where $not(t_3 \leq M_{start}$ or $M_{end} \leq t_1)$ $(M_{start} \leq M_{end})$.

2. Play back retrieved data.

   (a) Start playing back the retrieved movie and trigger the action Act_1. Let the start time be t.
   (b) Trigger the action Act_2 when it is $t + (t_2 - t_1)$.
   (c) Trigger the action Act_3 when it is $t + (t_3 - t_1)$ and stop playing back the movie.

This is very simplified algorithm for illustrating our basic idea. In practice there are some points to consider. For example, a type of actions must be triggered before another type of actions. The *mouse_button1_down* should be triggered before *mouse_button1_move*. Therefore if the first action in a retrieved action history is *mouse_button1_move*, the *mouse_button1_down* should be supplemented before the *mouse_button1_move* is triggered.

4   Related Work

The problem of retrieving continuous media objects such as videos has been studied before. Image understanding and speech recognition are very popular technology to realize it[7, 9, 2] and their cost of computation is high in general. In the Informedia Digital Video Library project at CMU[9] videos are indexed automatically with texts derived from audios by using speech recognition technology. Ariki Y. et al[2] propose a method to classify TV news using image processing techniques. In this paper we discuss another efficient way to retrieve continuous media objects using action histories recorded synchronizing with them. Zettsu K. et al.[10] focus on the description of contents of videos. In "time-stamped authoring graph"for it vertexes express time-stamp and annotations about contents of videos and undirected edges denote relationships between vertexes. Retrieving videos is realized by searching this graph.

Manohar et al.[6] studies about how to capture interactions between applications and users and play back them for asynchronous collaboration. Interactions, annotations using audio and references to resource data are recorded in a "session object". Recorded data in a session object are used to replay past interactions. One of the objectives of this research is similar to ours, that is, to enable users to search and retrieve events that they need, for example, a particular command is executed or a mouse is clicked. But a method to realize the search and its applications are discussed little.

The Classroom 2000 project[1] is initiated to record and reuse lectures by connecting each slide and annotations with the corresponding video automatically or manually. This connection is explicit while in our approach it is implicit, that is, a video is connected with a slide by using action histories.

5   Conclusions

In this paper we propose a new indexing method for continuous media objects. Action history views generated synchronizing with continuous media objects make it possible to provide a new kind of search mechanism. We expect that it works more useful and efficiently than methods using image processing technology. As this method cannot be used for continuous media objects without action histories, people need to use tools that can be generated continuous media objects with them. In VIEW Classroom which we are developing such facilities to record lectures as continuous media objects with action histories and retrieve the necessary parts for users by applying this method will be realized.

Acknowledgments

The authors are grateful to colleagues of our laboratory for discussions on the topics presented in this paper.

References

[1] Gregory D. Abowd, Christopher G. Atkeson, Ami Feinstein, Cindy Hmelo, Rob Kooper, Sue Long, Nitin "Nick" Sawhney, and Mikiya Tani. Teaching and learning as multimedia authoring: The classroom 2000 project. In *Proceedings of the Fourth ACM International Multimedia Conference (MULTIMEDIA'96)*, pages 187–198, Phoenix, Arizona, November 1996. ACM, Addison-Wesley Publishing Company.

[2] Y. Ariki, A. Shibutani, and Y. Sugiyama. Classification and retrieval of tv spots news by dct features. In *Proceedings of the IPSJ International Symposium on Information Systems and Technologies for Network Society*, pages 269–272, September 1997.

[3] Hajime Iwamoto, Chihiro Ito, and Yahiko Kambayashi. Design and implementation of action history view mechanisms for hypermedia systems. In *Proceedings of the Twenty-Second Annual International Computer Software and Applications Conference (COMPSAC'98)*, Vienna, Autstria, August 1998.

[4] Yahiko Kambayashi, Kaoru Katayama, Toshihiro Kakimoto, and Hajime Iwamoto. Flexible search functions for multimedia data with text and other auxiliary data. In *Proceedings of the 1998 ACM Symposium on Applied Computing*, pages 498–504, Atlanta, Georgia, February 27 - March 1 1998.

[5] Kaoru Katayama, Osami Kagawa, Yasuhiro Kamiya, and Yahiko Kambayashi. Flexible play back facilities for distance education. In Masatoshi Yoshikawa and Shunsuke Uemura, editors, *Proceedings of International Symposium on Digital Media Information Base (DMIB'97)*, pages 74–78, Nara, Japan, November 1997.

[6] N. R. Manohar and A. Prakash. The session capture and replay paradigm for asynchronous collaboration. In *Proceedings of the Fourth ECSCW Conference*, pages 149–164, September 1995.

[7] Behzad Shahraray and David C. Gibbon. Automated authoring of hypermedia documents of video programs. In *Proceedings of the 3th ACM International Multimedia Conference (ACM MULTIMEDIA95)*, pages 401–409, San Francisco, California, November 1995.

[8] Abdullar Uz Tansel, James Clifford, Shashi Gadia, Arie Segev, and Richard Snodgrass. *Temporal Databases: Theory, Design, and Implementation.* Series on Database Systems and Applications. Benjamin/Cummings, 1993.

[9] Howard D. Wactlar, Takeo Kanade, Michael A. Smith, and Scott M. Stevens. Intelligent access to digital video: Informedia project. *IEEE Computer*, 29(5), May 1996.

[10] Kohji Zettsu, Kuniaki Uehara, Katsumi Tanaka, and Nobuo Kimura. A time-stamped authoring graph for video databases. In *Proceedings of the 8th International Conference on Database and Expert Systems Applications (DEXA'97)*, pages 192–201, Toulouse, France, September 1997. Springer-Verlag.

Semantic Structures for Video Data Indexing

Koji Zettsu[1], Kuniaki Uehara[2], and Katsumi Tanaka[3]

[1] Kobe Research Center, Telecommunications Advancement Organization of Japan,
6-9-1 Kobe International Friendship Building, Chuo Kobe, 650-0046 JAPAN
Tel; +81-78-303-5521, Fax: +81-78-303-5519
zettsu@kobe-sc.tao.go.jp

[2] Research Center for Urban Safety and Security, Kobe University,
Nada Kobe, 657-0013 JAPAN
Tel; +81-78-803-1184, Fax: +81-78-803-1218
uehara@kobe-u.ac.jp

[3] Graduate School of Science and Technology, Kobe University,
Nada Kobe, 657-0013 JAPAN
Tel; +81-78-803-1211, Fax: +81-78-803-1217
tanaka@db.cs.kobe-u.ac.jp

Abstract. Video indexing based on contents annotations can fully explore semantic information of video data. However, the most difficult and time-consuming process in annotation-based indexing is to identify appropriate video intervals for various semantic contents manually. Thus, automatic discovering video intervals from video data will be helpful for the indexing work. For this purpose, we propose "semantic structures" of video data and a mechanism for discovering semantic structures. The basic concept of our approach is to (1) discover consecutive sequences of shots from video data, each of which represents a consistent action or situation, and (2) index each of the discovered video intervals based on its semantics. A semantic structure is a collection of discovered video intervals that are classified into three categories: "unchanged" (i.e. actors or backgrounds are unchanged throughout the interval), "gradually changing" (i.e. actors or backgrounds are changing shot by shot) and "multiplexing" (i.e. individual actors or backgrounds are appearing by turns). The mechanism discovers these types of video intervals by comparing and contrasting similarity between each shot, and indexes each of discovered intervals by using indexing algorithms prepared for each type. We show how well our approach works for identifying video intervals with some experimental results.

1 Introduction

With advances in computer technology, digital video data is becoming more and more common. Due to the huge data volume of video database, accessing and retrieving video data item is time-consuming effort. Indexing of the video data is needed to facilitate this process. Compare to the traditional text-based database systems, video indexing is more difficult and complex, because what to index on is not clear and difficult to determine in video data indexing. Video data indexing

S. Nishio, F. Kishino (Eds.): AMCP'98, LNCS 1554, pp. 356–369, 1999.
© Springer-Verlag Berlin Heidelberg 1999

is closely related to what kind of information is accessed and how the indices are derived.

Existing approaches on video indexing can be classified into two categories: feature-based indexing and annotation-based indexing. Feature-based indexing provides access to video data based on audio-visual information, such as color, texture, object motion and so on. Feature-based indexing can be done completely automatically by extracting key features from video data. However, due to the lack of semantics attached to the features, it causes inconveniences to users who are attempting to specify video database queries. On the other hand, annotation-based indexing provides access to video data based on semantic information, such as knowledge, impression and so on. Annotation-based indexing is usually a manual process that requires human intervention. An author or a librarian identifies semantic contents laid in video data during his/her indexing work. The advantages of annotation-based indexing are that it fully explores the richness of the information contained in the video data, and it is familiar with users. However, unlike feature based indexing, the fully automation of the video annotation process will remain impossible, due to the limitations of current image-processing techniques.

Let us consider accessing to video data based on annotation-based indexing. The most difficult and time-consuming process in annotation-based indexing work is to identify appropriate parts of video data for various semantic contents, such as actions or situations, because it depends on the temporal position and duration of what an author has seen. However, only few attempts have so far been made to this problem. The purpose of our research is to develop a mechanism for discovering appropriate parts for semantic contents, and help the annotation-based indexing work. For this purpose, we introduce the notion of "semantic structures" of video data in this paper.

The rest of this paper is organized as follows: Sect. 2 describes the motivation of our research. Section 3 describes our approach. Section 4 formally defines a semantic structure of video data and describes a mechanism for discovering semantic structures in detail. Section 5 presents experimental results. Finally, Sect. 6 concludes the paper.

2 Managing Video Data Based on Temporal Intervals

2.1 Structured Approach and Stratification Approach

Due to the temporal indeterminacy of semantic contents of video data, it is widely accepted to manage semantic contents based on temporal intervals[4,5]. Existing approaches for managing video data based on temporal intervals can be classified into two categories: structured approach[3], and stratification approach[1,2]. The structured approach first divides video data into a consecutive sequence of non-overlapping basic units called *shots* (Fig.1(a)), usually, based on audio-visual information of video data, and then describes the semantic contents of each shot. The advantage of structured approach is that video segmentation can be done automatically by an algorithm, such as a scene change detection algorithm[13]. However, it can only provide access to semantic contents within each shot. On the other hand, the stratification approach segments video data into overlapping intervals called *video intervals* (Fig.1(b)). Each video interval is defined for each of semantic contents, such as an action or a situation. Thus, it can fully explore semantic information of video data. However, video intervals are needed to be identified manually. That is a difficult and time-consuming, moreover, biased by the user doing the work.

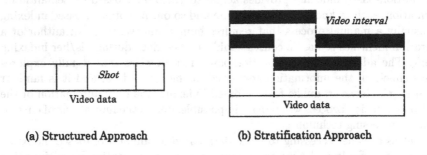

(a) Structured Approach   (b) Stratification Approach

Fig. 1. Video data management based on temporal intervals

Let us consider accessing video data by its semantic information. It is more preferable to managing video data based on stratification approach rather than structured approach. However, the most significant problem of stratification approach is how to identify video intervals easier.

2.2 Discovering Video Intervals

One important thing is that a video interval consists of a consecutive sequence of shots, each of which are semantically related with each other so as to represent a consistent action or situation. This leads to the idea that video intervals can be discovered automatically by comparing and contrasting the contents of shots. It will be helpful for an author to identify video intervals for various semantic contents based on stratified approach. Once video intervals are discovered, the author continues his/her indexing work as follows:

− Refines the annotations of discovered.intervals.
− Adjusts the temporal positions of discovered intervals.
− Aggregates some discovered intervals into a new interval.

A mechanism for discovering video intervals is expected to have the following characteristics:

− Segments video data into shots and annotate them based on simple structured approach.

- Defines mapping between video intervals and consecutive sequences of shots.
- Discovers appropriate video intervals from a shot sequence of whole video data, and indexes them.

Figure 2 illustrates the basic concepts of our approach.

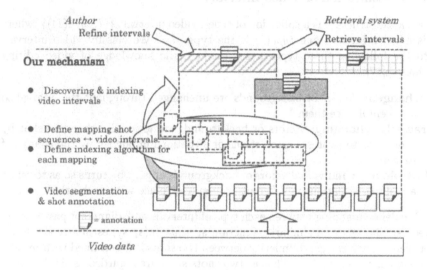

Fig. 2. Basic concepts of our approach

3 Semantic Structure of Video Data

Our approach identifies video intervals in accordance with shot annotation and mapping between shots and video intervals. A mechanism for discovering video intervals associates discovered intervals with these prerequisites. For this purpose, we define a "semantic structure" of video data. A semantic structure of video data is a set of video intervals identified by our mechanism under the current mapping and shot annotations. A user of our mechanism, such as an author, a librarian, or a retrieval system can obtain various semantic structures out of the same video data by adjusting the mapping and the shot annotations.

In the following sections, a semantic structure of video data is formally defined, and a mechanism for identifying video intervals is explained.

3.1 Basic Notations

Here are some basic definitions.

- A video sequence VD is a nonempty finite sequence of shots $s_1, s_2, \cdots, s_N$, which is temporally ordered with respect to '$<$', such as $s_1 < s_2 < \cdots < s_N$, where N is the number of shots.

- A video interval I is a consecutive subsequence of shots $s_i, \cdots, s_j$ of the video sequence VD, where $s_i, s_j \in VD$ and $i < j \leq N$ (denoted by $I[i,j]$). A video interval I of VD has a start shot $start(I) = s_i$ and an end shot $end(I) = s_j$.

3.2   Classification of Video Intervals

A semantic structure is a collection of typed video intervals $\{(I, type(I))\}$, where I is a video interval and $type(I)$ is the type of I. We classify video intervals into the following three types based on heuristic knowledge of video editing technique[17]:

Unchanged: Actors or backgrounds are unchanged throughout the interval so as to emphasize them.

Gradually changing: Actors or backgrounds are gradually changing shot by shot so as to express some actions or situations implicitly by the change of its contents.

Multiplexing: Individual actors or backgrounds appear by turns so as to relate actions or situations in different time and/or place with each other.

We define mapping between each type of intervals and alignment patterns between similar shots. A similarity function $similarity(s_i, s_j)$ estimates similarity between two shots s_i, s_j. Similarity between two shots is determined with respect to a similarity threshold θ. That is, two shots s_i, s_j are regarded as similar shots if $similarity(s_i, s_j) > \theta$. Each type of interval is defined as follows:

Unchanged: If $\forall s \in I$ $similarity(s, start(I)) > \theta$, the video interval I is classified into unchanged type.

Gradually changing: If $\forall s_i, s_{i+1} \in I$ $similarity(s_i, s_{i+1}) > \theta$, the video interval I is classified into gradually changing type.

Multiplexing: If $\forall s_i, s_{i+m} \in I$ $similarity(s_i, s_{i+m}) > \theta$, the video interval I is classified into multiplexing type, where m is a notion of multiplicity, that is, a similar shot appears at every m shots. With respect to multiplicity m, this type of interval $I[a,b]$ can be also defined as a set of m subsequences $\{I_i' | 1 \leq i \leq m\}$, where I_i' is one of the subsequences of I, defined as $I_i' = [s_{a+(i-1)}, s_{a+(i-1)+m}, s_{a+(i-1)+2m}, \cdots, s_{b-m+i}]$.

Figure 3 illustrates each type of video intervals.

The definitions described above are representing the following alignment patterns between similar shots. Here, each shot in a video interval is represented as an arbitrary symbol like A, B, and "a similar shot" of each shot is designated as dashed symbol like A', B'. The unchanged type defines the pattern sequence $[A, A', A', \cdots]$, that is all shots in the interval is similar to the first shot. The gradually changing type defines the pattern sequence $[A, A', A'', \cdots]$, where any adjacent two shots are similar. The multiplexing type defines the pattern sequence $[A, B, A', B', \cdots]$ (or $[A, B, C, A', B', C', \cdots]$), that is a set of consecutive shots A, B (or A, B, C) or the similar shots of them are repeated. We can obtain the sequences of A's and B's (or A's, B's and C's), each of which is called a

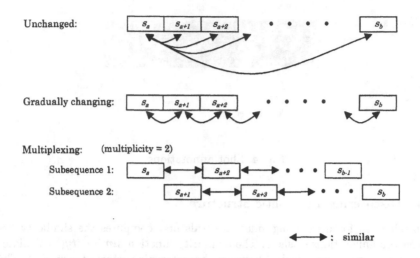

Fig. 3. Alignment patterns between similar shots for semantic structures

subsequence. In order to discover these types of video intervals, the mechanism for discovering a semantic structure searches for the sequences of shots ordered in these patterns (the mechanism is described in the following section).

3.3 Shot Annotation

Our approach assumes that video data is segmented into shots and every shot is annotated by its semantic contents. The video segmentation process can be done automatically by a scene change detection algorithm based on audio-visual information. An annotation of shot s consists of a set of keywords $\{w\}$, each of which describes an actor, a background or a simple action appearing in the shot. A video annotation database for a video data VD is defined as a set of pairs between shots and annotations $\{(s_1, \{w\}_1), (s_2, \{w\}_2), \cdots (s_N, \{w\}_N)\}$.

Figure 4 shows an example of shot annotation. Each shot is annotated by keywords representing actors (e.g. *red bean*), backgrounds (e.g. *farm, rice field*), and simple actions (e.g. *grow, appear*). Currently, the annotation process is done manually by an author, due to the limitations of current image-processing technique.

A characteristic vector $v(s)$ for each video annotation is defined in the video annotation database. A characteristic vector $v(s)$ is a K-dimensional vector, where K is the size of exclusive set of keywords in the video annotation database. In $v(s)$, absence of a keyword is indicated by 0, while presence of a keyword is indicated by 1.

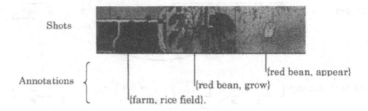

Fig. 4. Shot annotations

3.4   Discovering a Semantic Structure

A mechanism for discovering video intervals first computes the similarity between the annotations of shots. The similarity function *similarity*() calculates the cosine correlation value[10] between characteristic vectors of each shot. The mechanism computes the similarity of all pairs of shots.

Once the similarity between each shot is calculated, the mechanism searches for the sequences of shots in video data, each of which matches either of three patterns described above. With respect to a similarity threshold θ, a sequence of shots is represented as a pattern sequence between similar and dissimilar shots relative to an arbitrary shot. Starting from the first shot of video data, the mechanism searches for video intervals matching the alignment patterns between similar shots defined for video interval types.

Figure 5 illustrates how the mechanism discovers video intervals of video data. In Fig.5, the similarity between annotations of each shot is shown as a similarity matrix. In this matrix, each cell contains the similarity value between two shots, and is filled in gray color if the similarity value is greater than the threshold (= 0.2). Intuitively, the mechanism searches for the patterns between colored and non-colored cells, each of which represents either of video interval types. The discovered video intervals are also shown and labeled by their types. Note that the video intervals are stratified. This explains that the interval to which a shot belongs depends on where you start watching video data, since video data has temporal contexts.

Since each video interval type is defined based on semantic similarity between each shot, the mechanism can discover different video intervals dynamically by changing the threshold of similarity. This is the most significant characteristic of our mechanism.

3.5   Generating Indices for Discovered Intervals

After the discovering process, the mechanism generates indices for discovered video intervals. An index of interval also consists of a set of keywords, which suggests what is described in the interval. Thus, the author conducts his/her indexing work with referring to these indices.

Similarity between each shot   (threshold = 0.2)

	4	5	6	7	8	9	10	11	12	13
4		1	0	0	1	0	0	0	0	0	0
5	0	1	0.19245	0	0.57735	0	0	0.235702	0.19245	0.218218
6	0	0.19245	1	0	0	0.3849	0	0.272166	0.111111	0.125988
7	1	0	0	1	0	0	0	0	0	0
8	0	0.57735	0	0	1	0.288675	0	0.204124	0	0
9	0	0	0.3849	0	0.288675	1	0	0.471405	0	0
10	0	0	0	0	0	0	1	0	0.149071	0
11	0	0.235702	0.272166	0	0.204124	0.471405	0	1	0.272166	0.308607
12	0	0.19245	0.111111	0	0	0	0.149071	0.272166	1	0.377964
13	0	0.218218	0.125988	0	0	0	0	0.308607	0.377964	1
Shot No.	4	5	6	7	8	9	10	11	12	13

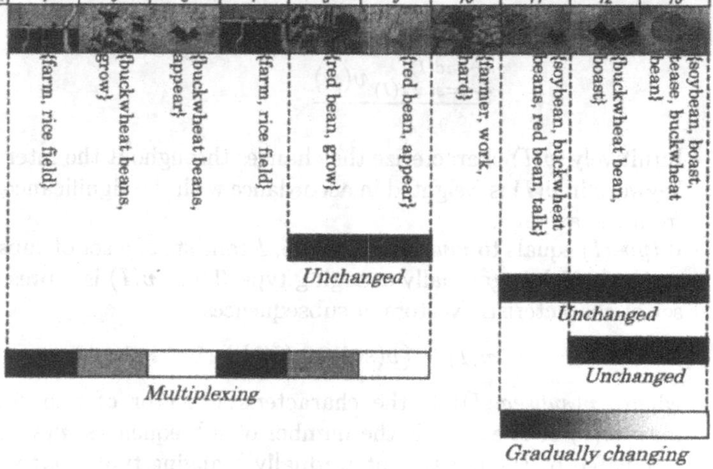

Shot annotations (left to right):
- {farm, rice field}
- {buckwheat beans, grow}
- {buckwheat beans, appear}
- {farm, rice field}
- {red bean, grow}
- {red bean, appear}
- {farmer, work, hard}
- {soybean, buckwheat beans, red bean, talk}
- {buckwheat bean, boast}
- {soybean, boast, tease, buckwheat bean}

Unchanged

Unchanged

Unchanged

Multiplexing

Gradually changing

Fig. 5. Semantic structures of video data

The mechanism generates an index for a discovered interval from annotations of shots included in the interval. An algorithm for generating an index consists of the following two steps:

1. The mechanism first calculates a characteristic vector $v(I)$ for the interval I. In $v(I)$, the importance of an keyword is indicated by the vector value. The characteristic vector is calculated by either of the following functions in accordance with the type of I. Here, n is the number of shots in I.
   - If $type(I)$ equals to $unchanged$ type, $v(I)$ is a weighted average of characteristic vectors of all shots in I. The characteristic vector of each shot is weighted by the similarity between the first shot of I and the shot.

$$v(I) = \frac{\sum_{i=start(I)}^{end(I)} similarity(s_i, s_{start(I)})v(s_i)}{\sum_{i=start(I)}^{end(I)} similarity(s_i, s_{start(I)})} \tag{1}$$

$v(I)$ implies the first shot dominates the semantic contents throughout the interval.

- If $type(I)$ equals to *gradually changing* type, $v(I)$ is an eigen vector for the maximum eigen value of the following matrix M:

$$M = \sum_{k=1}^{n} \delta_k \delta_k^T \tag{2}$$

where, δ_k is the differential vector between v_i and the average of characteristic vectors of all shots in I, and δ_k^T is the transposed matrix of vector δ_k.

$$\delta_k = v_i - \bar{v} \quad (start(I) \leq i \leq end(I), 1 \leq k \leq n)$$

$$\bar{v} = \frac{\sum_{i=start(I)}^{end(I)} v(s_i)}{n}$$

Intuitively, $v(I)$ characterize the changes throughout the interval. Every keyword in $v(I)$ is weighted in accordance with the significance of change from the average.

- If $type(I)$ equals to *multiplexing* type, I consists of a set of subsequences, each of which is gradually changing type. Thus, $v(I)$ is represented by a set of characteristic vectors of subsequences:

$$v(I) = \{v(subseq_i(I))\} \quad (1 \leq i \leq m) \tag{3}$$

where, $v(subseq_i(I))$ is the characteristic vector of i-th subsequence $subseq_i(I)$ of I, and m is the number of subsequences. $v(subseq_i(I))$ is calculated by the function of gradually changing type, that is described above.

2. Then, the mechanism ranks keywords in $v(I)$ in accordance with their vector values, and selects several keywords in high ranks as the index of I. Thus, the generated index contains limited number of keywords, which most describe the contents of the interval.

Figure 6 shows examples for generating indices for discovered video intervals. In Fig.6, each of characteristic vectors contains only several keywords in high ranks. The index of unchanged type emphasizes the keywords appearing in the first shot and also continuing in the following shots. The index of gradually changing type emphasizes the keywords changing its state of appearance significantly (i.e. changing its state between appearance and disappearance frequently, or at long intervals), which characterize the contents throughout the interval. In multiplexing type, each index represents individual plots embedded in the interval.

4   Experimental Results

In this section, we discuss the performance of the mechanism for identifying video intervals. We have tested the following matters:

- The performance of the algorithm for discovering video intervals.
- The stability for indexing discovered intervals.

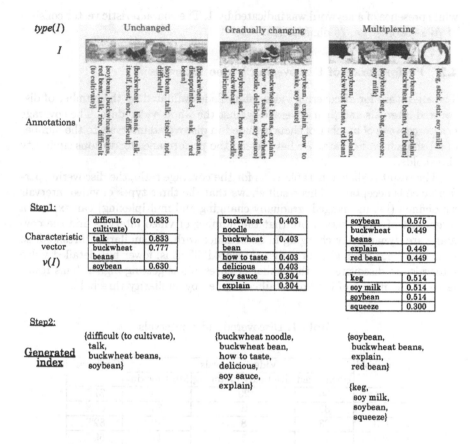

Fig. 6. Generating indices for discovered intervals

4.1 Preliminaries

Sample video data is digitized from a 10 minutes animation movie clip. Before processing sample video data with our mechanism, we first segment video data into shots. Sample video data is segmented into shots by using a scene change detection algorithm based on color distribution[15], which computes inter-frame difference based on χ^2 histogram[14]. We have got 79 shots from the sample video data.

After video segmentation process, we manually annotated every shot with a set of keywords. We described the following semantic contents within each shot: actors, backgrounds and simple actions. Then, each annotation was represented by a characteristic vector, in which absence of a keyword was indicated by 0, while presence of a keyword was indicated by 1. The characteristic vector consists of 76 keywords (i.e. 76 dimensions).

4.2   Performance of Discovering Video Intervals

The algorithm for discovering video intervals is evaluated by the number of discovered intervals and their coverage against the whole video data. The coverage is a percentage of number of shots included in discovered intervals to the number of all shots in video data. We have tested the performance for various similarity thresholds.

The result is shown in Table 1. As for the coverage ratio, the discovering performance is acceptable. This result shows that the three types of video intervals we defined (i.e. unchanged, gradually changing and multiplexing) can explore a large part of video data. Note that the number of video intervals and the coverage are dynamically changing in accordance with the similarity threshold, so that they become smaller in higher threshold. That is, fewer but detailed video intervals are discovered as the similarity threshold is getting higher. This result leads to the notion of level of detail controlled by similarity threshold.

Table 1. Discovered video intervals

Threshold	Video intervals			Converage
	Unchanged	Gradually changing	Multiplexing	
0.1	36	20	30	87%
0.2	35	19	30	86%
0.3	23	8	19	82%
0.4	19	5	11	69%
0.5	13	1	5	53%

4.3   Stability for Indexing Discovered Intervals

The algorithm for indexing discovered video intervals is evaluated by correctness of indices. There exists well-accepted measure for evaluating correctness of indices: precision ratio, which has been adjusted to our purpose as follows:

$$\text{precision ratio} = \frac{\left(\begin{array}{c} \text{number of discovered intervals whose} \\ \text{index correctly describes their contents} \end{array} \right)}{(\text{number of discovered intervals})} \quad (4)$$

Table 2 shows the precision ratios for various similarity thresholds. A user determines the correctness of each index with using several keywords highly ranked in the corresponding characteristic vector. The precision ratio ranges from 0.5

to 0.8, which remarks high value. In general, the higher the similarity threshold becomes, the higher the precision ratio becomes, even though the number of discovered intervals is decreasing.

Table 2. Precision ratio of discovered intervals

Threshold	Precision ratio		
	Unchanged	Gradually changing	Multiplexing
0.2	0.600	0.579	0.567
0.3	0.739	0.875	0.579
0.4	0.790	0.800	0.727

We have also evaluated the effectiveness of the algorithm for generating indices for discovered intervals. The effectiveness is also evaluated by the precision ratios of discovered intervals. This time, they are calculated by changing the functions for calculating characteristic vectors irrelevant to the types of discovered intervals. Table 3 shows the result. In table 3, "Function type" means the type of intervals for which each function is prepared, and "N/A" (Not Applicable) means the function cannot be applied to this type of intervals (i.e. the function for multiplexing type assumes that an interval is segmented into subsequences). The precision ratio remarks the best value when each type of function is applied to the same type of intervals (as shown at the diagonal cells in Table 3). This means that the algorithm for generating indices for discovered intervals is properly designed so as to reflect their semantics.

Table 3. Precision ratio of discovered intervals against functions for calculating characteristic vectors for index generation

Function type	Precision ratio		
	Discovered interval type		
	Unchanged	Gradually changing	Multiplexing
Unchanged	0.739	0.625	0.105
Gradually changing	0.696	0.875	0.316
Multiplexing	N/A	N/A	0.579

5 Conclusions and Future Work

We have described semantic structures of video data and a mechanism for discovering them for video indexing. The main purpose of our approach is to help the annotation-based video indexing work based on stratification approach. We have classified video intervals into three types (i.e. unchanged, gradually changing and multiplexing) based on alignment patterns between similar shots, and designed the mechanism for identifying video intervals for these types. We have shown that video intervals are successfully identified by using these classifications.

Currently, the mechanism assumes that every shot is annotated manually by an author. However, recent researches on image-processing techniques extend the ability to automate the annotation work based on audio-visual information. They have exploited automatic video annotation based on extracting textual information, speech recognition, face detection and so on[8,11,12]. We will explore using those approaches to automate our primary annotation work.

As for annotations, we assume that an annotation consists of a set of keywords. However, it is not enough to use only keywords to describe semantic contents of video data, due to the paraphrasing problem[16]. Remember that video data is continuous media, and a video interval consists of a consecutive sequence of shots. Thus, an annotation for the video interval is not only a simple collection of annotations of the shots, but also a completely new one paraphrasing (or summarizing) them. For example, considering a video interval which consists of the following sequence of shots: "taking a knife and fork", "cutting a meat", and "moving a meat into a mouth". The appropriate annotation for this interval is not a collection of these annotations, but a new keyword "eating", which paraphrases them. It is difficult to completely solve the paraphrasing problem, because it requires human knowledge and experiences. To overcome this problem, there are the following approaches we will exploit:

- Paraphrase process is done by an author, and the mechanism supports the process. That is, the mechanism computes suggestive annotations, each of which represents the most significant characteristics of the interval, and then an author refines the annotations.
- The mechanism emulates paraphrase with using AI techniques. In [6,7], their authoring system supports inheritance, similarity, or inference between descriptors (i.e. icons, instead of keywords) required for paraphrase by predefining descriptors and relationships between them in a hierarchical structure.

Finally, we will work on a more sophisticated version of our mechanism, which can handle more complex semantics than "unchanged", "gradually changing" and "multiplexing". We will explore combining those primitive types in our future work.

Acknowledgements

This project is supported in part by Kobe Research Center, Telecommunications Advancement Organization of Japan under the Project "Research and Development for Advanced Digital Video Transmission System". This project is also supported in part by the Japanese Ministry of Education Grant-in-Aid for Scientific Research on Priority Area: "Research and Development of Advanced Database Systems for Integration of Media and User Environments." This work is supported in part by Research for the Future Program of Japan Society for the Promotion of Science under the Project "Researches on Advanced Multimedia Contents Processing." The authors would like to thank Kazuaki Tanizawa of Kobe University for providing a scene change detection program.

References

1. Thomas, G., Smith, A. and Davenport, G.: The stratification system: A design environment for random access video, Proc. of Workshop on Networking and operating System Support for Digital Audio and Video, ACM (1992).
2. Davenport, G., Thomas, G., Smith. A. and Pincever, N.: Cinematic primitives for multimedia. Proc. of IEEE Computer Graphics & Applications, pp.67–74 (1991).
3. Tonomura, Y.: Video handling based on structured information for hypermedia systems, Proc. of Intl. Conf. on Multimedia Information Systems, pp.333–344 (1991).
4. Weiss, R., Duda, A. and Gifford, D.: Content-Based Access to Algebraic Video, Proc. of IEEE Multimedia, pp.140–151 (1994).
5. Allen, J. F.: Maintaining Knowledge about Temporal Intervals, C. ACM, Vol.26, pp.832–843 (1983).
6. Davis, M.: Media Streams: An iconic visual language for video annotation, Proc. of IEEE Symposium on Visual Languages, pp.196–202 (1993).
7. Davis, M.: Knowledge representation for video, Proc. of Workshop on Indexing and reuse in Multimedia Systems, pp.19–28 (1994).
8. Smith, M. A. and Kanade, T.: Video Skimming for Quick Browsing based on Audio and Image Characterization, Tech-Report CMU-CS-95-186 (1995).
9. Hampapur, A., Jain, R. and Weymouth, T.: Digital video indexing in multimedia systems, Proc. of the Workshop on Indexing and Reuse in Multimedia Systems (1994).
10. Salton, G.: The SMART Retrieval System - Experiments in Automatic Document Processing. Prentice-Hall Inc, Englewood Cliffs: New Jersey (1971).
11. Lienhar, R.: Automatic Text Recognition for Video Indexing. Proc. of the 4th ACM Multimedia, pp.11–20 (1996).
12. Ariki, Y., Iwanari, E. and Motegi, Y.: Detection and Description of TV News Article, Proc. of the 47th FID, pp.198–202 (1994).
13. Boreczky, J., S. and Rowe, L., A: A comparison of Video Shot Boundary Detection Techniques, Strage & Retrieval for Image and Video Databases IV, Proc. of SPIE 2670, pp.170–179 (1996).
14. Zhang, H., J., Kankanhalli, A. and Stephen, W., S.: Automatic parsing of full-motion video, Multimedia Systems, 1:10–28, July (1993).
15. Tanizawa, K.: Video Clustering and Scene Detection based on Visual Information, Graduation thesis, Kobe University (1998).
16. Schank, R.: Dynamic Memory, Cambridge University Press: Cambridge (1982).
17. Arijon, D.: Grammar of the Film Language, Silman-James Press (1991).

A Study of Emergent Computation of Life-like Behavior by Indefinite Observation

Michita Imai and Tsutomu Miyasato

ATR Media Integration & Communications Research Labs.
Address: 2-2,Hikaridai,Seikacho,Sourakugun,
Kyoto Pref.,619-0288,Japan
Phone:+81.774.95.1491
Fax:+81.774.95.1408
{michita,miyasato}@mic.atr.co.jp

Abstract. This paper describes the action generation of an autonomous animated agent. The agent behaves as if it has subjective motivation that is affected by its experiences. A typical study of this generation uses an ad-hoc method for the subjective action because it focuses on the action selection to solve some task. This paper proposes an indefinite communication architecture (ICA) to generate a subjective action. ICA keeps the experiences of an agent using indefinite observation. In particular, since the indefinite observation gives some chaotic feature to the experience, ICA generates various actions in response to an environment. This paper simulates and discusses the behavior of ICA in our interactive system.

Keywords: Life-like Behavior, Autonomous Agent, Indefinite Observation, Emergent Computation, Subjective Experience

1 Introduction

Recently, there has been a lot of research examining a subjective action for believable agents [6] and robot entertainment [3]. The subjective action is a response triggered by a physiological phenomena. For example, "what should a robot do if it is hungry?" or "what is best if a robot is in pain or is injured ?". This paper describes a subjective action of an autonomous animated agent. We define the word "subjective" as something existing only in an agent's internal state.

Typical studies use behavior sets [5], which are distributed representations [2] of the action to generate a subjective action. The approach[1] of Blumberg and his colleagues added Level of Interest variables to the behavior sets to handle a subjective action. The Level of Interest variables represent the strength of a subjective motivation. In addition, the variables are adjusted using a learning mechanism. However, such research [3][6] almost always deals with subjective action in an ad hoc manner.

The main problem when dealing with subjective action is to prepare a definitive tool for measuring the quality of subjective motivation. Unfortunately, there

S. Nishio, F. Kishino (Eds.): AMCP'98, LNCS 1554, pp. 370–385, 1999.

is no such a tool. However, someone may insist that this quality can be measured statistically, but statistics cannot explain why each quality emerges. People have the same problem because they cannot give an accurate report about the degree of a subjective quality with definitive descriptions in things like the report of a feeling, the quality of color, or the quality of pain. Even though a person can use various words as definite descriptions to represent the quality, these words cannot adequately assess a degree of the quality. The absence of a definitive description indicates that there is no accurate representation of a subjective quality. For example, the designer of an agent cannot represent an accurate degree of hunger or of pain. The quality of a subjective motivation also has the same difficulty in representations. Due to the representation difficulty, the designer cannot give the agent a subjective motivation as well as an action in response to the motivation. Since this fact suggests the impossibility of designing a subjective action by a tool that measures subjective qualities, a significant problem for an agent designer is how to prepare a mechanism instead of the measuring tool.

In this paper, our contribution is to show that there is a possible mechanism for generating a subjective action without the Level of Interest variables or an ad hoc implementation. The main mechanism is indefinite observation[4]. Our idea is to use indefinite observation as an emergent computation to generate a subjective action. In our approach, indefinite observation forms a contradiction between the generation and the perception of an agent's action. Since the contradiction takes apart an old standard and generates a new standard, indefinite observation has an advantage for emerging a new agent's action based on its experience.

We named the mechanism of the emergent computation Indefinite Communication Architecture (ICA). ICA changes the quality of a subjective action according to the observation of an action contradiction. The action generated by ICA is in contrast to the one generated by ordinary systems that use the selection of behavior. In particular, the action of ICA depends more on a subjective experience of an agent than an ordinary system using a learning mechanism. This is because the subjective experiences immediately affect the action generation through an observation log.

In this paper, ICA gives a subjective action to an autonomous mobile agent like that in figure 1. An example in the virtual world is the process where the agents and a user gather at one place to have their photographs taken. This virtual world appears in the interactive system, we developed, FurintClub. The agent behaves in the virtual world affected by its subjective motivation like real people. For example, many actions emerge when real people gather in a place. A person may stop to go forward in spite of the fact that there is more space between him and other people. A person may or may not also proceed much closer to others in response to their movement. The motivation of these movements is in the subjective point of view because the decision of the movement depends on a single person's experiences. FurintClub provides such a subjective motivation to agents by using ICA.

Fig. 1. Snapshot of user and autonomous agents in FurintClub

We hope that ICA can generate lifelike behavior, which sometime obeys a rule and sometimes does not. Moreover, this research tries to make a model of emergent computation as the origin of lifelike behavior.

In this paper, we introduce an interactive system FurintClub. In section 2, we give formal descriptions for the experience-based action of an agent and the goal of the spontaneous generation of an action standard in the virtual world. We propose an Indefinite Communication Architecture crucial for the generation of an agent's action in section 3, and show an example of an interaction between a user and these agents in section 4. We evaluate our architecture in section 5 with simulation of an agent's behavior that is generated by ICA. Finally, we conclude this paper and discuss the problems of ICA in section 6.

2 Interactive System FurintClub

Figure 2 shows the structure of the interactive system FurintClub. FurintClub consists of Pfinder[7][1] and a CG generator. Pfinder is used to track a user in front of the display, and the user can interact with an agent in the virtual world. When the user moves toward the left or the right in the real world, an avator of the user moves in the virtual world with the same movements.

The action of gathering or the arrangement of agents for the photograph emerges from an interaction in FurintClub. The response to a user is not a static one like an ordinary behavior set defined prior to the interaction. The response action is based on an agent's experiences. Since the experiences of agents vary, the actions of each agent differ from each other. As a result, the user is attracted to the movements of the agent, which responds to the user in a different way. Moreover, the user feels that the agent has a kind of subjective motivation.

---

[1] Pfinder is a vision system developed at MIT media labs. This system can find a person from the camera view point.

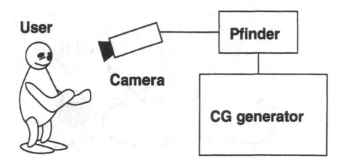

Fig. 2. Structure of FurintClub

2.1 Experience-Based Action of Autonomous Agent

An agent takes an experience-based action. Figure 3 indicates the structure of an action generation mechanism for the autonomous agent. The structure has three components: a perception, a generation, and a context. The perception part perceives an action generated by other agents as well as the action of itself. The generation part generates an action for the agent according to the results of the perception. The context part gives a standard for the perception to judge an action and for the generation to generate an action. Moreover, the contents of the context are also generated by the perception. In short, the context is based on the experience of each agent. Therefore, we call the action generated by these components an experienced-based action. We explain, in section 2.2, the reason why the generation of context must depend on the perception.

There are two movement features of an agent in the virtual world: the speed V of the agent and an angle Θ (the direction the agent proceeds). A representation of the movement is a pair of the feature F and a value X: (F, X). For example, when an agent moves at a speed of 10, the representation is $(V, 10)$. A user's movement is also represented in the same form.

We formalize the experience-based action in terms of the perception and the generation. The perception part is defined in the following function:

$$\phi_o(X_F) = True \ or \ False, \phi_o \in O(F, C), \tag{1}$$

where $O(F, C)$ is a set of a function ϕ_o, which observes a movement X_F of a feature F according to a judgment standard C. The function ϕ_o returns a truth-value according to a perception standard C. For example, under the standard of speed $C : 5 < x < 10$, the function decides that the speed of an agent $X_V = 2$ is false.

The generation part is defined in the following function:

$$\phi_g(T) = X_F, \phi_g \in G(F, X), \tag{2}$$

where $G(F, X)$ is a set of a function ϕ_g which gives an agent a movement X of feature F. T is a truth-value generated from a result of the perception. For

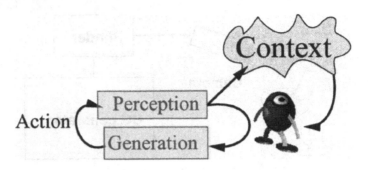

Fig. 3. Action generation mechanism of autonomous agent.

example, in $\phi_g \in G(V, 10)$ and $T = True$, the agent's speed is given a value of 10.

We must prepare a rule for a goal before formalizing ICA. The following rule represents the goals rule that gives an agent a subjective action,

$$X_{other} \to C, X. \tag{3}$$

Here, X_{other} is the variable that does not fit in the standards. The rule means that the action generation mechanism obtains a new action standard C, X in the function (1), (2) from the movement X_{other} of the other agent.

2.2 Indefinite Observation

The implementation of rule 3 is crucial for providing an agent with a subjective experience. The origin of the rule is indefinite observation [4]. Indefinite observation is proposed to explain the subjective behavior of a creature. It suggests that a subjective behavior emerges from an observation and a generation. The function generating the subjective behavior comes from the results of the action observation. For example, when a person divides oranges into two groups like big or small, he/she will do according to his/her experiences gained through past divisions. In the example, the action of the division develops a standard by dividing an orange, and people also divide it according to an acquired standard; however, in comparison, a computer divides oranges firmly based on a threshold value given by a programmer prior to the division.

There are some chaotic features based on an observation slip in the process of indefinite observation like the example of oranges. The slip is attributed to an observer. The observer of an indefinite observation has a local viewpoint of a behavior. Since a local observer cannot immediately perceive all of the world or the behavior himself, the slip occurs between experiences and actions[2]. Rule 3 emulates such a observation slip by adding X_{other} to the standards C, X.

---

[2] Since there are many parameters in complex system, an observer cannot completely detect the parameters related to an event in finite time.

3   Indefinite Communication Architecture

ICA observes the actions of its own agent and of the two nearest agents (or a user), and generates an action according to the observation. The function between the observation and the generation of the action is defined in the following form:

$$F(a_1, a_2, a_3) \rightarrow m. \tag{4}$$

Here, $a_i, m = \{True\ or\ False\}$. a_1 and a_3 represent the results of the observations of the nearest agents' movements, and a_2 is also the result of ICA's own movements. In another notation, a_i forms an equation $\phi_o(A_i) = a_i$ by function (1) where A_i is a movement of an agent i, and m in function (4) is input to the action generation. As a result of function (2), the equation between m and A_i is $\phi_g(m) = A_i$.

ICA has the following features to realize a subjective behavior:

- "Indefinite observation", which makes an agent's action similar to the another agent.
- "Order generation function", which produces an original action for the agent.

Although indefinite observation results in rule (3), there is lack of a spontaneous aspect when generating a behavior. The order generation function provides the agent with the spontaneous aspect.

Indefinite observation incorporates a movement observed as false into a standard of perception C and generation X in some probability. In short, putting X_{other} in its standards, this observation carries out rule (3) in some probability. Since the indefinite observation adapts the standards to the movements of others, the actions of the agents become similar to each other.

An order generation function F is function (4) itself, which controls the order of the action generation $\phi_g(m)$. In the process of an action generation, if $m = True$, then $\phi_g(m) = A_i$. If $m = False$, then $\phi_g(m) = \neg A_i$. $\neg A_i$ indicates an action that does not exist in an action generation standard X. Since $\neg A_i$ generates an action outside of the standard, it prevents the movements of agents from becoming similar to each other.

Figure 4 indicates the ICA structure of agent A2. There are two agents A1 and A3 near agent A2. The ICA structure of A2 is also shown at the top of the figure.

A perception (observation) part, a generation part of an action, and a context part are on the left-side, on the right-side , and in the middle at the top of figure 4, respectively. The perception observes the three actions of agents A1, A2, and A3, and gives the result of the observation to the order function. The order function decides the order of an action and gives the result to the generation. The generation generates an action A_2 according to the order from the order function, and gives the action A_2 to agent A2.

The context is a table which consists of several movement values (F, X). The context's contents develop through the observation of the perception. The perception ϕ_o writes down an observed action (F, X) in the context when it

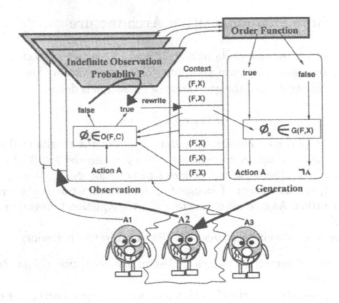

Fig. 4. ICA structure of agent A2

judges that the action is true. For example, when $X_v = 2$ and $\phi_o(2) = True$, the perception writes down the value $(V, 2)$ in the context. According to the change of the context, both standard C and X get close to the value $(V, 2)$. Here, the standard of judgment C is made from inequalities of the maximum and minimum value of the context, and the standard of generation X is selected from the context at random. The more the perception perceives the same value, the more the contents of the context approach the value.

3.1   Contribution of Indefinite Observation

Indefinite observation contributes to the development of a subjective experience with the contradiction of an observation because the contradiction performs rule (3), which adds X_{other} to the standards C and X. Since the perception adds a value judged as true to the context, the result of the perception must come to the same thing as $\phi_o(X_{other}) = True$ to add X_{other}. However, since X_{other} does not suit the standard C, the actual result is $\phi_o(X_{other}) = False$. ICA uses the contradiction of the observation to turn the judgment of $\phi_o(X_{other})$ into true. For example, when the speed of an agent is 4 and the standard of the perception is $C : 5 < x < 10$, the indefinite observation decides that speed 4 is true. The standard changes to $C : 4 < x < 10$ after the observation.

ICA uses the contradiction in some probability. The contents become an agent's own experiences because the probable observation is a unique experience

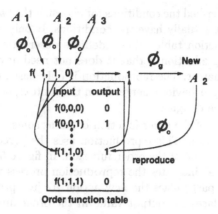

Fig. 5. Order function usage and reproduction

to each agent. This is reason why we uses the word "subjective" for an agent's experience.

3.2 Contribution of Order Function

Order function contributes to the adaptation of the degree of spontaneity for an agent's behavior. The degree means how often generated actions depend on the context. The more the action depends on a context, the more it becomes ordered. In contrast, when the action depends less on a context, it becomes disordered. The dependency on the context occurs when a true value is given to the generation part ϕ_g. Since the output value of the order function is given to the generation part ϕ_g, the degree of spontaneity is arranged by the ratio of a true value to a false one in output values of the order function.

ICA spontaneously generates the output values of order function (m in function 4). Figure 5 shows the outline of usage and reproduction of the order function. The order function consists of a table (order function table), which suggests a relation between the inputs and outputs of the function. In this table, 1 indicates a true value and 0 a false value. The order function is given input values from the perception part ϕ_o. Then, it seeks an output value appropriate for the input values and gives the generation part ϕ_g its value.

There are two functions that are opposite to each other in reproducing the order function: one makes an agent's behavior depend on a context when the behaviors of near agents become disordered; the other makes the behavior disordered when the behaviors of near agents depend on the context of the agent. The ordered behavior means that the agent has a behavior similar to the other agents' behaviors according to indefinite observation. The disordered behavior means that the agent has a spontaneous behavior. The order function can adapt the degree of behavioral spontaneity with these two functions.

Although we described the conditions for choosing the two reproduction functions, ICA does not actually have the conditions. It only spontaneously reproduces the order function table using indefinite observation. The advantage of the spontaneous reproduction is that it does not need any evaluation function for good or bad behavior. The reproduction is only based on the context of an agent. The value of a behavior emerges from the context, not from an evaluation function given by the designer.

The reproduction of the order function contains some complicated processes by indefinite observation. This reproduction uses the perception part ϕ_o to reproduce the output value of the order function. In figure 5, the arrow with the comment "reproduce" indicates the reproduction process with perception part ϕ_o. The perception part judges the new action $A2$ just produced by the generation part, and replaces the output value in the order function table with the judgment result of the new $A2$.

The judgment result of the new $A2$ is sometimes not equal to the value m given to the generation part $\phi_g(m)$ to produce a new $A2$. This is because the context used by the reproduction is sometimes different from that used by the generation when producing a new $A2$. There are two mechanisms connected with this contextual difference. The first mechanism is where the generation part uses the same context used by the perception part when observing the old $A2$, $A1$, and $A3$. The second mechanism is where the perception part of the reproduction uses the same context generated by the observation part when observing the old $A2$, $A1$, and $A3$. Since indefinite observation sometimes makes the context vary before and after the observation of the old $A2$, $A1$, and $A3$, the difference of context occurs between the generation in generating a new $A2$ and the perception in observing a new $A2$ for reproduction.

As a result, the reproduction process has the ability to adjust the ratio of a true value to a false one in the order function table. If the behavior of the near agents, $A1$ and $A3$, depends on the context of agent A2, the reproduction process increases the ratio of a false value on the order function table. Since the context of A2 is restricted to the same behavior as the other agent's under this condition, the context is very sensitive to strange behavior. It is easy for the context to change with indefinite observation when observing strange behavior. Once the context changes, a false value increases in the order function table. If the agent's behavior becomes disordered, the reproduction process increases the ratio of true values on the order function table. Since indefinite observation adapts the context for the spontaneous action, the reproduction process increases the ratio of true values on the order function table with the new context.

3.3   Task of Taking Photographs

ICA gives constraints to the context part for the agents when they gather to have their photographs taken. There are two constraints: one is where an agent turns in the direction of the user, and the other is where the agent near the user

Fig. 6. Independent movement agent

turns in the direction of the camera[3]. Since the constraints are provided to ICA through the context, it only indirectly affects the agent's behavior. As a result, ICA can sometimes generate various actions obeying the constraints.

4 Example of Agent's Action in FurintClub

Figures 6, 7, and 8 show an example of an interaction between a user and agents.

Figure 6 shows that the agent in the black circle moved alone in the virtual world in spite of the fact that a user and two agents gathered at one place. This independent behavior occurs when the contents of the context part differ from those of other agents. The difference is a result of the observation part ϕ_o outputting a false value in the judgment of other agents' movements.

However, the agent that moved alone began to participate in the group (the user and two agents) according to the change in the ICA context's contents. The reason for the change in the contents is that an indefinite observation inserts the movements of the other agents into the ICA context.

The upper left side of figure 7 shows the agent on the way to join the group of the user and the agents. As a result of this participation, the agent that joined the group moves with the others like so in the upper right side of figure 7. The contents of the context parts in the group members are similar to each other during the group behavior.

For a while, ICA generates a new type of action that does not exist in the contents of the context part because the order function inputs a false value to the action generation part ϕ_g. If indefinite observation takes the new actions into the context, its contents gradually become different from other members in the group. Such an agent drops out of the group like so in the bottom of figure 7.

---

[3] this is not the real camera for Pfinder.

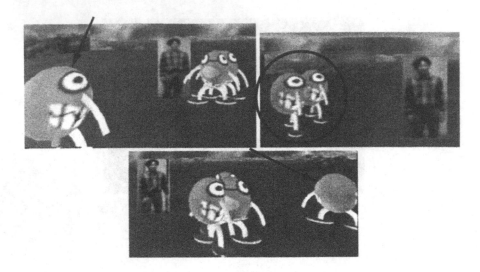

Fig. 7. Agent participation in group and dropping out of group

When constraints for the photograph exist in the context part, the behavior for the photograph dominates the agent's movement. As a result, the agent moves in the direction of the user like so in the left side of figure 8. Furthermore, when the agent arrives at a place near the user, it turns in the direction of the camera. The two agents in the foreground of figure 6 were situated for a photograph. However, the behavior resulting from the constraints disappears according to a disassembly of the contents by the false output of the order function. As a result, the agent moves away from the user like so in the right side of figure 8.

The main factor in a series of these behaviors is a change in the relation in each agent's context. The contexts change by becoming similar or opposite to each other. Indefinite observation is the main mechanism for such similar or opposite characteristics.

5   Evaluation of ICA

5.1   Simulation of Agents

We evaluate ICA by simulating the movements of agents. Ten agents are only given a speed feature (V, v) to easily determine the properties of ICA. Hence, these agents only proceed forward. The movement speed for each agent is decided by ICA. We plot the distance of each agent from its initial position.

Moreover, we evaluate ICA by two parameters. One is λ, which is the ratio of the true value to the false value in the order function table. The range of value of λ is $0 \leq \lambda \leq 1$. The closer the value λ is to 1, the more true values exist in the context part. The other parameter is P, which is a probability of contradiction

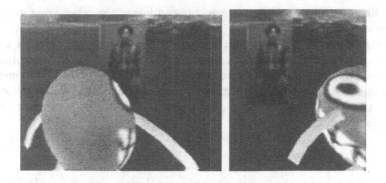

Fig. 8. Agent moves toward user and away from user

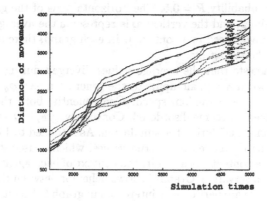

Fig. 9. Plot of agents' movements

by indefinite observation. We conducted two simulations to evaluate the effect of these parameters on ICA. The first simulation examines the behavior of the context with respect to static λ by recording the average value of the contents in the context. The other simulation examines the diversity of λ with respect to static probability P.

5.2   Movements of Agents

Figure 9 represents the forward movement of the ten agents. The horizontal axis of the graph represents the simulation times. The vertical axis represents the distance of the agents' movements from their initial positions. Each line represents the movement of one agent. Figure 9 suggests that several agent groups emerged from the simulation when the agents moved forward. Between the 2000 and the 2500 step stage in the simulation, one of the agent groups sped up. Near

382   Michita Imai and Tsutomu Miyasato

the 4300 step stage some agents sped up and joined the other agent groups at
the newly generated speed.

This simulation suggests that a group behavior emerged through mutual
observation and the groups also collapsed. The groups suggest the existence of
a stable structure of context parts between agents; the agent maintained the
same speed as other agents in its group. Furthermore, the construction and
deconstruction of the groups also indicates ICA's ability to produce various
behaviors for a group activity.

5.3   Behavior of ICA

Figure 10 represents the diversity of a context during a simulation. In figure 10,
each graph represents a different simulation in terms of λ of the order function
table: in graph A, $\lambda = 0$; in B, $\lambda = 0.5$; in C, $\lambda = 0.9$; and in D, $\lambda = 1.0$. In the
simulation the probability $P = 0.5$. The horizontal axis of the graph represents
the simulation times, and the vertical axis represents the average of values in a
context(the average speed in a context). In each graph, there are ten plots for
the averages of the ten agents.

Graph A suggests that the average values diverged during the simulation.
This is because when $\lambda = 0$, all inputs to the generation part ϕ_g are false, which
causes ϕ_g to generate a random speed. Consequently, both the behavior and
context of the agents became disordered. Conversely, graph D suggests that the
average value converged during the simulation. As λ tended to 1.0, all the inputs
to generation part ϕ_g tended to be true values, which caused ϕ_g to generate a
speed only from a context. As a result, the context of each agent became similar
through mutual observation. Graph B shows behavior between those of graphs A
and D. However, we are particularly interested in graph C because it exhibits the
kind of lifelike behavior in ICA that we hope to achieve. By the 1600 step stage,
the average of the contexts was stable at a speed of 3.0, although some diversity
could be seen in graph C. Some kind of catastrophe occurred near 1600 steps.
The average of the agent's context rose to the speed of 8.0 and the two groups
emerged after this catastrophe. Consequently, ICA exhibited both construction
and destruction of behaviors at $\lambda = 0.9$

Figure 11 represents the diversity of λ throughout the length of the simulation. In figure 11, each graph exhibits a different simulation of probability P for
an observation part ϕ_o: in graph A, $P = 0.5$; in B, $P = 0.25$; and in C, $P = 0.01$.
The horizontal axis of the graph represents the simulation times and the vertical
axis represents λ of each agent's order function table. In each graph, there are
ten plots for each agent.

We pay particular attention to graph A because the λ of graph A has suggested values between 0.7 and 0.9. Moreover, λ approaches 0.9 from an initial
value of 0.5. This property suggests that ICA with $P = 0.5$ has the capacity
to recover from disorder even when ICA has lost its order of context. In the
other graphs, both B and C had a wide range of values for λ. The range is far
form $\lambda = 0.9$. In terms of $\lambda = 0.9$, ICA presents an unappealing property for
probability P in graphs B and C.

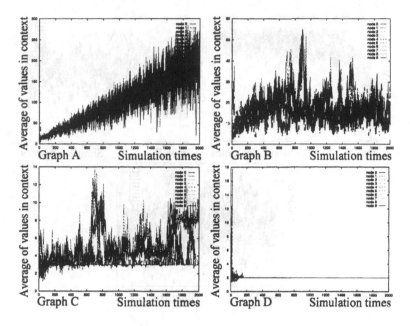

Fig. 10. Diversity of context in response to λ: in graph A, $\lambda = 0$; in B, $\lambda = 0.5$; in C, $\lambda = 0.9$; and in D, $\lambda = 1.0$

5.4 Discussion of Simulation

Both graph C in figure 10 and graph A in figure 11 demonstrate that the context has features of stability and liquidity in its structure when $P = 0.5$. The groups of the agents appeared and collapsed like figure 9 because of the features.

The behavior of each agent does not come from an evaluation function provided by the designer. It certainly comes from the experiences of each agent. In particular, the collapse and the appearance of the groups is different from a search for the optimum grouping. The worthiness of a behavior is estimated by the context a group keeps, and is transformed according to the collapse and the appearance of the group. Evaluation functions are unnecessary because the group has experience-based values. This is an advantage of ICA compared to other generation mechanisms of subjective behavior that use some evaluation mechanism.

6 Conclusion

This paper proposed a new action generator named ICA, where the action of an agent emerges from its calculation as if the action is a result of a subjective motivation. ICA generates an action standard by indefinite observation and order function. The action generated by ICA has two opposite features: one is an

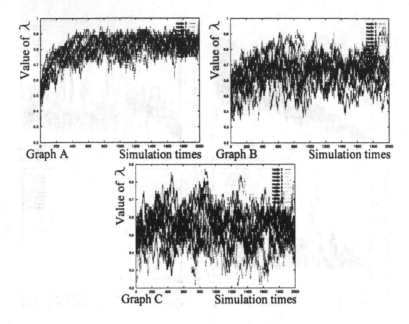

Fig. 11. The change of λ in response to Probability P: in graph A $P = 0.5$, in B $P = 0.25$, and in C $P = 0.01$

individual behavior, and the other is a group-based behavior. While the individual behavior is a result of the subjective experience of the agent by observation, the group behavior is also a result of the observation because the contents of the observation are generated in a selfish manner and are also affected by the action of the other agents. These features play a significant role for ICA in generating a subjective action.

Moreover, we constructed an interactive system, FurintClub, using ICA. The agents in FurintClub generate various actions such as an individual action, a group action, movement toward a user, and away from a user in response to the user's or another agent's movements. As a result of the various actions, FurintClub attracts the user's attention by the movement of the agents.

The action generated by the existing ICA is a very simple action such as the movement of an agent in the virtual world. For our next step, we will implement more action features in ICA to support more powerful interaction. Presently, ICA has a problem of controllability because it is still unclear as to how experience leads to some kind of an action. We have not captured such a behavioral tendency driven by the environment. Fortunately, since the experience depends on the environment, we will be able to control the behavior of ICA by giving an agent a particular environment if the tendency becomes clear. In the future, we will challenge the problem of controlling ICA with environmental parameters.

References

[1] B. M. Blumberg, P. M. Todd, and P. Maes. No bad dog: Ethological lessons for learning in hamsterdam. In *SAB'96*, 1996.

[2] R. A. Brooks. Intelligence without reason. In *IJCAI-91*, pages 561–595, 1991.

[3] M. Fujita and K. Kageyama. An open architecture for robot entertainment. In *Proc. of Autonomous Agent'97*, pages 435–442, 1997.

[4] Y. P. Gunji. Life, time, protozoan-calculation, and ontological observation. *Revue de la pensée d'aujourd'hui*, Sep, 1996.

[5] P. Maes. How to do the right things. In *Connection Science*, volume 1(3), pages 291–323, 1989.

[6] W. S. N.Reilly. Believable social and emotional agents. Technical Report CMU-CS-96-138, School of Computer Science, Carnegie Mellon University, May 1996.

[7] C. R. Wren, A. Azarbayejani, T. Darrell, and A. Pentland. Pfinder: Real-time tracking of the human body. In *IEEE Tranc. on Pattern Analysis and Machine Intelligence*, volume 19(7), pages 780–785, July 1997.

An Interactive Digital Fishtank
Based on Live Video Images

Toshihiro Masaki, Tetsuya Yamaguchi, and Yoshifumi Kitamura

Department of Electronic and Information Systems Engineering,
Graduate School of Engineering, Osaka University
2-1, Yamada-Oka, Suita, Osaka 565–0871, Japan,
{masaki, tetsuya, kitamura}@eie.eng.osaka-u.ac.jp

Abstract. This paper describes an interactive manipulation technique
for objects extracted from video images by an efficient algorithm in order
to produce multimedia contents. An object contained in a video image
must be able to be extracted at a reasonable speed and a moving object
able to be tracked with sufficient accuracy. Moreover, real object interactions that meet such criteria must be established using techniques
for manipulating objects in virtual worlds. The proposed technique has
potential to generate novel multimedia contents without using shape
modeling. An interactive digital fishtank is described as one example application of the proposed technique. In this application, the user is able
to obtain various video sequences of fishtanks from the original video
sequence.

1 Introduction

Recently, multimedia data can be treated more and more readily in computer
systems, owing to remarkable advances in electronics, computer science, and
communication systems have made remarkable advances. Therefore, the processing of multimedia data has been a main focus of the development of applications
of these systems. In order to develop the advanced techniques for processing multimedia contents, very diverse approaches have been proposed; virtual reality is
regarded as the promising technology for multimedia contents.

Various kinds of techniques to construct realistic virtual environments have
been used to process multimedia contents. In particular, progressive research in
3-D computer graphics and 3-D interaction techniques has played an important
role in the evolution of virtual reality. At present, a user can manipulate "virtual"
objects created by computer graphics in the same way he/she would do in a
real environment with a sophisticated user interface [3]. However, such virtual
objects are mainly artifacts. Living things or plants have been, and continue to
be, difficult to introduce into a virtual environment because of the difficulty of
generating geometry-based realistic computer simulations of them.

Image-based rendering techniques have recently been used to provide photo-
realistic images without using objects' geometric models [1,2]. Moreover, live

S. Nishio, F. Kishino (Eds.): AMCP'98, LNCS 1554, pp. 386–396, 1999.
© Springer-Verlag Berlin Heidelberg 1999

video of real scenes taken by a camera has been used together with virtual objects in advanced virtual environments (the so-called "augmented virtuality" [4]). In such cases, virtual objects can be manipulated by using an intuitive user interface. However, it must be emphasized that it is difficult to manipulate objects in video sequences because sequences of individual frames are typically involved. Accordingly, in order to manipulate objects in video sequences in the same way as one would manipulate virtual objects in a consolidated manner, it is necessary to develop reasonable storage/retrieval/editing/transmission techniques for objects contained in video scenes.

2   Interactive Manipulation of Objects in Video Images

Research in augmented reality [5,6], augmented virtuality, and mixed reality [4] has been seeking a visual space having an effective combination of a virtual world and the real world. A virtual world is usually generated by computer graphics. The real world, in contrast, has mainly involved two basic approaches. The first is the optical see-through technique in which a user observes the real environment directly through a half-silvered mirror on which a virtual world is combined. The second is the video see-through technique in which video images of the real world captured by cameras are combined with graphic images of a virtual world. With either approach, the development of reasonable storage/retrieval/editing/transmission techniques will enable a user to observe the combination of the real world and a virtual world over time and distance.

Our interactive digital fishtank is an example application that involves manipulating objects in video images. This fishtank is set in a virtual environment in which live video of real fish is displayed together with virtual fish. A user can manipulate various attributes of this fishtank. Examples of such interaction include cutting/copying/pasting real fish, moving virtual objects (e.g., rocks) into the real environment, and interacting with the real fish (e.g., catching, feeding, and communicating with them).

For these purposes, robust algorithms for the extraction of video contents must be developed. Algorithms for the retrieval of and interaction with the extracted video contents must also be developed. In particular, each object contained in a video sequence must be able to be extracted at a reasonable speed and a moving object able to be tracked with sufficient accuracy. Then, a user can manipulate real objects using a consolidated interaction method for real and virtual objects and create various multimedia contents from one video scene by using this tool.

Figure 1 shows the flow chart of the interactive digital fishtank. The algorithm is separated into two parts: the first (process 1 in Fig. 1) is the process for extracting a moving object, and the second (process 2 in Fig. 1) is the process for interacting and controlling behavior of the video contents. The first process hands the moving vector and the boundary of a moving object over to the second. The processes are described in sections 3 and 4, respectively.

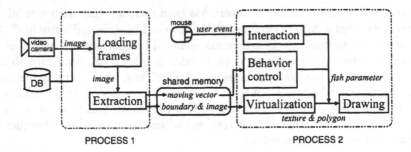

Fig. 1. Flow chart of an interactive digital fishtank.

3   Extraction of Moving Objects

In this section, the first process (process 1 in Fig. 1) for extracting moving objects from video images is described. The proposed extraction algorithm is able to find the boundaries of objects with a low computational complexity. This process can be mainly divided into the two steps shown in Fig. 2. In the first step, coarse areas of moving objects are extracted by using the differences among three video frames. Then in the second step, a pixel sequence of the boundary of an accurate object is determined based on the edge detection. Here, the extracted objects can be manipulated in a manner similar to virtual objects.

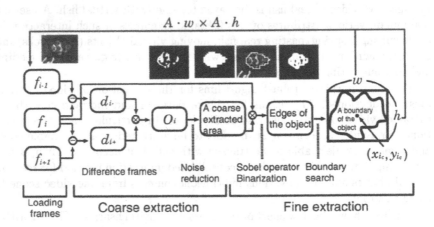

Fig. 2. Flow chart of extraction algorithm.

3.1 Coarse Extraction of Moving Objects

Three frames f_{i-1}, f_i, and f_{i+1} in a video sequence taken by a stationary camera are loaded to extract the coarse areas of moving objects (where f_i denotes the i-th frame of the video sequence).

The first step for this coarse extraction is to obtain difference frames d_{i-} and d_{i+}. Here, d_{i-} and d_{i+} are difference frames in terms of the luminance between f_{i-1} and f_i, and between f_i and f_{i+1}, respectively. d_{i-} and d_{i+} are given by

$$
\begin{aligned}
d_{i-}(x,y) &= |f_{i-1}(x,y) - f_i(x,y)|, \\
d_{i+}(x,y) &= |f_i(x,y) - f_{i+1}(x,y)|,
\end{aligned}
\tag{1}
$$

where x and y are the coordinates of the frame and $X(x,y)$ denotes the luminance pixel at point (x,y) of frame X. Then, we can obtain differential frame O_i, where the areas of the moving objects come to the front, as follows:

$$
O_i(x,y) = d_{i-}(x,y) \times d_{i+}(x,y).
\tag{2}
$$

O_i, a gray image, contains noise caused by light, shadows, and so on, so noise reduction is introduced next as the second step of the coarse extraction. The noise reduction used here is block-based noise reduction. This noise reduction method reduces the whole noise in the video images and also extracts the coarse area without gaps which sometimes appear in the case of monotonous texture.

Before the noise reductions, O_i is converted into a binary image Ob_i with an appropriate threshold. Those pixels that are possibly in the areas of the moving objects are set equal to "1". In the block-based noise reduction, the binary image is divided into 4×4 blocks, and at the same time, the value of each block is set. This value is determined not to be narrower than the essential areas of the moving objects. When the number of pixels having the value "1" is under two in each block, the block is set equal to "0", otherwise, the block is set to "1". The block-based noise reduction is determined by the 8-neighboring blocks, excluding the block itself. When the number of blocks with the value "1" is under three in the 8-neighboring blocks, the value of the block is set equal to "0". On the contrary, if the number of blocks having the value "1" is three or more, the value of the block is set equal to "1". The thresholds of these processes are found by way of experimentation.

The procedure outlined in this section enables the coarse extraction of moving objects without noise. The next step, fine extraction, involves only the extracted areas being processed.

3.2 Fine Extraction – Searching for Pixel Sequences of Moving Objects

In order to detect the boundaries of moving objects, first of all, a sobel operator is used to obtain the edges of objects. In order to reduce the computational cost, this operator is applied only to each area of original frame f_i obtained by the coarse extraction. The edge image in this area is converted into a binary image, and then the pixel-based noise reduction is applied. Also employed is a border-following process, which traces the boundaries of the moving objects, since only the borders of real objects are required in order to manipulate them.

The border-following is started from the scanning line at the center of balance (x_{i_c}, y_{i_c}) of the coarse extracted area in the i-th frame. The computation time for this border following process is faster than if started from the top or bottom scanning line. Moreover, the area to search coarse extraction of moving objects in the frame f_{i+1} is limited to

$$f_{i+1}(x, y), \quad (x_{i_c} - \tfrac{A}{2} \cdot w \le x \le x_{i_c} + \tfrac{A}{2} \cdot w),$$
$$(y_{i_c} - \tfrac{A}{2} \cdot h \le y \le y_{i_c} + \tfrac{A}{2} \cdot h), \tag{3}$$

where w and h are the maximum width and height of the object's bounding box, respectively. A is a constant which is determined by the average speed of fish and sampling rate of video scenes.

The proposed extraction algorithm is able to detect the moving objects within a sufficient accuracy, and these objects are tracked in each frame in the video sequence. The first process outputs a moving vector and a boundary of the extracted area to the second.

4   Virtualization, Interaction Technique, and Control of Behavior of Video Contents

In this section, the second process (process 2 in Fig. 1) for virtualization, controlling behavior, and interacting with these moving objects is described.

The second process converts the extracted video image into the object which consists of polygons and textures at the start. The interactive manipulations of moving real objects in video sequences are achieved by using the image data of extracted area. Manipulations like feeding, size reductions, movements, and changes of color/size/direction/background are implemented as combinations of cutting, copying, pasting, and changing attributes. The user can interact with the moving real objects by using a mouse and function menu.

In the case of an interactive digital fishtank, the behavior of virtual fish is under the control of the system in order to show them realistically. The motion of the virtual fish are determined by using the position of the real fish in original images (Fig. 3). If a virtual fish has the same size and direction as an original real fish, the moving vector (v_{x_i}, v_{y_i}) defined by

$$v_{x_i} = x_{i_c} - x_{(i-1)_c},$$
$$v_{y_i} = y_{i_c} - y_{(i-1)_c}, \tag{4}$$

is simply apply to the motion of the virtual fish. In case of a scaled down (up) fish, vector $(K \cdot v_{x_i}, K \cdot v_{y_i})$ is used, where K is the scale factor of the fish . If the direction of a virtual fish has been reversed, the fish is flipped by using vector $(-v_{x_i}, v_{y_i})$. On the other hand, virtual fish are prevented from leaving the fishtank by using constraints on x and y. The center of a virtual fish in the i-th frame (x_{i_c}, y_{i_c}) is given by,

$$
x_{i_c} = \begin{cases} \frac{w}{2} & (x_{(i-1)_c} + v_{x_i} < \frac{w}{2}), \\ x_r - \frac{w}{2} & (x_{(i-1)_c} + v_{x_i} > x_r - \frac{w}{2}), \\ x_{(i-1)_c} + v_{x_i} & (otherwise), \end{cases} \tag{5}
$$

$$
y_{i_c} = \begin{cases} \frac{h}{2} & (y_{(i-1)_c} + v_{y_i} < \frac{h}{2}), \\ y_t - \frac{h}{2} & (y_{(i-1)_c} + v_{y_i} > y_t - \frac{h}{2}), \\ y_{(i-1)_c} + v_{y_i} & (otherwise), \end{cases} \tag{6}
$$

where x_r and y_t denote the coordinates of a right wall and the surface of the water, respectively. In case $(x_{(i-1)_c} + v_{x_i} < \frac{w}{2})$ or $(x_{(i-1)_c} + v_{x_i} > x_r - \frac{w}{2})$, the sign of $v_{x_{i+1}}$ should be turned over so that the fish turns back at the wall of a fishtank. Thus, the user can obtain various scenes from one video sequence by interacting with the fishtank.

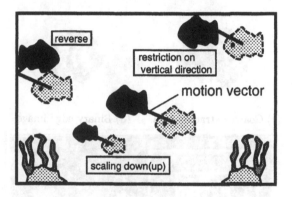

Fig. 3. Behavior of virtual fish.

5   Implementation Results

The proposed approach is implemented in an interactive digital fishtank application. The fishtank implemented on a SGI Onyx2 (two CPUs) using OpenGL enables users to make desirable scenes from one fixed video sequence. The performance of the extraction algorithm and interaction are evaluated as follows.

5.1   Extraction Algorithm

The proposed algorithm is evaluated on a video scene featuring dummy coral, rocks, and a swimming toy fish in a real fishtank. The scene consists of 200 frames, and the size of each video frame is 320 dots × 240 dots. Figures 4 (a), (b), (c), (d) and (e) show the original source image, differential image, coarse extracted area, binary edge, and extracted boundary, respectively. The swimming fish is extracted within a sufficient accuracy in almost all of the frames. However, some extraction errors can be observed when the fish swims in front of the dummy coral, since the coarse extracted area is not sufficiently cramped so as to exclude the edge of the coral.

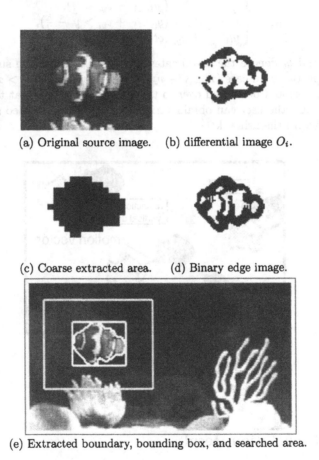

(a) Original source image.   (b) differential image O_i.

(c) Coarse extracted area.   (d) Binary edge image.

(e) Extracted boundary, bounding box, and searched area.

Fig. 4. Processes of the proposed extraction algorithm.

Table 1 lists the calculation time per frame of each step to extract the toy fish in Fig. 4 (a). The constant A, which denotes the area in which to search for moving objects, is set equal to two.

The whole execution time of the proposed algorithm is determined by the slower process (process 1), since the Onyx2 has two processors, and two processes

Table 1. Calculation time of the algorithm ($A = 2$).

process		time [msec/frame]
process 1	loading frames	34.2
	coarse extraction	22.8
	fine extraction	18.9
	total	75.9
process 2	virtualization	33.3
	interaction and behavior control	0.3
	drawing	7.8
	total	41.4

are able to run asynchronously. The first process hands the parameters, such as moving vectors and boundary information, to the second with the use of a shared memory (Fig. 1). The average frame rate is 13.2 frame/sec. The rate varies from 11.0 frame/sec to 23.2 frame/sec in the video sequence, since the searched area in the frame f_{i+1} depends on the bounding box of the extracted objects ($w \times h$) in the frame f_i.

Figure 5 shows the processing time of process 1, varying the searched area. The extraction time T in sec is approximated as:

$$T = 6.64 \times 10^{-6} \times A^2 \cdot h \cdot w + 1.97 \times 10^{-2}. \qquad (7)$$

The proposed algorithm is able to track objects which do not move more than $\frac{A-1}{2} \cdot w$ dots horizontally, or $\frac{A-1}{2} \cdot h$ dots vertically in T sec. For example, if the size of the extracted area is 50 dots $\times$ 40 dots (i.e. the number of pixels in the searched area is 8,000), the maximum horizontal and vertical speeds of an extractable object are approximately 340 dots/sec and 270 dots/sec (which are wider than the size of frame) respectively. If a tuna fish, two meters long, (the size is 15 dots $\times$ 6 dots) is swimming in the video sequences, maximum speed of an extractable object is over 160km/h, which overcomes the swimming speed of the tuna fish.

5.2 Interaction with Moving Objects

Figures 6 (b), (c), (d), (e), and (f) show examples of manipulations in the interactive digital fishtank; all of them generated from an original source image (see Fig. 6(a)). Figure 6(b) shows a result of cutting and pasting. The left fish is the original fish, and the right one is the copied and pasted fish. A copied fish can be pasted anywhere the user wants to put it. Figures 6(c) and (d) are example images of attribute changes. The left fish in Fig. 6(c) is copied and pasted with a change of direction. In Fig. 6(d), on the other hand, the user changes the size and color. Figures 6(e) and (f) are examples of combinations of interactions. They show that a user can create desired fishtanks from one video sequence.

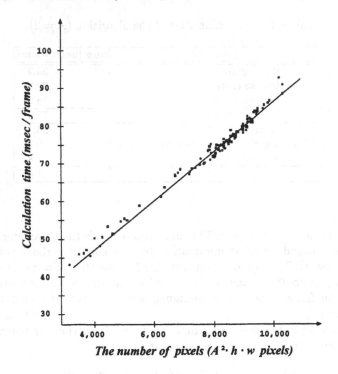

Fig. 5. Calculation time of process 1 against the size of searched area.

6   Conclusions

This paper has described a means of interactive manipulation using processed video contents, placing stress mainly on moving object extraction and interaction with real objects in real time. The proposed low-cost extraction algorithm can track objects at 13 frame/sec within a sufficient accuracy. Moreover, a number of interactive manipulation techniques have been introduced into our interactive digital fishtank so as to enhance the capabilities of augmented virtuality. The user is able to obtain various video sequences of fishtanks from one video sequence. The interactive digital fishtank is an example of the novel creation of multimedia contents by using the technologies of virtual reality.

Future work includes reducing the time and making the algorithm more robust in extracting coarse areas of moving objects. A 3-D interactive environment using virtual reality techniques will be established by expanding the proposed tracking and interaction method to include depth information obtained by multiple cameras.

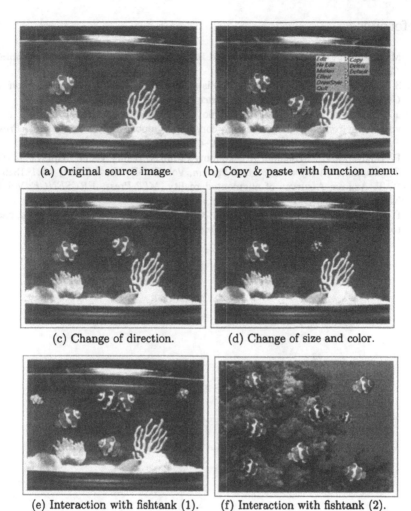

(a) Original source image.   (b) Copy & paste with function menu.

(c) Change of direction.   (d) Change of size and color.

(e) Interaction with fishtank (1).   (f) Interaction with fishtank (2).

Fig. 6. Example frames of an interactive digital fishtank.

Acknowledgments

The authors would like to express sincere gratefulness to Professor Fumio Kishino who gave us valuable comments. This research was supported by "Research for the Future" Program of Japan Society for the Promotion of Science under the Project "Advanced Multimedia Content Processing" (Project No. JSPS-RFTF97P00501).

References

1. M. Levoy and P. Hanrahan: Light field Rendering. in *Computer Graphics, Annual Conference Series* (1996) 31–42
2. S. J. Gortler, R. Grzeszczuk, R. Szeliski, and M. F. Cohen: The lumigraph. in *Computer Graphics, Annual Conference Series* (1996) 43–54
3. Y. Kitamura, A. Yee, and F. Kishino: A sophisticated manipulation aid in a virtual environment using dynamic constraints among object faces. *MIT Press, PRESENCE*, Vol. 7, No. 5 (1998) 460–477
4. P. Milgram and F. Kishino: A taxonomy of mixed reality visual displays. *IEICE Transactions on Information and System*, Vol. E77-D, No. 9 (1994) 1321–1329
5. R. T. Azuma: A survey of augmented reality. *MIT Press, PRESENCE*, Vol. 6, No. 4 (1997) 355–385
6. P. Wellner, W. Mackay, and R. Gold: Computer-augmented environments: back to the real world. *Special Issue of Communications of the ACM*, Vol. 36, No. 7 (1993)

Contents Creation for Interactive Media

Ryohei Nakatsu

ATR Media Integration & Communications Research Laboratories
2-2, Hikaridai, Seika-cho, Soraku-gun, Kyoto, 619-0288 Japan
Tel:+81 774 95 1400, Fax:+81 774 95 1408
E-mail:nakatsu@mic.atr.co.jp

Abstract. Multimedia technology is deeply related with contents. Therefore, it is an important issue to consider contents in the development process of multimedia technologies. This paper summarizes how research and development aim at various types of services, proposes an approach to researching and developing interactive media, and presents some specific examples of research.

1 Introduction

With the coming of the multimedia era, the importance of "contents" in addition to that of technology has come to be recognized. While various reasons can be given for this development, certainly one is that multimedia in itself has a deep relationship with contents. In other words, engineers that work on multimedia technology must always consider how contents will be used, i.e., work must now be approached from a broader point of view compared to the conventional way of focusing on technology only. At the same time, the field of multimedia has provided engineers more freedom and possibilities. As a result, how exactly to proceed with the research and development of multimedia technology while considering contents is now an important issue. The author and his colleagues have been researching interactive media as one form of multimedia, and it is here that the connection between contents and technology takes on special significance. In this paper, we summarize how research and development aim at various types of services, propose an approach to researching and developing interactive media, and present some specific examples of research now being performed at ATR in this field.

2 Contents and Technology

Here we examine the relationship between contents and technology. At this time, the people involved in these areas can be classified into three main types: engineers, contents creators, and users, and that which is provided to users is

S. Nishio, F. Kishino (Eds.): AMCP'98, LNCS 1554, pp. 397-405, 1999.
© Springer-Verlag Berlin Heidelberg 1999

generally referred to as "services."   With the above in mind, the following categories can be established with respect to contents and technology and their use (Fig. 1).

(1) Service = existing media (no contents)

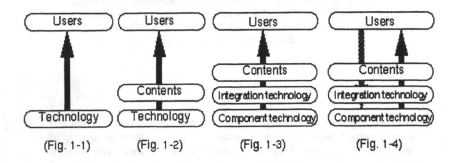

(Fig. 1-1)   (Fig. 1-2)   (Fig. 1-3)   (Fig. 1-4)

Fig. 1 Technologies, contents, and users

In this scheme, engineers provide a service and users use that service to create contents.   A good example of such a service in the communications field is the telephone, where the user is responsible for creating all contents transmitted by the telephone set.   This example represents the most basic relationship between technology and people. In this case, the user is the same as the contents creator (Fig. 1-1).

(2) Service = existing media (with contents)
In this scheme, engineers provide a system and contents creators use that system to create contents. Typical examples of such contents are movies and television programs.   Users in this case become users and recipients of contents (Fig. 1-2).   A major feature of this scheme is that information flows in the direction from technology-->contents creation-->use. (Of course, as contents may or may not be accepted by the user, a strong feedback loop exists that drives change and improvement in both contents and technology.)   Such services and associated technology and contents creation have been well established and recognized as stable media.

(3) Service = multimedia
This scheme requires technology that can integrate various types of media.   It would be difficult to say, however, that media integration technology has already been established.   It would be more correct to say that both engineers and contents creators are still groping for the ultimate multimedia service.   As a consequence, media integration at this time depends on contents.   In other words, service development must proceed in the order of component technology-->integration technology-->contents creation while keeping the overall service in mind.   This kind of service development, however, is highly inefficient.   At the least, however, if

engineers involved with integration technology work closely with contents creators in this process, we may see the birth of the ultimate multimedia service (Fig. 1-3)

(4) Service = interactive media
The situation becomes even more complicated in the case of interactive media. A major feature of this form of multimedia is that contents itself can be changed through user interaction.   In current examples of interactive media such as games, the contents creator determines the overall framework of contents while the user is left to determine contents on a partial basis (for example, the user can select one of several stories).   In the future, however, we can imagine the appearance of interactive media that provides the user with an even greater degree of freedom.   This kind of media will give birth to a whole new situation in which information flows from the user to the service-providing side in real time (Fig. 1-4).   A completely different approach compared to past forms of media is therefore needed to develop services that include contents creation.

3 Approach to Contents Creation

As described above, interactive media not only requires an overlapping structure of media component technology, media integration technology, and contents creation on the service-providing side, it also introduces a complicated situation in which user interaction results in changes to contents.   As a consequence, the one-way development flow of component technology development-->media integration technology development-->contents creation does not necessarily work well here. Some solutions to this problem include having engineers also perform contents creation and developing technology having contents-creation functions.   These types of solutions, however, are "seeds" (or "means") oriented, and a situation results in which engineers can easily become dogmatic during service development.   In contrast, our approach to developing new interactive media is to have contents creators and engineers cooperate in development work. The advantages of this approach are as follows.
(1) Research and development by engineers only can easily become seeds oriented. Contents creators, on the other hand, are sensitive to "needs."   By therefore having contents creators and engineers cooperate, needs-oriented research and development becomes possible.
(2) Due to the nature of interactive media, the entire service including contents is indefinite, which actually unifies the development side and makes for more efficient development.
With regards to (2) in particular, we have invited computer-graphic artists as guest researchers to be contents creators, and are having them work together with engineers to research and develop new interactive media [1].   However, as the contents of specific interactive media that we wish to develop and the personalities of the artists and engineers involved all tend to become interrelated, we have avoided establishing a standardized mechanism of cooperation.   We have instead adopted a flexible system that allows selection of any of the following possibilities depending on the project in question (Fig. 2).

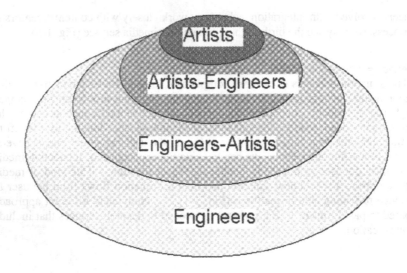

Fig. 2 Collaboration between engineers and artists

(1) Artist-only format: The artist himself/herself possesses engineering knowledge and ability and creates art that incorporates new technology.
(2) Artist-led format: The artist conceives a new concept and the engineer cooperates with the artist by providing technology to realize that concept.
(3) Engineer-led format: The engineer presents the entire concept and the artist cooperates with the engineer by creating the artistic part of the concept.
(4) Engineer-only format: The engineer performs standard engineering research.

The next section presents examples of specific research projects now being performed at our research laboratories in the field of interactive media. Please note that these projects do not exactly correspond to any one of the above formats, and that the format that a project most closely resembles may change over the course of research.

4 Research Project Examples

4.1 Interactive Art

We have constructed a system called "MIC Exploration Space" [2] whose environment changes in response to communication occurring between two people at a distance. The concept behind this system is shown in Fig. 3. Here, each person stands before a screen unit located at each of the two locations. A video image of each person is obtained by cameras installed on top of each screen, and the figure of the person is extracted from the background in this image. At the same time, the camera above is used to recognize the position and simple movements of the person.

Fig. 3 MIC exploration space
(by Laurent Mignonneau & Christa Sommerer)

Then, based on the results obtained here, computer graphics are used to construct virtual trees, grass, flowers, etc. The video of the two persons are then superimposed on these virtual objects, the result of which is projected onto the two screens. In this way, two people at remote locations can experience a sense of face-to-face communication in virtual space, and in addition, a virtual environment can be dynamically produced in response to communication signals like body shifting, hand waving, etc.

Most of the development work in the project was performed by two computer-graphic artists. These artists also had an engineering background, which they used to develop the entire system. Although this particular system is essentially interactive art, it shows promise as a future video communications system. Accordingly, interesting issues like research and development of a communications system that makes use of this concept and evaluation of contents other than virtual plants are being presented to engineers for consideration. Figure 4 illustrates an example of the interaction.

4.2 Interactive Movies

Since their invention over 100 years ago, movies have become firmly established as a major medium of entertainment. The addition of an interactive function however, will broaden the range of possibilities by allowing a viewer to enter and experience a virtual world in the role of a "hero." Research of interactive movies is now being performed along these lines [3]. An interactive movie consists of an interactive story whose development changes through interaction; a participant who experiences the world of the interactive story in the role of a hero; and characters that participate in story development while interacting with the hero. The hardware configuration of a prototype interactive-movie system is shown in Fig. 5. The video output subsystem shown here makes use of a high-speed computer-graphics

Fig. 4 Example of an interaction in the MIC exploration space

workstation to output video.   Character images are stored beforehand as computer-generated animation data and character graphics are generated in real time. In the same way, background graphics are stored as digital data and background images are generated in real time.   Processing to overlap multiple character graphics, background graphics, and background video is performed on a video board.   The speech and emotion recognition subsystem consists of one workstation, and this machine performs speech recognition by HMM and emotion recognition by a neural net.   The motion recognition subsystem also consists of one workstation to perform recognition of simple body movements in real time.   The sound output subsystem consists of multiple personal computers to perform simultaneous output of background music, audio effects, and character dialog.

Research work for such interactive movies is divided among engineers, who develop the system, and artists, who create contents.   However, the performance of the system as a whole including contents can begin to be evaluated by having users engage in actual interaction, and by performing ongoing evaluation, results can be fed back to contents creators and engineers to continuously enhance the entire system. Figure 6 illustrates an example of the interaction.

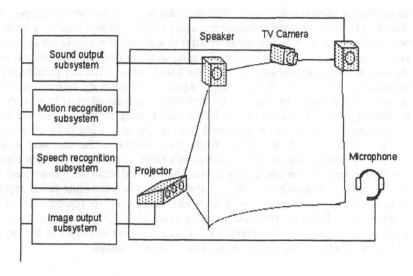

Fig. 5 Hardware configuration of the interactive movie system

4.3 Communication by Human Metamorphosis

In conventional video communications, much effort has been spent in developing technology that can transmit the image of a remotely located person as faithfully as possible.   One factor, however, in why video telephones have not penetrated the market is that such a device is inconvenient at times when one does not want to show one's image to the other party.   In short, instead of showing how one actually looks, there are times when it might be better to send a transformed image of oneself

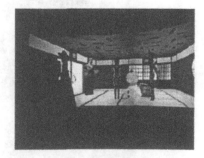

Fig. 6 Example of an interaction in the interactive movie system

depending on current conditions.   With this idea in mind, we have developed Virtual Kabuki Theater," a human metamorphosis system[4].   In this system, a model of a person's face and body is first transformed into a model of a kabuki actor.   Then, by recognizing facial expressions and body movements in real time, the current state of the person is reflected onto the model of the kabuki actor.   The key point here from a technical point of view is that facial expressions and body movement must be detected in real time.   In addition, detected facial expressions must be reproduced as a different facial expression on the kabuki actor using metamorphosis technology. For this part of the system, an artist will employ artistic techniques to create a model of the kabuki actor's face.   Thus, while the objective in the past was to simply achieve an accurate reproduction of a person's facial expressions, the application of artistic elements in the above way means that anybody can be transformed into a kabuki actor.   Although a number of studies are now examining facial-expression recognition and reproduction, all of these emphasize accurate reproduction of facial expressions.   There is apparently no other research like that described here that is working on metamorphosis technology.   Figure 7 shows examples of the generated Kabuki actor's facial expressions along with a full body image.

This project, in contrast to faithful transmission of a person's image over a future communications network, is based on an idea conceived by an engineer that it would be better to send out a transformed image according to the current situation.   While this project therefore takes on an engineer-led format, the involvement of artists in the development of the kabuki actor's model has made a major contribution to the appearance of the overall system.   At the same time, as this system might also be applied to games, Karaoke, and other uses outside of communications, this project has demonstrated how cooperation between engineers and artists can be truly effective.

Fig. 7 Metamorphosis into a Kabuki actor

5 Conclusion

This paper has described methods for creating contents in interactive media, a future form of multimedia. It was shown that interactive media exhibits both an overlapping structure consisting of component technology, integration technology, and contents creation, and complicated characteristics by which contents undergo change due to user interaction. Moreover, as a technique for researching and developing such incomplete media, we have proposed that engineers and contents creators cooperate in the overall development of media. Several examples of research projects at the ATR Media Integration & Communications Research Laboratories based on this idea were presented. The author and his colleagues hope that the information provided here becomes a useful reference for researchers and contents creators working on multimedia technology.

References

[1] Ryohei Nakatsu, "Virtual Reality Technology Adopting an Artistic Approach," Transactions of the Virtual Reality Society of Japan, Vol. 1, No. 1, pp 1-9 (1996. 9).
[2] Christa Sommerer and Laurent Mignonneau, " Intro Act & MIC Exploration Space," Visual Proceedings of SIGGRAPH96, pp.17 (1996.8).
[3] Ryohei Nakatsu and Naoko Tosa, "Toward the Realization of Interactive Movies - Inter Communication Theater: Concept and System," Proceedings of IEEE International Conference on Multimedia Computing and Systems, pp. 71-77 (1997.6).
[4] Jun Ohya, et al., "Virtual Kabuki Theater: Towards the Realization of Human Metamorphosis Systems," Proceedings of RO-MAN '96, pp.416-421 (1996.11).

Visual Modeling for Multimedia Content

Demetri Terzopoulos

Department of Computer Science, University of Toronto
10 King's College Road, Toronto, ON M5S 3G4, Canada
dt@cs.toronto.edu

Abstract. This paper reviews research that addresses the challenging problem of modeling living systems for multimedia content creation. First, I discuss the modeling of animals in their natural habitats for use in animated virtual worlds. The basic approach is to implement realistic artificial animals (in particular, fish) and to give them the ability to locomote, perceive, and in some sense understand the realistic virtual worlds in which they are situated so that they may achieve both individual and social functionality within these worlds. Second, I discuss the modeling of human faces. The goal is to develop facial models that are capable of synthesizing realistic expressions. At different levels of abstraction, these hierarchical models capture knowledge from psychology, facial anatomy and tissue histology, and continuum biomechanics. The facial models can be "personalized", or made to conform closely to individuals, once facial geometry and photometry information has been captured by a range sensor.

1 Introduction

Multimedia content creation for animation and virtual reality has made significant progress in the last decade. We are presently witnessing the transition from an earlier generation of purely geometric models of inanimate objects, to the creation of virtual worlds populated by much more sophisticated models—models of objects that are *alive*. The modeling and simulation of living systems for multimedia content creation resonates with the field of scientific inquiry called *Artificial Life*, conceptually a discipline that transcends the traditional boundaries of computer science and biological science.[1] This paper reviews two research projects which relate to the visual modeling of living systems for multimedia content creation.

The first project involves the modeling of complete animals of nontrivial complexity; in particular, teleost fishes in their natural habitats (Fig. 1). The basic approach is to implement realistic *artificial animals* and to give them the ability to locomote, perceive, and in some sense understand the realistic virtual worlds in which they are situated so that they may achieve both individual and

---

[1] For an engaging introduction to the field, see, e.g., S. Levy, *Artificial Life* (Pantheon, 1992). Journals such as *Artificial Life* and *Adaptive Behavior* document the state of the art.

S. Nishio, F. Kishino (Eds.): AMCP'98, LNCS 1554, pp. 406–421, 1999.

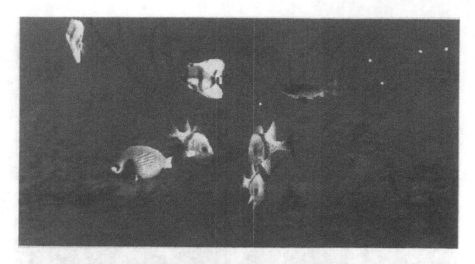

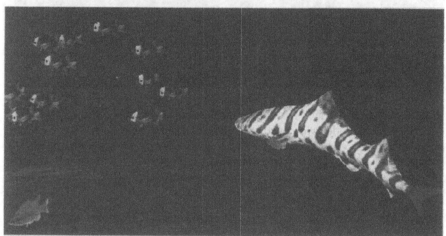

Fig. 1. Artificial fishes in their physics-based virtual world as it appears to an underwater observer (monochrome version of original color images). Top: The 3 reddish fish (center) are engaged in a mating ritual, the greenish fish (upper right) is a predator hunting for small prey, the remaining 3 fishes are feeding on plankton (white dots). Dynamic seaweeds grow from the ocean bed and sway in the current. Bottom: A predator shark stalking a school of prey.

social functionality within these worlds. To this end, each artificial animal is an autonomous agent possessing a muscle-actuated body that can locomote and a brain with motor, perception, behavior, and learning centers.

The second project, involves the modeling of faces, a vitally important communicative medium of the most highly evolved living systems known—human

Fig. 2. An artificial face (monochrome version of original color images). The functional face model was constructed automatically from an RGB/range laser scan of an individual, "George". Artificial George is shown here engaged in synthesizing various facial expressions and pretending to read a technical paper about how he was constructed.

beings (Fig. 2). Our goal has been to develop *artificial faces* that are capable of synthesizing realistic facial expressions. At different levels of abstraction, these hierarchical models capture knowledge about facial expression from psychology, facial anatomy and tissue histology, and continuum biomechanics. I will show that a generic facial model of this sort can be "personalized", or made to conform closely to individuals once the geometry and photometry of their faces has been captured by a range sensor. I will also describe how muscle-actuated facial models can be used for model-based facial image analysis/synthesis.

2   Artificial Fishes

Imagine a virtual marine world inhabited by a variety of realistic fishes (Fig. 1). In the presence of underwater currents, the fishes employ their muscles and fins to swim gracefully around immobile obstacles and among moving aquatic plants and other fishes. They autonomously explore their dynamic world in search of food. Large, hungry predator fishes stalk smaller prey fishes in the deceptively

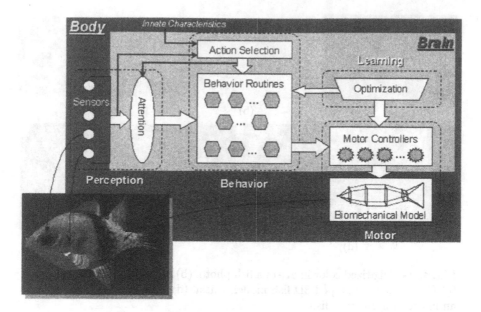

Fig. 3. Control and information flow in artificial fish.

peaceful habitat. The sight of predators compels prey fishes to take evasive action. When a dangerous predator appears in the distance, similar species of prey form schools to improve their chances of survival. As the predator nears a school, the fishes scatter in terror. A chase ensues in which the predator selects victims and consumes them until satiated. Some species of fishes seem untroubled by predators. They find comfortable niches and feed on floating plankton when they get hungry. Driven by healthy libidos, they perform enticing courtship dances to attract mates.

Each artificial fish is an autonomous agent with a deformable body actuated by internal muscles. The body also harbors eyes and a brain with motor, perception, behavior, and learning centers, as Fig. 3 illustrates. Through controlled muscle actions, artificial fishes are able to swim through simulated water in accordance with hydrodynamics. Their functional fins enable them to locomote, maintain balance, and maneuver in the water. Thus the artificial fish model captures not just 3D shape and appearance in the form of a display model, but it also captures the basic physics of the animal (biomechanics) and its environment (hydrodynamics). Though rudimentary compared to real animals, the brains of artificial fishes are nonetheless able to learn some basic motor functions and carry out perceptually guided motor tasks. In accordance with their perceptual awareness of the virtual world, their minds arbitrate a repertoire of piscine behaviors, including collision avoidance, foraging, preying, schooling, and mating.

The details of the artificial fish model are presented in [1, 2]. I will summarize its main features in the remainder of this section.

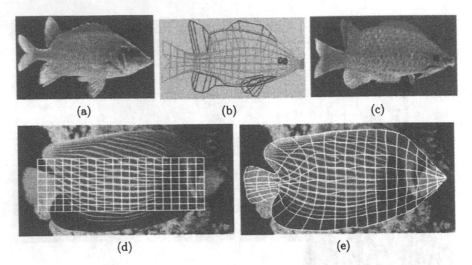

Fig. 4. (a) Digitized color image of a fish photo. (b) 3D NURBS surface fish body. (c) Color texture mapped 3D fish model. Initial (d) and final (e) snake-grid on an image of a different fish.

2.1 Modeling Form and Appearance

Artificial fish display models, such as the one illustrated at the lower right of Fig. 3, should capture the form and appearance of real fishes with reasonable visual fidelity. To this end, photographs of real fishes, such as the one shown in Fig. 4(a), are converted into 3D spline (NURBS) surface body models (Fig. 4(b)). Using an interactive image based-modeling strategy, the digitized photographs are analyzed semi-automatically using a "snake-grid" tool which is demonstrated in Fig. 4(d–e) on a different fish image. The grid of snakes [3] floats freely over an image. The border snakes adhere to intensity edges demarcating the fish from the background, and the remaining snakes relax elastically to cover the imaged fish body. This yields a smooth, nonuniform coordinate system (Fig. 4(e)) for mapping the texture onto the spline surface to produce the final texture mapped fish body model (Fig. 4(c)).

2.2 Motor System

The motor system comprises the dynamic model of the fish including its muscle actuators and a set of motor controllers (MCs). Fig. 5(a) illustrates the biomechanical body model which produces realistic piscine locomotion using only 23 lumped masses and 91 elastic elements. These mechanical components are interconnected so as to maintain the structural integrity of the body as it flexes due to the action of its 12 contractile muscles.

Artificial fishes locomote like real fishes, by autonomously contracting their muscles. As the body flexes it displaces virtual fluid which induces local reaction

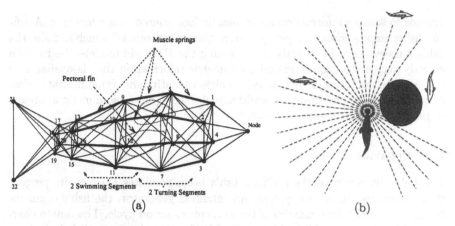

Fig. 5. Biomechanical fish model (a). Nodes denote lumped masses. Lines indicate uniaxial elastic elements (shown at natural length). Bold lines indicate muscle elements. Artificial fishes perceive objects (b) within a limited field view if objects are close enough and not occluded by other opaque objects (only the fish towards the left is visible to the one at the center).

forces normal to the body. These hydrodynamic forces generate thrust that propels the fish forward. The model mechanics are governed by Lagrange equations of motion driven by the hydrodynamic forces. The system of coupled second-order ordinary differential equations are continually integrated through time by a numerical simulator.

The model is sufficiently rich to enable the design of motor controllers by gleaning information from the fish biomechanics literature. The motor controllers coordinate muscle actions to carry out specific motor functions, such as swimming forward (swim-MC), turning left (left-turn-MC), and turning right (right-turn-MC). They translate natural control parameters such as the forward speed or angle of the turn into detailed muscle actions that execute the function. The artificial fish is neutrally buoyant in the virtual water and has a pair of pectoral fins that enable it to navigate freely in its 3D world by pitching, rolling, and yawing its body. Additional motor controllers coordinate the fin actions.

2.3 Perception System

Artificial fishes are aware of their world through sensory perception. The perception system relies on a set of on-board virtual sensors to gather sensory information about the dynamic environment. As Fig. 5(b) suggests, it is necessary to model not only the abilities but also the limitations of animal perception systems in order to achieve natural sensorimotor behaviors. The perception center of the brain includes a perceptual attention mechanism which allows the artificial fish to sense the world in a task-specific way, hence filtering out sensory information superfluous to its current behavioral needs. For example, the artificial fish attends to sensory information about nearby food sources when foraging. Artificial fishes employ *simulated* perception, a "perceptual oracle" which satisfies the fish's sensory needs by directly interrogating the 3D world model—the fish can directly access the geometric and photometric information that is available to the graphics rendering engine, as well as object identity and dynamic state information about the physics-based world model. (See [4] for a biomimetic approach to perception using computer vision.)

2.4   Behavior

The behavior center of the artificial fish's brain mediates between its perception system and its motor system. An intention generator, the fish's cognitive faculty, harnesses the dynamics of the perception-action cycle. The innate character of the fish is established by a set of habits that determine if it is male or female, predator or prey, etc. At each simulation time step, the intention generator takes into account the habits of the fish and the incoming stream of sensory information to generate dynamic goals for the fish, such as to avoid an obstacle, to hunt and feed on prey, or to court a potential mate. It ensures that goals have some persistence by exploiting a single-item memory. Persistence makes sustained behaviors such as foraging, schooling, and mating more robust. The intention generator also controls the perceptual attention mechanism. At every simulation time step, the intention generator activates behavior routines that attend to sensory information and compute the appropriate motor control parameters to carry the fish one step closer to fulfilling its current intention. The behavioral repertoire of the artificial fish includes primitive, reflexive behavior routines, such as obstacle avoidance, as well as more sophisticated motivational behavior routines such as schooling and mating whose activation depends on the dynamic mental state of the fish as represented by hunger, fear, and libido mental variables (see [1] for the details).

2.5   Learning

The learning center of its brain enables the artificial fish to learn how to locomote through practice and sensory reinforcement. Through optimization, the motor learning algorithms discover muscle controllers that produce efficient locomotion. Muscle contractions that produce forward movements are "remembered". These half-success then form the basis for the fish's subsequent improvement in its swimming technique. Their brain's learning center also enable these artificial animals to train themselves to accomplish higher level sensorimotor tasks, such as maneuvering to reach a visible target (see [1, 5] for the details).

2.6   Real-Time Virtual Marine World

We have recently developed a real-time, interactive version of our virtual marine world [6]. Our approach was to replace the computationally expensive biome-

Fig. 6. A user is enjoying the ride stereoscopically, wearing a pair of CrystalEyes stereo glasses.

chanical fish models with fast kinematic replicas that preserve as much as possible the lifelike appearances, locomotions, and behaviors of the fully dynamic originals. In particular, we capture motion data through the systematic biomechanical simulation of locomotion in the original models. We refer to this technique as *synthetic motion capture* since it is in principle not unlike natural motion capture applied to real animals, particularly human actors. We appropriately process the recorded data and compile the captured actions into *action repertoires*. The action repertoire implements motion synthesis in a kinematic creature, and it is rich enough to support natural looking locomotion and complex behavior.

Our virtual marine world runs at interactive rates on a deskside graphics workstation (Silicon Graphics, Inc., 1×194 MHz R10000 InfiniteReality system). The user pilots a submarine, navigating in a 3D virtual world populated by lifelike marine animals. The user may maneuver the submarine into a large school of fishes, chase a fleeing fish, or simply look around and observe colorful marine life unfold. Our interactive virtual marine world is inhabited by 60 artificial fishes of 7 different species. Fig. 6 shows a user enjoying the ride, wearing a pair of CrystalEyes stereo glasses to obtain a compelling depth perception and a quasi-immersive 3D experience.

Fig. 7. The virtual undersea world experienced on the panoramic display in a Trimension Reality Theater.

We have furthermore developed a large scale version of our virtual undersea world in a "Relocatable Reality Theater" marketed by Trimension, Inc., which combines a Silicon Graphics $8 \times R10000$ CPU Onyx2 system and multichannel PRODAS projection technology from SEOS Displays, Ltd. The system features three InfiniteReality graphics pipelines, each feeding video to an overhead projector. This system animates and renders our virtual world at a sustainable rate of at least 30 frames per second. It renders through the three projectors a seamless image of approximately 4000×1000 pixel resolution across a 18×8-foot curved screen, producing a large panoramic display that fills the peripheral view. Fig. 7 shows the theater.

3 Artificial Faces

Next, I will discuss the modeling of humans, focusing on the important and challenging problem of modeling faces. The human face has attracted much attention in several disciplines, including psychology, computer vision, and computer graphics. Psychophysical investigations clearly indicate that faces are very special visual stimulii. Psychologists have studied various aspects of human face perception and recognition. They have also examined facial expression—the result of a confluence of voluntary muscle articulations which deform the neutral face into an expressive face. The facial pose space is immense. The face is capable of generating on the order of 55,000 distinguishable facial expressions with about 30 semantic distinctions. Studies have identified six primary expressions that communicate anger, disgust, fear, happiness, sadness, and surprise in all cultures.

Ekman and Friesen's "Facial Action Coding System" (FACS) provides a quantification of facial expressions [9]. The FACS quantifies facial expressions in terms of 44 "action units" (AU) involving one or more muscles and associated activation levels.

3.1 A Functional Facial Model

We have developed a hierarchical model of the face which provides natural control parameters and is efficient enough to run at interactive rates [10]. Conceptually, the model decomposes into six levels of abstraction. These representational levels encode specialized knowledge about the psychology of human facial expressions, the anatomy of facial muscle structures, the histology and biomechanics of facial tissues, and facial skeleton geometry and kinematics:

1. *Expression.* At the highest level of abstraction, the face model executes expression (or phoneme) commands. For instance, it can synthesize any of the six primary expressions within a specific time interval and with a specified degree of emphasis.
2. *Control.* A muscle control process, a subset of Ekman and Friesen's FACS, translates expression (or phoneme) instructions into a coordinated activation of actuator groups in the facial model.
3. *Muscles.* As in real faces, muscles comprise the basic actuation mechanism of the model. Each muscle model consists of a bundle of muscle fibers. When fibers contract, they displace their points of attachment in the facial tissue or the jaw.
4. *Physics.* The face model incorporates a physical approximation to human facial tissue. The tissue model is a lattice of point masses connected by non-linear elastic springs. Large-scale synthetic tissue deformations, subject to volume constraints, are simulated numerically by continuously propagating through the tissue lattice the stresses induced by activated muscle fibers.
5. *Geometry.* The geometric representation of the facial model is a non-uniform mesh of polyhedral elements whose sizes depend on the curvature of the neutral face. Muscle-induced synthetic tissue deformations distort the neutral geometry into an expressive geometry.
6. *Images.* After each simulation time step, standard visualization algorithms implemented in dedicated graphics hardware render the deformed facial geometry in accordance with viewpoint, light source, and skin reflectance information to produce the lowest level representation in the modeling hierarchy, a continuous stream of facial images.

The hierarchical structure of the model encapsulates most of the complexities of the underlying representations, relegating the details of their computation to automatic procedures. At the higher levels of abstraction, our face model offers a semantically rich set of control parameters which reflect the natural constraints of real faces.

Our synthetic facial tissue model is motivated by histology and tissue biomechanics. Human skin has a nonhomogeneous and nonisotropic layered structure consisting of the epidermis, dermis, subcutaneous fatty tissue, fascia, and muscle layers. The synthetic tissue is a deformable lattice, an assembly of discrete finite elements (see [10] for the details).

3.2   Personalizing the Functional Model

We have developed a highly automated approach to constructing realistic, functional models of human heads [11]. These physics-based models are anatomically accurate and may be made to conform closely to specific individuals. Currently, we begin by scanning a subject with a laser sensor which circles the head to acquire detailed range and reflectance information. Next, an automatic conformation algorithm adapts a triangulated face mesh of predetermined topological structure to these data. The generic mesh, which is reusable with different individuals, reduces the range data to an efficient, polygonal approximation of the facial geometry and supports a high-resolution texture mapping of the skin reflectivity.

The conformed polygonal mesh forms the epidermal layer of a physics-based model of facial tissue. An automatic algorithm constructs the multilayer synthetic skin and estimates an underlying skull substructure with a jointed jaw. Finally, the algorithm inserts synthetic muscles into the deepest layer of the facial tissue. These contractile actuators, which emulate the primary muscles of facial expression, generate forces that deform the synthetic tissue into meaningful expressions. To increase realism, we include constraints to emulate tissue incompressibility and to enable the tissue to slide over the skull as real skin does.

Fig. 8 illustrates the aforementioned steps. The figure shows a 360° head-to-shoulder scan of a woman, "Heidi," acquired by a Cyberware Color 3D Digitizer. The data set consists of a radial range map (Fig. 8(a)) and a registered RGB photometric map (Fig. 8(b)). The range and RGB maps are high-resolution 512×256 arrays in cylindrical coordinates, where the x axis is the latitudinal angle around the head and the y axis is vertical distance. Fig. 8(c) shows the generic mesh projected into the 2D cylindrical domain and overlayed on the RGB map. The triangle edges in the mesh are elastic springs, and the mesh has been conformed automatically to the woman's face using both the range and RGB maps [11]. The nodes of the conformed mesh serve as sample points in the range map. Their cylindrical coordinates and the sampled range values are employed to compute 3D Euclidean space coordinates for the polygon vertices. In addition, the nodal coordinates serve as polygon vertex texture map coordinates into the RGB map. Fig. 8(d) shows the 3D facial mesh with the texture mapped photometric data.

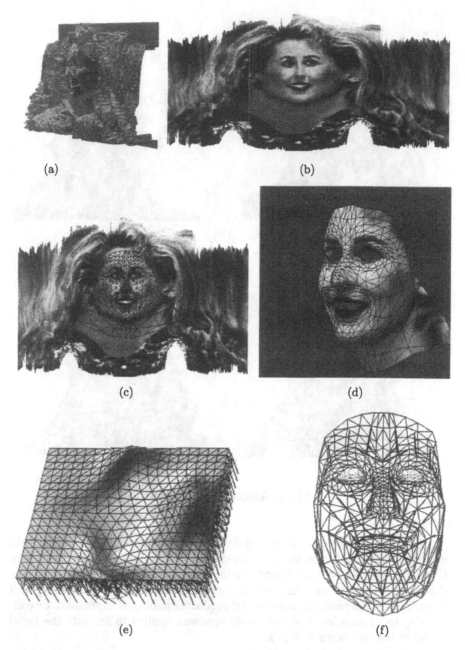

Fig. 8. Facial modeling using scanned data. (a) Radial range map. (b) RGB photometric map. (c) RGB map with conformed epidermal mesh overlayed. (d) 3D mesh and texture mapped triangles. (e) Physics-based skin model. (f) Muscles under facial mesh.

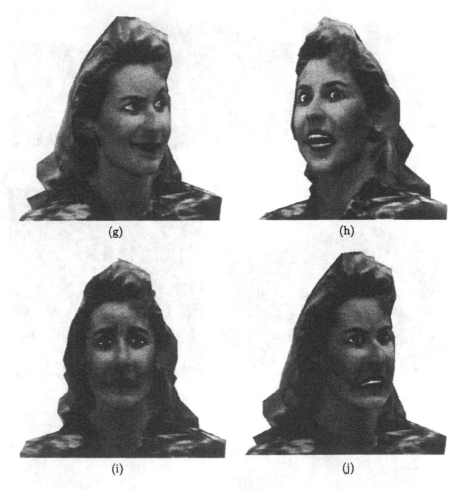

(g)   (h)

(i)   (j)

Fig. 9. Animated face model.

Once we have reduced the scanned data to the 3D epidermal mesh of Fig. 8(d), we can assemble a physics-based face model of Heidi, including the synthetic skin (Fig. 8(e) shows a skin patch undergoing deformation) and muscles (Fig. 8(e) shows the contractile muscles (vectors) underneath the epidermal mesh). Fig. 9 demonstrates the resulting facial model producing animated expressions by contracting facial muscles. The same technique was applied to animate the facial model of George shown in Fig. 2.

3.3   Model-Based Facial Image Analysis

Facial image analysis and synthesis is necessary for numerous applications. Among them is low bandwidth teleconferencing which may involve the real-time extraction of facial control parameters from live video at the transmission site and the reconstruction of a dynamic facsimile of the subject's face at a remote receiver. Teleconferencing and other applications require facial models that are computationally efficient and also realistic enough to accurately synthesize the various nuances of facial structure and motion. We have argued that the anatomy and physics of the human face, especially the arrangement and actions of the primary facial muscles, provide a good basis for facial image analysis and synthesis [12].

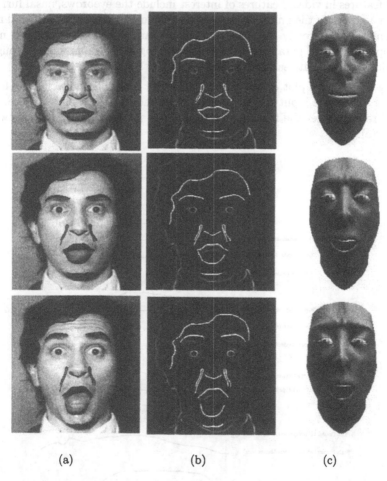

(a)   (b)   (c)

Fig. 10. Dynamic facial image analysis and expression resynthesis. Sample video frames with superimposed deformable contours tracking facial features; (a) intensity images with black snakes, (b) image potentials with white snakes. (c) Facial model resynthesizes surprise expression from estimated muscle contractions.

The physics-based anatomically motivated facial model has allowed us to develop a new approach to the analysis of dynamic facial images for the purposes of estimating and resynthesizing dynamic facial expressions [12]. Part of the difficulty of facial image analysis is that the face is highly deformable, particularly around the forehead, eyes, and mouth, and these deformations convey a great deal of meaningful information. Techniques for tracking the deformation of facial features include "snakes" [3]. Motivated by the anatomically consistent musculature in our model, we have considered the estimation of dynamic facial muscle contractions from video sequences of expressive faces. We have developed an analysis technique that uses snakes to track the nonrigid motions of facial features in video. Features of interest include the eyebrows, nasal furrows, mouth, and jaw in the image plane. We are able to estimate dynamic facial muscle contractions directly from the snake state variables. These estimates make appropriate control parameters for resynthesizing facial expressions through a generic face model at real-time rates.

Fig. 11 shows a plot of the estimated muscle contractions versus the frame number. They are input to the physics-based model as a time sequence. The model resynthesizes the facial expression. Three rendered images are shown in Fig. 10(c).

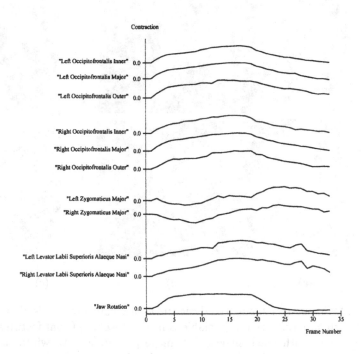

Fig. 11. Estimated facial muscle contractions plotted as time series.

Acknowledgements

I thank my students whose research work is highlighted in Sections 2 and 3; in particular, Xiaoyuan Tu developed the artificial fish model, and Qinxin Yu developed the real-time virtual marine world simulation, and Yuencheng Lee developed the realistic facial model. My thanks also go to Keith Waters for his collaboration on the problem facial modeling. Funding was provided in part by the Natural Sciences and Engineering Research Council of Canada. The author is a Killam Research Fellow of the Canada Council for the Arts.

References

[1] D. Terzopoulos, X. Tu, and R. Grzeszczuk. Artificial fishes: Autonomous locomotion, perception, behavior, and learning in a simulated physical world. *Artificial Life*, 1(4):327–351, 1994.

[2] X. Tu and D. Terzopoulos. Artificial fishes: Physics, locomotion, perception, behavior. In *Computer Graphics* Proceedings, Annual Conference Series, Proc. SIGGRAPH '94 (Orlando, FL), pages 43–50. ACM SIGGRAPH, July 1994.

[3] M. Kass, A. Witkin, and D. Terzopoulos. Snakes: Active contour models. *International Journal of Computer Vision*, 1(4):321–331, 1988.

[4] D. Terzopoulos and T.F. Rabie. Animat vision: Active vision in artificial animals. *Videre: Journal of Computer Vision Research*, 1(1):2–19, 1997.

[5] R. Grzeszczuk and D. Terzopoulos. Automated learning of muscle actuated locomotion through control abstraction. In *Computer Graphics* Proceedings, Annual Conference Series, Proc. SIGGRAPH '95 (Los Angeles, CA). ACM SIGGRAPH, August 1995.

[6] Q. Yu and D. Terzopoulos. Synthetic motion capture for interactive virtual worlds. In *Proc. IEEE Computer Animation 98 Conference*, pages 2–10, Philadelphia, PA, June 1998.

[7] G. M. Davies, H. D. Ellis, and G. M. Shepherd. *Perceiving and Remembering Faces*. Academic Press, New York, 1981.

[8] V. Bruce. *Recognizing Faces*. Lawrence Erlbaum, Hillsdale, 1988.

[9] P. Ekman and W. V. Friesen. *Manual for the Facial Action Coding System*. Consulting Psychologist Press, Palo Alto, CA, 1977.

[10] D. Terzopoulos and K. Waters. Physically-based facial modelling, analysis, and animation. *The Journal of Visualization and Computer Animation*, 1(2):73–80, 1990.

[11] Y. Lee, D. Terzopoulos, and K. Waters. Realistic facial modeling for animation. In *Computer Graphics* Proceedings, Annual Conference Series, Proc. SIGGRAPH '95 (Los Angeles, CA), pages 55–62. ACM SIGGRAPH, August 1995.

[12] D. Terzopoulos and K. Waters. Analysis and synthesis of facial image sequences using physical and anatomical models. *IEEE Transactions on Pattern Analysis and Machine Intelligence*, 15(6):569–579, 1993.

Automatic Generation of Moving Crowds
in the Virtual Environment

Naoki Saiwaki, Toshiaki Komatsu, and Shogo Nishida

Graduate School of Engineering Science, Osaka University
1-3, Machikaneyama, Toyonaka, Osaka 560 JAPAN
E-mail : {Saiwaki, komatsu, nishida}@nishilab.sys.es.osaka-u.ac.jp

Abstract. In this paper, we are concerned with modeling and generation of
the moving crowd in a three dimensional virtual city to increase reality. In the
virtual city, the habitants are usually controlled by some strict rules. To
actualize more natural atmosphere, many data and rules are needed.

In such case, we can easily control the shape of the crowd by using the
characteristics of chaos. Although random numbers are often applied to such
purposes, it is difficult to create and control such patterns. We constructed the
models of the crowd in the street and the folk dance. The prototype system
was developed for evaluation and it was confirmed that users felt easy to
generate and control the crowd.

1 Introduction

Recent development of VR (Virtual Reality) technology enables the real-time
interaction in the 3D virtual space on computers, which can be used as the
quasi-experience of the real world.

The construction of virtual cities is one of the typical examples of the
utilization of VR technology. However, in previous research the number of people
who move around in the VR space tends to be relatively small. Moreover, the
behavior of people is often directly controlled by the user or is programmed
beforehand only to produce typical responses. Thus, there are few approaches to
reproduce the realistic motion of crowds in cities[1]. One of the reasons for making
the behavior of people in the VR space unrealistic is that the load required for the
realistic motion is too huge for the designer and/or user of the system[2][3]. For
instance, to enable the "natural" appearance of the direction and speed of the motion
of crowds, it is necessary to program the behavior of each person appropriately as
well as to provide enormous data.

We focus on the motion of crowds which is observed in cities at the macro
level, not on that of each person at the micro level as in the other approaches. Our
approach aim to construct the behavior model of crowds which appears to have a
pattern as a whole while each person moves around independently. We propose
several basic methods for the construction of systems which easily produce the
characteristic behavior of crowds in the VR space. In our approach the behavior of
crowds is represented at the macro level, and the chaos equation is used to describe

S. Nishio, F. Kishino (Eds.): AMCP'98, LNCS 1554, pp. 422-432, 1999.

both the behavior of crowds as a whole and the relationship between persons. Thus, it is possible to represent the global behavior of crowd without specifying the behavior of each person nor the relationships between them. It also enables to create various behavioral patterns of crowd through simple parameter tuning. Various experiments with the computer simulation were carried out to examine the effectiveness of our method[4][5]. Our method is expected to contribute to reducing the load in the creation of CG (Computer Graphics) when it is used as the module to produce the CG data.

2 Modeling of Behavior with Chaos

2.1 Behavior of Crowds

The issue of describing the behavior of flocks as a whole has been pursued since the beginning of the research in ALife (Artificial Life). Several methods have been developed to represent the behavior of flocks with as little data as possible by furnishing each individual with autonomy[6][7]. For instance, there are three basic rules for each member of a flock to create the behavioral pattern for the flock of birds and/or fish, as follows:

1)   maintain a minimum distance from other ones in the environment, including other members
2)   match velocities with others in its neighborhood
3)   move toward the perceived center of mass of others in its neighborhood

The flock with these rules moves around by itself simply being provided with the initial configuration of each member and the initial speed. With the right setting for the parameters, a collection of birds will collect into a dynamic flock, which flies around obstacles in a fluid and natural manner, occasionally breaking up into sub-flocks to avoid obstacles. However, there are few researches which try to control the general behavior of crowds in human cases with small number of parameters, in contrast to the ones which deal with the collective behavior of animals.

There are several reasons for the lack of research on human cases. First, it is difficult to find out remarkable behavioral patterns, in contrast to the case in animals. Second, even if such patters are found, it is necessary to provide complex descriptions to produce the realistic behavior. In some cases it is necessary to provide so many data, the amount of which is almost the same to that of data required to specify the individual behavior in crowds. These reasons suggest the importance of the selection of behavioral patterns and their representation to deal with the crowd in human cases.

This paper considers the case in which people flows realistically along the street and the one in which people dances as a group by combining the circular motion which can be observed in folk dances. We propose to utilize the characteristics of chaos to represent the above cases. For instance, it is possible to exploit the fact that small disturbance in the initial condition greatly affects the subsequent trajectory in chaos. It is also possible to utilize the fact that chaos constitutes the pattern such as a strange attractor at the macro level even if it appears to move randomly at first. These features enable to represent the behavior of crowds with small number of description models, their initial conditions, and parameters.

2.2 Application of Chaos

There are three characteristics in chaos, each of which can be utilized to model the behavior of crowds. First, it has non-regularity, which can be used to represent the fuzziness in the behavior of people. Second, it is sensitive to the initial condition, which enables the completely different behavior as time goes by when the initial condition for each person differs. Third, the parameter value in chaos equations greatly affects the possible range and characteristic of the trajectory, which is useful to change the feature of the behavior of crowds as a whole by simple tuning of parameters.

Suppose that we try to model the case in which people walk along a straight street. It is possible to represent the random walk with probability (e.g., random number), however, it is impossible to control the behavior of crowds as a whole. In contrast, with chaos it is possible for the crowd to divide spontaneously into the sub-crowd which gets away swiftly and the one which wonders about. It is also possible to control the circular motion and the direction of dance when chaos is used to represent a dance. Furthermore, it is possible to locally modify the behavior of crowds as a whole by supplementing the above ones with electric charges. The details of these methods are explained in the following sections.

3 Modeling and Creation of Crowds with Linear Motion

3.1 Characteristics in Linear Motion

People primary notice the crowd which moves toward the place with lots of coming and going, even at the location where many people with various objectives move around freely. In this case it is almost impossible to walk straightly toward the destination due to the random influence from the surrounding people, and they tend to walk in the wandering manner to avoid the touching and colliding with others. As described in the previous section, although it is difficult to describe the wandering behavior of each person directly, it is possible to represent it with chaos easily.

3.2 Chaos and Wandering

Chaos can be considered as the non-regular vibration which is controlled by relatively simple rules[8]. The typical rule for chaos is the logistic function:

$$X[n+1] = R\ X[n](1 - X[n]); \quad R_\infty < R \le 4 \tag{1}$$

It is known that this function creates the chaotic progression with the initial value $(0 < X[0] < 1)$ as long as the coefficient R is. The value of R also affects the trajectory of progression (see Fig. 1 and 2 with $X[0] = 0.3$).

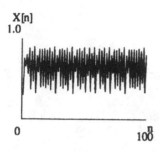

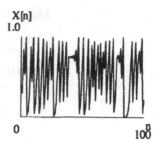

Fig. 1. $X[n]$ (R=3.6)   **Fig. 2.** $X[n]$ (R=4.0)

Each member of a crowd is provided with the motion vector at each time to move the crowd as a whole. Utilizing the non-regular behavior in chaos to calculate the motion vector realizes the behavior of crowds, in which each member appears to wander at random but the global pattern is still observed.

3.3 Calculation of Motion Vector

The motion vector is expressed as in Fig. 3 and is calculated with the following expressions:

$$\vec{v} = r\vec{b} \begin{pmatrix} \cos\theta & -\sin\theta \\ \sin\theta & \cos\theta \end{pmatrix} \qquad (2)$$

$$\begin{cases} r = 1.0 + X[n] - Avx \\ \theta = Y[n] - Avy \end{cases} \qquad (3)$$

$$\begin{cases} X[n+1] = Rx\, X[n](1-X[n]) \\ Y[n+1] = Ry\, Y[n](1-Y[n]) \\ R_\infty < Rx \le 4.0, R_\infty < Ry \le 4.0 \end{cases} \qquad (4)$$

In the above expressions b is the basic motion vector with magnitude 1 and is given as the initial condition. r and theta are the parameters to represent the wandering behavior and are determined as in expression (3) with the logistic function $X[n]$ and

Y[n]. Note that Avx and Avy are the constants which are determined by Rx and Ry. The initial value X[0] and Y[0] in the logistic functions are given at random. Our modeling make it possible to represent the behavior of crowds with two parameters Rx and Ry.

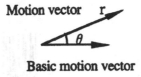

Basic motion vector

Fig. 3. Motion Vector

3.4 Control of Behavior through Parameter

Fig. 4 and 5 show the trajectory of people in our model with the motion vector in expression (2). Fig. 4 shows the case in which only Ry is changed to model the behavior of single person. It can be observed that the amount of wandering grows as Ry gets larger.

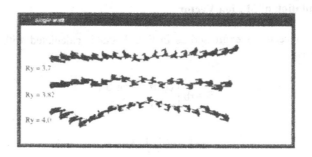

Fig. 4. Locus of wandering
(Ry = 3.7, 3.82, 4.0)

Fig. 5 shows the example behavior of crowds with various settings for Rx and Ry by extending the above model to the multiple person case. In Fig. 5(a) the crowd flows smoothly along the street.

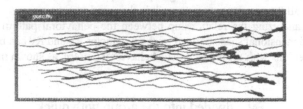

(a) behavior of crowds (Rx=3.7, Ry=3.7)

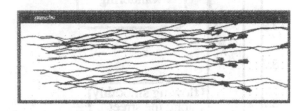

(b) behavior of crowds (Rx=3.6, Ry=3.7)

(c) behavior of crowds (Rx=3.7, Ry=3.6)

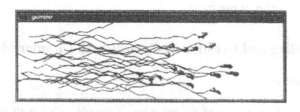

(d) behavior of crowds (Rx=4.0, Ry=4.0)

Fig. 5. Examples of behavior of crowds

In Fig. 5(b) the crowd is getting to be divided laterally into two groups. In Fig. 5(c) the crowd is divided into the group with quick steps and that with slow ones. In Fig. 5(d) all the members seem to walk in the wandering manner.

By analyzing the relationship between the behavioral pattern and parameter, the two dimensional space defined by the parameter Rx and Ry can be classified into several regions with the characteristic behavioral pattern, as shown in Fig. 6.

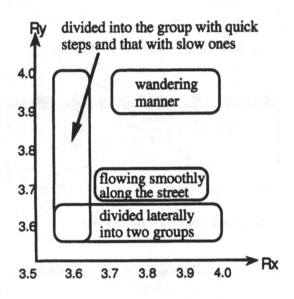

Fig. 6. Relation between the parameters and the characteristic behavioral patterns

The simulation has confirmed that one of the advantages of our modeling method with chaos is to allow the handy creation of complex behavior of crowds by tuning only small number of parameters.

4 Modeling and Creation of Crowds with Circular Motion

4. 1 Attractor

We discuss the modeling of the behavior of crowds with the attractor in chaos by focusing on the circular motion, which is often observed in folk dances. Some trajectories with the chaos equation keep the complex and non-regular behavior confined within some region, and these are called attractors.

$$\begin{cases} \dfrac{dx}{dt} = (K_0(y-x) - g(x))/K_1 \\[2mm] \dfrac{dy}{dt} = K_0(x-y) + z \\[2mm] \dfrac{dz}{dt} = -K_2 y \end{cases} \tag{6}$$

$$g(x) = K_3 x + \frac{1}{2}(K_4 - K_3)|x + K_5|$$
$$+ \frac{1}{2}(K_3 - K_4)|x - K_5|$$
$$(K_0 = 0.68, K_1 = 9.0, K_2 = 7.0, K_3 = -0.5, K_4 = -0.8, K_5 = 1.0)$$

The equations (6) are example equations for creating the attractor, whose trajectory is shown in Fig. 7. The 4th order Runge-Kutta method is used to solve the first order differential equations with high accuracy.

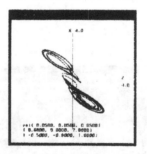

Fig. 7. Example of Attractor

4.2 Representation of Circular Motion

It is possible to reproduce the circular motion along the trajectory of attractor by disposing people on that trajectory. This modeling also scales up to the multiple person case as in the modeling of linear motion in section 3. The small difference at the initial condition amounts to produce a different trajectory and motion vector for each person, which can be used to represent a subtle difference in each behavior while showing the similarity as a whole. It is also possible to modify the shape of attractor by tuning parameters. For instance, it is possible to control the magnitude of the radius at will and to make the trajectory similar to the shape of number 8. Fig. 8 shows the trace of the trajectories by modifying the parameter K0 to 0.68, 0.67, and 0.65. It is possible to observe from these figures how trajectories vary with different parameter values.

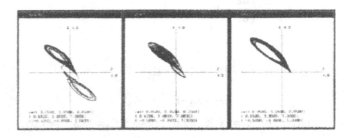

Fig. 8. Trace of the trajectories

Suppose the case in which the pairs of two persons perform the folk dance or social dance. The motion can be modeled and represented as the combination of one large attractor trajectory and multiple small attractor ones. The former determines the global behavior and the latter determines the motion of the pairs, as shown in Fig.9.

Fig. 9. Folk dance

5 Creation of Local Events

Besides the global pattern of motion described in the previous sections, the change of flow can occur as local events in the real behavior of crowds. For instance, some portion of a crowd can be attracted by a street performance. After a while those who have seen it go away from the performance bit by bit and new audience come to see the performance. When the performance ends, such a local accumulation of crowds disappears and the global behavior comes to dominate the motion. Not only this kind of behavior is often observed in reality, but also it can be considered as a crucial element to represent the realistic appearance of crowds. We extend the model of linear motion in section 3 by utilizing the electric attractive and repulsive forces. Each member of crowds receives the former one by attractive events and the latter by obstacles.

Suppose there are persons who play a street performance. Negative electric charges are given to them periodically while small positive ones are given to the surrounding people for a short period of time. This modeling enables to represent the people around the performance. It is also possible to represent the passengers who avoid the crowd with the repulsive force by the positive charges on people, as shown in Fig. 10. This kind of behavior is realized by calculating the electric field and adding up the force from it into the motion vector explained in section 3.

Fig. 10. Street performer and spectators

6 Conclusion

We have shown that the characteristics of chaos can be utilized to reproduce the behavior of crowds and have presented the creation the behavior of crowds as computer animation in the case of linear and circular motion. We also have shown that the local accumulation of crowds and the avoidance of collision can be realized by adding electric charges. The user who used our prototype system commented that the system presented the interesting and realistic behavior of crowds. Especially, it was highly evaluated on the aspect of controlling the global behavior with a few parameters and that of the accumulation of people with electric charges.

We have used only small portion of the characteristics of chaos to model the behavior of crowds. Thus, it is possible to extend our method by utilizing other ones. It is necessary to investigate which characteristics of chaos are easy and/or hard to use when they are introduced into our system. Furthermore, at present the evaluation of the system relies on the subjective judgment by the user. We plan to define the evaluation criteria based on the analysis of real behavior of crowds and its quantitative evaluation so that the system can produce more realistic behavior in terms of the criteria. We also plan to extend the method with electric charges so that it can be applied to avoid the collision with obstacles such as buildings and cars. We hope that the behavior of crowds in our system can be made more realistic by combining the approach in Artificial Life[9].

References

1. Unuma, T., Yasuo, K., Takeuchi R. : "Modeling of moving crowd" T.IEE Japan, Vol. 115-C, No2, pp.212-pp.221 (1995)
2. Lasseter, J.: "Principles of traditional animation applied to 3D computer animation", Proc.SIGGRAPH'87, p.35 (1987)
3. Glenn, S.: "Vector Animation Creation System", SIGGRAPH'93 Visual Proceedings, p.223 (1993)
4. Komatsu, T., Saiwaki, N., Nishida, S.: "Automatic Generation of Moving Crowd using Chaos Model", HIP96-44, pp.31-36 (1997)
5. Komatsu, T., Saiwaki, N., Nishida, S. : "Automatic Generation of Moving Crowd using Electric Charge Model", Proceedings of the 13th Symposium on Human Interface, pp.475-480 (1997)
6. Uchiki, T., Maruichi, T., Tokoro, M. : "A method of making animation with active characters", Proc. of the paper contest of the 2nd NICOGRAPH , p. 197 (1986)
7. Langton, C. G. : "Artificial Life", (1987) Addison-Wesley
8. Aihara, K. : "Chaos", (1986) Science Co. Ltd
9. Langton, C. G. (Eds.) : "Artificial Life II", (1992) Addison-Wesley

also being discussed for MPEG4, the next image compression standard. Once the standard is fixed, an application will be available to allow the user to easily manipulate these parameters. However, it has been a difficult problem to extract these parameters from the human face without placing markers on the face or using a special device attached on the face. In this paper, we discuss a method for extracting facial motion parameters using only image processing techniques.

In computer vision research, several techniques have been proposed to extract facial motion for facial expression recognition [3,4,5,6,7]. Terzopoulos and Waters were the first to demonstrate the tracking of facial features such as the eyes, eyebrows and mouth, as well as to estimate facial muscle actions from tracked data [3,4]. However, their tracking method is based on snakes, so real-time performance is not feasible. Mase has used a method in which optical flow is averaged over the window for the maximum motion of AU to estimate the magnitude of a limited number of AUs. Essa and Pentland have extended Mase's method using a Newtonian equation to model the motion of each point of the face and solving that equation by the finite element method. This method is accurate in estimating the magnitude of AUs, but the computational cost is so high that it cannot be executed in real time. DeCarlo and Metaxas have applied a deformable template [8] to estimate the shape of facial features, such as the eyes, nose, and mouth. Even though these models are simpler than a 3-D model, the processing speed is not fast enough (10 seconds for each frame on an SGI Indigo 2).

In this paper, we propose an algorithm that has a lower computational cost than the above optical flow and the 3-D face model methods. This will achieve a real-time execution. The algorithm consists of three steps. First, feature points of the face that are automatically selected in the first frame are tracked in the successive frames. Then, the set of feature points is triangulated so that the motion of each point relative to the surrounding points can be easily computed. Finally, the motion of each muscle is estimated based on the motion of feature points in the region around the muscle.

The proposed method can be applied to multimedia content production such as facial animation. FaceWorks [9], developed in digital's Cambridge Research Laboratory, is a program that produces facial animation in which a characters' mouth motions are synchronized with their speech. Although facial expressions can be added manually, the categories of expressions and facial motions for each category are fixed. Our proposed method can be used to control facial motions more concisely. In addition, the motion patterns extracted from actors and actresses can be used on the characters' face.

This paper is structured as follows. In Section 2, the proposed algorithm is explained. In Section 3, the experimental results are described. Finally, the paper is concluded in section 4.

2   Algorithm

The flowchart of the proposed algorithm is shown in Fig. 1. In the first step, feature points of the face are selected automatically and tracked for each frame.

Extracting Facial Motion Parameters by Tracking Feature Points

Takahiro Otsuka and Jun Ohya

ATR Media Integration & Communications Research Laboratories
Seika-cho, Soraku-gun, Kyoto, 619-0288, JAPAN
{otsuka, ohya}@mic.atr.co.jp

Abstract. A method for extracting facial motion parameters is proposed. The method consists of three steps. First, the feature points of the face, selected automatically in the first frame, are tracked in successive frames. Then, the feature points are connected with Delaunay triangulation so that the motion of each point relative to the surrounding points can be computed. Finally, muscle motions are estimated based on motions of the feature points placed near each muscle. The experiments showed that the proposed method can extract facial motion parameters accurately. In addition, the facial motion parameters are used to render a facial animation sequence.

1 Introduction

The human face displays various kinds of information in face-to-face communications, i.e., emotions via facial expressions and the category of speech sounds via the shape of the mouth. By adding the facial image to the telephone, the video phone has provided an environment for incorporating such information. However, the communication capability of the video phone is limited because the narrow bandwidth of the telephone line results in a low quality facial image. On the other hand, a method using Computer Graphics (CG) for rendering the face is not be affected by bandwidth if the number of parameters transmitted is low. A similar CG approach has been applied in a text-to-speech system that makes the mouth movement appear to be synchronous to the speech sound. Psychological experiments have demonstrated that human subjects have a higher speech recognition rate using the CG face image and speech than using speech alone [1]. Therefore, it seems natural that understanding between speakers can improve by using facial motions other than mouth shape, which only gives information on speech sound.

The number of parameters for facial motions can be determined from the Facial Action Coding System (FACS) [2] proposed by the psychologist Paul Ekman. According to his theory, any facial expression can be represented as a combination of atomic facial motions called Action Units (AUs). The whole set of AUs contains only 44 units, so the data load is very low even if the parameters are transmitted at video rate. Parameters for facial motions are

S. Nishio, F. Kishino (Eds.): AMCP'98, LNCS 1554, pp. 433–444, 1999.
© Springer-Verlag Berlin Heidelberg 1999

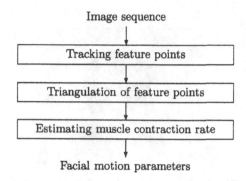

Fig. 1. Flowchart of proposed algorithm.

In addition, the head motion between successive frames is estimated. In the second step, the feature points are connected with Delaunay triangulation. This is intended to reduce the number of variables that represent facial motion over the entire face. In the final step, the motion of each muscle is estimated based on the motion of feature points in the region around the muscle. Details of the algorithm are described in the following subsections.

2.1   Tracking Feature Points

To track the feature points of a head, we apply the Kanade-Lucas-Tomasi tracking algorithm [10]. First, the feature points are selected from local extrema or saddle points of luminance distributions. For each pixel, the following matrix is computed.

$$\mathbf{Z} = \begin{bmatrix} g_x^2 & g_x g_y \\ g_x g_y & g_y^2 \end{bmatrix}, \tag{1}$$

where g_x and g_y denote the average of the gradient in the x and y directions, respectively, over the square-shaped window centered at the pixel. If the matrix $\mathbf{Z}$ has two non-zero eigenvalues, the point of the pixel can be interpreted as a local extremum or saddle point; here, the eigenvalues are related to the rate of change of the gradient value along the steepest direction and along its perpendicular direction. Feature points with a larger second eigenvalue are selected to be tracked.

Feature points as selected above are tracked by solving the following equation.

$$\mathbf{Z}\mathbf{d} = \int \int_{\mathbf{W}} [I(\mathbf{x}) - J(\mathbf{x})] \begin{bmatrix} g_x \\ g_y \end{bmatrix} d\mathbf{x}, \tag{2}$$

where $I(\mathbf{x})$ and $J(\mathbf{x})$ denote the luminance of successive frames.

If the determinant of the matrix $\mathbf{Z}$ is zero, the vector $\mathbf{d}$ is not defined. In this case, the feature point is replaced by the new feature point with the larger second eigenvalue. An example of selected feature points is shown in Fig. 2.

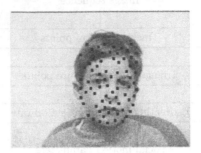

Fig. 2. Example of selected feature points.

Then, the head motion is estimated from the coordinates of feature points in successive frames using the epipolar geometry of a weak perspective projection model [11].

2.2 Triangulation of Feature Points

The feature points are connected with Delaunay triangulation using the algorithm proposed by Lawson [12,13]. In this algorithm, Delaunay triangulation is obtained by increasing the number of feature points one at a time. When a new point is added, the triangle including the point is computed using the data structure mentioned below. Then, the triangle is divided into three new triangles with the added point as a common vertex. Finally, the legality of the new triangles is determined, i.e., when there is a point that lies in the interior of the circle through the vertex of the new triangles, then, the triangle is illegal. The illegal triangle can be made legal by flipping the diagonal edge of the quadrilateral formed from the three vertexes of triangle and the point in the circle (Fig. 3). The data structure is maintained so that each triangle is represented by a node

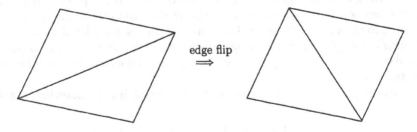

Fig. 3. Flipping an edge to remove an illegal triangle.

that has links from the divided triangle to the three new triangles and from the illegal triangles to the legal triangles.

This algorithm is proved to be an order of $n \log n$ for the expected computation time and an order of n for the expected storage area in the case of triangulation of n points. Similarly, the triangle, that includes a given point, can be found in an order of $\log n$.

The Delaunay triangles are computed for every frame of an image sequence, because the feature points cannot be accurately tracked over all the frames. For example, if the feature points are occluded by the head motion, feature points can not be tracked. Also, the feature points will disappear when facial change is large, i.e., the teeth suddenly appear while the mouth is opening.

The Delaunay triangulation is used to estimate the motion of facial points from the motion of near-by feature points by linear interpolation. In particular, the motion of points at the end of the linear muscles and on circular muscles can be used to estimate the contraction rate of the corresponding muscles. The velocity $\mathbf{u}$ of the point $\mathbf{p}$ in a triangle $\mathbf{abc}$ is derived from the velocity of the vertices of the triangle $\mathbf{a}$, $\mathbf{b}$ and $\mathbf{c}$ by linear interpolation by the following equation.

$$\mathbf{u} = \alpha \mathbf{u}_a + \beta \mathbf{u}_b + \gamma \mathbf{u}_c, \tag{3}$$

where $\mathbf{u}_a$, $\mathbf{u}_b$, and $\mathbf{u}_c$ are the velocity on the vertices of the triangle, and α, β, and γ are the barycentric coordinates of the point $\mathbf{p}$ as shown in Fig. 4

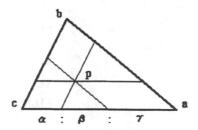

Fig. 4. Barycentric coordinates of the point p.

2.3   Estimating Muscle Contraction Rate

The facial motion parameters are determined based on the FACS as shown in Fig. 6. Table 1 lists the facial motion and the activated muscle for each AU. As these 19 AUs are the main components in six principal facial expressions (Table 2 [14]), it is probable that almost any facial expression can be generated by combining these AUs. In Fig. 5, an example of the Delaunay triangulation is shown.

The magnitude of the facial motion parameters is defined as the contraction rate of the corresponding muscles. The value of the contraction rate is one when the muscle is relaxed, and decreases toward zero when the muscle contracts. As there are three kinds of muscles, linear muscles, circular muscles and the mandible muscle, three methods for estimating their contraction rates are derived.

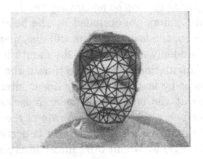

Fig. 5. Example of a Delaunay triangulation of feature points.

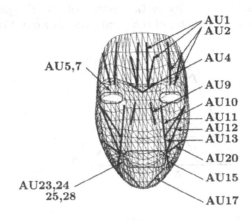

Fig. 6. Facial muscles corresponding to principal Action Units.

a. Linear muscles

In the linear muscles, one end attaches to a bone and the other end attaches to skin that moves to the other end while contracting. Here, we model a linear muscle as a line segment (Fig. 6) and the contraction rate of the muscle as a proportion of the length of the line to the original length. The displacement of the ends of linear muscles is interpolated from the displacement of the vertices of the triangle that includes the ends of the linear muscles. The positions of the line segments are manually adjusted at the first frame in x and y-coordinates (Fig. 7). The z-coordinates of the positions can be determined from the 3D geometry of the face. In the experiments mentioned below, however, we used the same z-coordinate value for the positions of the line segments. Therefore, the positions of the template for the linear muscle was deviated from the exact position when the the angle of head rotation was large.

AU1	Inner brow raiser	Frontalis, pars medialis
AU2	Outer brow raiser	Frontalis, pars lateralis
AU4	Brow lowerer	Depressor supercilii
AU5	Upper lid raiser	Levator palebrae superioris
AU7	Lid tightener	Orbicularis oculi
AU9	Nose wrinkler	Levator labii superioris alaeque nasi
AU10	Upper lid raiser	Levator labii superioris major
AU11	Nasolabial furrow deepener	Zygomatic minor
AU12	Lip corner puller	Zygomatic major
AU13	Cheek puffer	Levetor anguli oris
AU15	Lip corner depressor	Depressor anguli oris
AU16	Lower lip depressor	Depressor labii inferioris
AU17	Chin raiser	Mentalis
AU20	Lip stretcher	Risorius
AU23	Lip tightener	Orbicularis oris
AU24	Lip presser	Orbicularis oris
AU25	Lips part	Orbicularis oris
AU26	Jaw drop	Massetter (jaw rotation)
AU28	Lip suck	Orbicularis oris

Table 1. Facial motion for each AU and its activated muscle.

Expression	AU Combinations
Anger	AU2+4+5+10+23+24+28
Disgust	AU4+9+10+17
Fear	AU1+2+4+5+7+15+20+25
Happiness	AU1+2+5+6+10+12+13+20+25
Sadness	AU1+4+7+15
Surprise	AU1+2+5+26

Table 2. AU coding for principal facial expressions.

b. Circular muscles

The second kind of muscle is a circular muscle, which is placed around either the eyes or the mouth, which contracts toward the center of the muscles. We model a circular muscle as an ellipse (Fig. 6) and the contraction rate of the muscle as a proportion of the perimeter of the ellipse to the original length.

In the case of the mouth, we assume that the circular muscle lies under the edge of the upper and lower lips and that the shape of muscle changes similarly

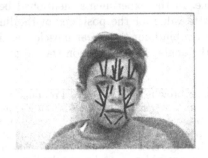

Fig. 7. Example of muscle positions.

in shape. Then, the displacement of the points on the ellipse is estimated by interpolation, and its normal component is averaged around the ellipse.

In the case of the eyes, the motion of the upper and lower eyelids is not similar unlike the motion of the mouth. Therefore, the displacement of the points on the major axis of the ellipse is estimated and averaged over the axis.

c. Mandible muscle

The mandible muscle connects the mandible to the skull, and opens the mouth while contracting. Here, we estimate the angle of the jaw rotation from the displacement of the lowest part of feature points on the face from Eq. 4.

$$\rho = \frac{\Delta V_\perp}{V_z}, \tag{4}$$

where $\Delta V_\perp$ is a component of the displacement vector ΔV parallel to the symmetric axis of the face, and V_z is the z-coordinate of the feature point derived from the assumption of cylindrical head shape (Fig. 8).

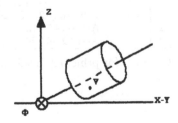

Fig. 8. Cylinder model for heads.

Therefore, the contraction rates of muscles for the 19 AUs are estimated. Of these 19 AUs, the linear muscles and the circular muscle of the eyes are placed at symmetric positions about the symmetric axis of the face. As the same muscles at symmetric positions can be activated independently, 38 (16×2+6) independent parameters are obtained.

3   Experimental Results

To evaluate the proposed method, we executed experiments using image sequences displaying facial expressions. In these experiments, the number of feature points was set as high as possible to get a large amount of information. Fig. 9 shows a sample of experimental results.

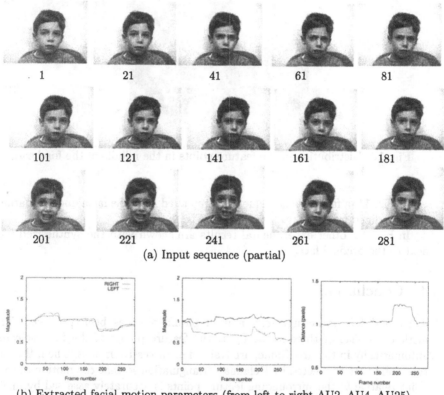

(a) Input sequence (partial)

(b) Extracted facial motion parameters (from left to right AU2, AU4, AU25)

Fig. 9. Sample of experimental results.

The figure shows the magnitude of AUs that are the main components of the displayed expressions such as 'Anger' and 'Fear.' In the left graph, the magnitude increased when the eyebrows was lowered, and decreased when the eyebrows was raised. As AU2 is defined as the magnitude of the eyebrow raising, this magnitude can be interpreted as the difference of the eyebrow raising (AU2) and the eyebrow lowering (AU4). As the motion of AU2 and AU4 occur simultaneously, this interpretation may be valid. Similarly, the right graph can be interpreted as the difference of the magnitude of the mouth tightening and the mouth opening. In the middle graph, the magnitude is not inversely related to the left graph as expected from that interpretation. This can be explained by the sparseness of feature points in the region around the muscles for AU4 (Fig. 10). As the motion of AU4 is accompanied with the wrinkles in the middle of the forehead, the detection of the wrinkles may be helpful in estimating the magnitude of AU4.

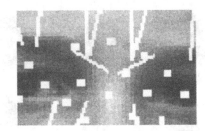

Fig. 10. Distribution of the feature points in the middle of the forehead.

In Fig. 11, a facial image sequence is rendered from the facial motion parameters obtained from the results shown in Fig. 9.

In Fig. 12, other experimental results are shown with the synthesized face next to the original face.

4   Conclusions

A method for extracting facial motion parameters has been proposed. The method consists of three steps. First, the feature points of the face, selected automatically in the first frame, are tracked in successive frames. Then, the feature points are connected by Delaunay triangulation so that the motion of each point relative to the surrounding feature points is accurately assessed by linear interpolation. Finally, the motion of a muscle is calculated based on the motion of the feature points placed near the muscle. The experiments showed that the proposed method can accurately extract facial motion parameters. Future work will aim to improve the computation speed in order to execute the process in real time.

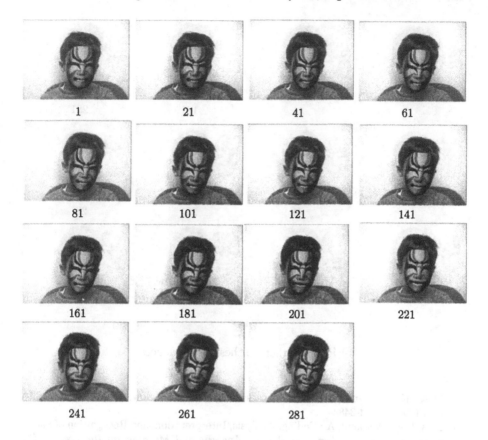

Fig. 11. Synthesized sequence from the results of Fig. 9.

Acknowledgments

We thank Dr. Yaser Yacoob and the Computer Vision Laboratory of Maryland University for making their facial expression sequences available to us.

References

1. Massaro, D, W.: *Perceiving Talking Faces*, MIT Press (1998).
2. Ekman, P., Friesen, W. V.: *The Facial Action Coding System*, Consulting Psychologists Press, Inc., (1978).
3. Terzopoulos, D., Waters, K.: Physically-based facial modeling, analysis, and animation, *The J. of Visualization and Computer Animation*, **1** (2) (1990) 73-80.
4. Terzopoulos, D., Waters, K.: Analysis and synthesis of facial image sequences using physical and anatomical models, *IEEE Trans. on Pattern Analysis and Machine Intelligence*, **15** (6) (1993) 569-579.

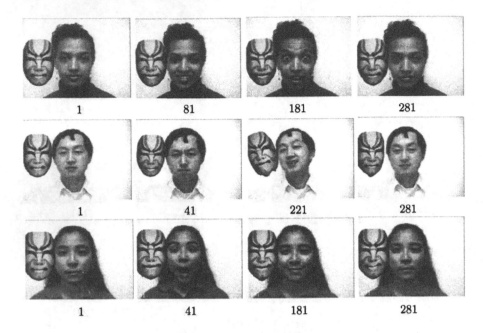

1	81	181	281
1	41	221	281
1	41	181	281

Fig. 12. Other synthesized sequences.

5. Mase, K.: Recognition of facial expression from optical flow, *IEICE Trans.*, **E74** (10) (1991) 3474-3483.
6. Essa, I. A., Pentland, A.: Coding, Analysis, Interpretation, and Recognition of Facial Expressions, *IEEE Trans. on Pattern Analysis and Machine Intelligence*, **19** (7) (1997).
7. Otsuka, T., Ohya, J.: Recognizing Multiple Persons' Facial Expressions Using HMM Based on Automatic Extraction of Significant Frames from Image Sequences, *ICIP'97*, vol. II (1997) 546-549.
8. DeCarlo, D., Metaxas, D.: Deformable Modes-Based Shape and Motion Analysis from Images using Motion Residual Error, *Proc. ICCV'98* (1998) 113-119.
9. Cambridge Digital Research Laboratory: FaceWorks, URL http://www.interface.digital.com/.
10. Shi, J., Tomasi, C.: Good features to track, *Proc of the IEEE Conf. on Computer Vision and Pattern Recognition* (1994) 593-600, 1994..
11. Shapiro, L., Zisserman, A., Brady, M.: 3D motion recovery via affine epipolar geometry, *Int. J. Computer Vision*, **16** (2) (1995) 147-182.
12. Lawson, C. L., Transforming triangulations, *Discrete Math.*, **3** (1972) 365-372.
13. de Berg, M., van Kreveld, M., Overmars, M., Schwarzkopf, O.: Computational Geometry, Chapter 9. Springer-Verlag (1997).
14. Pelachaud, C., Badler, N. I., Steedman, M.: Generating facial expressions for speech, Cognitive Science, **20** (1) (1996) 1-46.
15. Otsuka, T., J. Ohya, J.: Converting Facial Expressions Using Recognition-Based Analysis of Image Sequences, *ACCV'98*, vol. II (1998) 703-710.

Immersion Reconsidered
Virtual Reality Projects of Art+Com

Steffen Meschkat

Art+Com AG, Kleiststrasse 23-26, D 10787 Berlin, Germany
mesch@artcom.de

Abstract. This paper presents two application research and development projects in the field of virtual reality by Art+Com. The focus in this presentation as well as of the projects themselves is on the adaequate communication of information through virtual worlds. We particularly emphasize the medial and communication aspect of virtual reality applications and the role of the user interface in supporting collective communication processes based on the application of virtual reality.

1 Introduction

The developing technology of virtual reality matures and reaches ubiquity. Applications become essentially content driven and user focused. Virtual reality develops from a visualization tool to a medium of communication. While the experience of virtual reality merges into everday culture, the way of integration of the virtual within the real world becomes central to the success of a virtual reality application. Two applications of virtual reality are discussed in the following. They emphasize communication over simulation and thus exemplify the particular interdisciplinary understanding of Virtual Reality as is has been developed over the past decade by Art+Com.

The article proceeds as follows: The historical background is presented in the next section. Section 3 defines the understanding of virtual reality as it underlies the work of Art+Com. The Terravision telecommmunication research project is discussed in section 4, the Virtual Car application is discussed in section 5. We summarize and conclude in section 6.

2 History and Background

Art+Com was founded in 1988 as a research institute for computer-based design and visualization by architects affiliated with the Academy of Arts (HdK) in Berlin. The purpose was to facilitate interdisciplinary cooperation between artists, scientists and engineers, with the common basis of computer visualization. The motivating assumption was that the computer had developed from a technology which manipulates information to a medium of visual communication, suitable for expression of contents from different domains with its own

S. Nishio, F. Kishino (Eds.): AMCP'98, LNCS 1554, pp. 445–450, 1999.
© Springer-Verlag Berlin Heidelberg 1999

characteristics. At the very beginning, virtual reality was explored as a particular visual mode of computer based expression. First subjects of applications were architecture and town planning, scientific visualization, medical visualization. The subject of Art+Com may be described as application research, as a major part of the project work consists in figuring out what possible purposes an application of the new technology could have. However, the essence of this work is to develop the principles of visual expression within the new medium.

3   Virtual Reality

Virtual reality: *A medium that facilitates communication through the viewpoint dependent interactive presentation of spatial or spatialized content.*

We emphasize the communication medium and spatialization. Spatial representation provides a generic way to organize, navigate and manipulate information. It is crucial, however, to recognize the situations in which these advantages of spatial representation are important and why. Under the assumption that virtual reality indeed constitutes a new medium, this translates to the necessity of discovering and formulating its expressive language.

The definition above abstracts from some aspects which usually are considered crucial to virtual reality, namely immersion and multisensuality. We found classical immersive user interfaces to disrupt any collective communication process, both temporally and spatially, and hence to be precluded in our emphasis on the communication aspect. From the observation of the particular advantages of spatial organization of content follows a dominant role of visual perception in information systems. Hence we concentrate on the visual sense in virtual reality and consider the other sensual perceptions as auxiliary, albeit important.

The definition altogether ignores the dichotomy of real and virtual world, which is the actual subject of this paper. Virtual reality is well an increasingly important part of the reality, and mediated perception needs and creates real and material artefacts within the directly perceived world.

4   Terravision

Terravision is a VR model of the planet Earth in 1:1 scale. Geographic information is organized and represented within a virtual model of the planet Earth. The globe model of the Earth consists of a basis from satellite images and aerial photographs combined with digital elevation models at a wide range of resolutions. The system allows interactive visualization of this database with support for the wide range of resolution with which information are provided by the system. It is possible to interactively and continuously fly and zoom from an orbit with a view of the whole planet, down to Europe, to Germany, to Berlin, to the Breitscheidt square, and to enter the buildings where the Art+Com offices are located.

Terravision provides a uniform presentation of information of different types and origins. The range of combination of hetrogeneous information includes, but is not limited to, terrain elevation data at different resolutions, archived satellite images, online satellite images, and architectual CAD models.

The size of data which is available to the presentation within Terravision precludes the ability to load the whole model into memory at once. It is impossible to even concentrate all data onto secondary storage of one single computer. Instead, the data is distributed across a global network of computers. Only the data which is pertinent to the current field of view are loaded over the network and integrated in the representation of the model. The model in the memory of the computer is continuously modified and adapted according to the current field of view of the user.

It is worth noticing how the particular properties of spatial representation of information allow for a meaningful interaction with models as large as the whole planet Earth (modelled at 10cm resolution), which are impossible to be stored locally at one single point. Spatial representation places information in a hierarchy of contexts, it allows for synopsis and focus, and it provides generalization. These properties are desirable for information systems in general in order to provide access to and overview of very large amounts of information. They are present in a natural way in spatial information systems. Thus an essential connection between virtual reality and telecommunication is revealed.

To navigate this virtual world, a particular physical user interface is provided: The earth tracker implements the interfaces for global and local navigation. It integrates different input devices which serve in different aspects of navigation and manipulation: The big 3DOF Trackball for global navigation, a 6DOF Spacemouse for local navigation, and a touch display for interaction and manipulation. This user interface thus creates a physical representation of the virtual within the real world.

5 Virtual Car

The virtual car is a fully configurable model of an automobile (the A-Class car by Mercedes-Benz) within a virtual show room. It is intended to assist the communication between sales staff and the customer, and to support the customer's decision of their desired configuration of the car. Because of this intention, the interface to this system must not disrupt the natural flow of communication between the customer and the sales person. Collective and shared interaction processes among multiple persons must be supported by the system. The visual and spatial realism must allow for a life-like impression of the car.

The VR model of the car is presented through a user interface, called Holoflex, which makes any detail of the virtual car accessible at its position in real space: A flat panel is suspended within a mechnical mount. It can be moved freely in space at 5 degrees of freedom by holding handles to the right and left of the display and moving the display. When releaseing the grip of the handles, the display is arrested by mechanical brakes and thus keeps its position. The display is touch sensitve, allowing for interaction with the objects and components in view through the panel. The user interface creates thus a direct experience of the actual size of the represented object. Real space and virtual space are aligned and connected through the user interface. The Holoflex as the interface immerses the virtual world in the real world, which gives the interesting spin to the concept of immersive user interface which was hinted at earlier in this article.

6   Conclusion

This article presents two recent virtual reality applications realized by Art+Com. The unifying assumption underlying the concepts of this projects is that virtual reality has developed from a technology to a medium of communication in its own right which possesses its own intrinsically adeaquate language of visual expression.

In the rapidly developing information society, spatialization potentially harnesses the amount of information which is available through digital communication networks. There is an essential interrelationship of virtual reality and digital telecommunication networks.

As with every telecommunication medium, there appear particular physical artefacts within the real material world which serve as anchors to the mediated perception. In the case of virtual reality, it is the role of the user interfaces to seamlessly embed the objects of the virtual world into the real world.

7   Acknowledgements

The explanations in this paper were not to be made without the actual work done with the colleagues of Art+Com. Terravision was funded within the application research program of the german Telecom, Deutsche Telekom AG, and its research subsidiary Berkom GmbH. The virtual car was commissioned by Mercedes-Benz AG.

References

[1] Art+Com AG, http://www.artcom.de/
[2] Terravision, http://www.artcom.de/projects/terra/
[3] Virtual Car http://www.artcom.de/projects/vrf/
[4] Mercedes-Benz AG, http://www.mercedes-benz.com/
[5] Berkom GmbH, http://www.berkom.de/

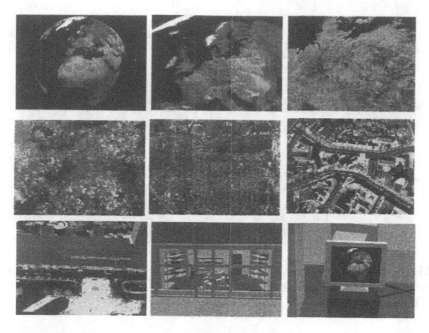

Fig. 1. The virtual world of Terravision. The figure shows 9 stages of a continuous flight and zoom from an orbit down into the offices of Art+Com in Berlin.

Fig. 2. Terravision within the real world. The trackball of the earth tracker provides a handle of virtual earth.

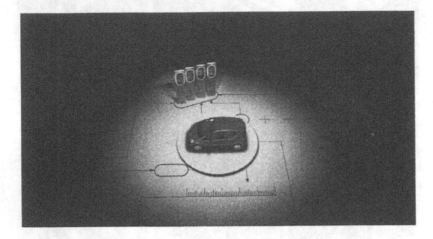

Fig. 3. Virtual Car within the virtual show room. Notice that a virtual show room does not need walls as necessary as a real show room does. A floor, however, is essential.

Fig. 4. Virtual Car within the real show room. Beware, the image is digitally composed. The car in the center is from the virtual world; only the images of the car on the both displays are "real".

Synthetic Characters: Behaving in Character

Bruce M. Blumberg

Asahi Broadcasting Corporation Career Development
Assistant Professor of Media Arts and Sciences
The Media Lab, MIT

E15-311, 20 Ames St Cambridge Ma. 02139
Bruce@media.mit.edu

Abstract.

Digital pets such as the Tamagotchi or Creatures are harbingers of a new kind of interactive character. They are characters that must maintain the perception of sentience over extended periods of un-scripted interaction in dynamic and noisy environments. This perception of sentience is created by the character's behavior over time (i.e. what it does and how it does it) as a coherent, character-specific, expression of its motivational and emotional state and reflecting its acquired "knowledge" of its environment. Ultimately this means we must understand how to build complete, if simple characters. In this talk I will present the approach the Synthetic Characters Group at the MIT Media Lab is taking to address this problem. Borrowing ideas from sources as wide ranging as animal behavior to classical animation, I will describe our approach to modeling drives, emotions, perception and behavior in an integrated architecture, as well as how we address the problem of expressive motor control. But I will also argue that the issues of camera control and interaction need to be more than an after-thought and I will present our approach to each of these problems. Finally, I will show videos of our SWAMPED! project, an interactive virtual cartoon world in which the participant uses a wireless, instrumented stuffed chicken to direct a semi-autonomous animated chicken as it interacts with a fully autonomous animated raccoon in a manner reminiscent of a Warner Brothers cartoon. SWAMPED! was part of the Enhanced Realities venue of Siggraph '98 and was used by over 400 participants.

S. Nishio, F. Kishino (Eds.): AMCP'98, LNCS 1554, pp. 451-451, 1999.
© Springer-Verlag Berlin Heidelberg 1999

Author Index

Aksoy, D.	194	Lee, S. H.	131
Akutsu, A.	17	Li, K. F.	177
Altinel, M.	194		
Apers, P.	119	Masaki, T.	386
Arikawa, M.	313	Mase, K.	161
Ariki, Y.	75	Matsuyama, T.	252
		Meschkat, S.	445
Blumberg, B. M.	451	Mille, A.	328
Bose, R.	194	Miyabe, Y.	236
		Miyahara, H.	221
Cetintemel, U.	194	Miyasato, T.	370
Cho, K. H.	131		
Choi, K. S.	131	Nabeshima, S.	236
Chu, C.-X.	145	Nakajima, M.	301
Chua, T.-S.	145	Nakatsu, R.	397
		Nakazawa, M.	103
Endo, T.	103	Niikura, Y.	17
		Nishida, M.	75
Fels, S.	161	Nishida, S.	422
Franklin, M.	194	Nishimura, G.	289
Fuchs, H.	30	Nomura, Y.	221
		Ogata, J.	75
Guibaly, F. E.	177	Ohya, J.	433
		Oka, R.	103
Hur, D. Y.	131	Okada, Y.	208
		Okamura, K.	236
Ide, I.	87	Otsuka, T.	433
Ikeuchi, K.	44		
Imai, M.	370	Paulin Carlos, R.	1
		Pinon, J.-M.	328
Kagawa, O.	344	Prié, Y.	328
Kakiuchi, T.	236		
Kamahara, J.	221	Saiwaki, N.	422
Kambayashi, Y.	344	Sakamoto, Y.	208
Kamiya, Y.	344	Sato, I.	44
Kamiyama, T.	313	Sato, Y.	44
Kandori, K.	221	Shimojo, S.	221
Kaneji, K.	208	Shin, M. K.	131
Kataoka, R.	289	Sumiya, K.	236
Katayama, K.	344		
Kersten, M.	119	Takahashi, H.	301
Kihara, T.	289	Takao, N.	236
Kitamura, Y.	386	Tanaka, H.	87
Komatsu, T.	422	Tanaka, K.	356

Taniguchi, Y.	17	Watanabe, Y.	208
Terzopoulos, D.	406	Wedlake, M.	177
Tokai, S.	252		
Tonomura, Y.	17	Yamaguchi, T.	386
Tsukamoto, M.	59	Yamamoto, K.	87
Tsushima, H.	344	Yoo, J. S.	131
		Yoshihiro, T.	344
Ueda, K.	221		
Uehara, K.	1, 356	Zdonik, S.	194
		Zettsu, K.	356
Wada, T.	252	Zhang, J.	103
Wang, J.	194	Zhang, X.	301

Springer
and the
environment

At Springer we firmly believe that an international science publisher has a special obligation to the environment, and our corporate policies consistently reflect this conviction.

We also expect our business partners – paper mills, printers, packaging manufacturers, etc. – to commit themselves to using materials and production processes that do not harm the environment. The paper in this book is made from low- or no-chlorine pulp and is acid free, in conformance with international standards for paper permanency.

Springer

Lecture Notes in Computer Science

For information about Vols. 1–1496
please contact your bookseller or Springer-Verlag

Vol. 1497: V. Alexandrov, J. Dongarra (Eds.), Recent Advances in Parallel Virtual Machine and Message Passing Interface. Proceedings, 1998. XII, 412 pages. 1998.

Vol. 1498: A.E. Eiben, T. Bäck, M. Schoenauer, H.-P. Schwefel (Eds.), Parallel Problem Solving from Nature – PPSN V. Proceedings, 1998. XXIII, 1041 pages. 1998.

Vol. 1499: S. Kutten (Ed.), Distributed Computing. Proceedings, 1998. XII, 419 pages. 1998.

Vol. 1500: J.-C. Derniame, B.A. Kaba, D. Wastell (Eds.), Software Process: Principles, Methodology, and Technology. XIII, 307 pages. 1999.

Vol. 1501: M.M. Richter, C.H. Smith, R. Wiehagen, T. Zeugmann (Eds.), Algorithmic Learning Theory. Proceedings, 1998. XI, 439 pages. 1998. (Subseries LNAI).

Vol. 1502: G. Antoniou, J. Slaney (Eds.), Advanced Topics in Artificial Intelligence. Proceedings, 1998. XI, 333 pages. 1998. (Subseries LNAI).

Vol. 1503: G. Levi (Ed.), Static Analysis. Proceedings, 1998. IX, 383 pages. 1998.

Vol. 1504: O. Herzog, A. Günter (Eds.), KI-98: Advances in Artificial Intelligence. Proceedings, 1998. XI, 355 pages. 1998. (Subseries LNAI).

Vol. 1505: D. Caromel, R.R. Oldehoeft, M. Tholburn (Eds.), Computing in Object-Oriented Parallel Environments. Proceedings, 1998. XI, 243 pages. 1998.

Vol. 1506: R. Koch, L. Van Gool (Eds.), 3D Structure from Multiple Images of Large-Scale Environments. Proceedings, 1998. VIII, 347 pages. 1998.

Vol. 1507: T.W. Ling, S. Ram, M.L. Lee (Eds.), Conceptual Modeling – ER '98. Proceedings, 1998. XVI, 482 pages. 1998.

Vol. 1508: S. Jajodia, M.T. Özsu, A. Dogac (Eds.), Advances in Multimedia Information Systems. Proceedings, 1998. VIII, 207 pages. 1998.

Vol. 1510: J.M. Zytkow, M. Quafafou (Eds.), Principles of Data Mining and Knowledge Discovery. Proceedings, 1998. XI, 482 pages. 1998. (Subseries LNAI).

Vol. 1511: D. O'Hallaron (Ed.), Languages, Compilers, and Run-Time Systems for Scalable Computers. Proceedings, 1998. IX, 412 pages. 1998.

Vol. 1512: E. Giménez, C. Paulin-Mohring (Eds.), Types for Proofs and Programs. Proceedings, 1996. VIII, 373 pages. 1998.

Vol. 1513: C. Nikolaou, C. Stephanidis (Eds.), Research and Advanced Technology for Digital Libraries. Proceedings, 1998. XV, 912 pages. 1998.

Vol. 1514: K. Ohta, D. Pei (Eds.), Advances in Cryptology – ASIACRYPT'98. Proceedings, 1998. XII, 436 pages. 1998.

Vol. 1515: F. Moreira de Oliveira (Ed.), Advances in Artificial Intelligence. Proceedings, 1998. X, 259 pages. 1998. (Subseries LNAI).

Vol. 1516: W. Ehrenberger (Ed.), Computer Safety, Reliability and Security. Proceedings, 1998. XVI, 392 pages. 1998.

Vol. 1517: J. Hromkovič, O. Sýkora (Eds.), Graph-Theoretic Concepts in Computer Science. Proceedings, 1998. X, 385 pages. 1998.

Vol. 1518: M. Luby, J. Rolim, M. Serna (Eds.), Randomization and Approximation Techniques in Computer Science. Proceedings, 1998. IX, 385 pages. 1998.

1519: T. Ishida (Ed.), Community Computing and Support Systems. VIII, 393 pages. 1998.

Vol. 1520: M. Maher, J.-F. Puget (Eds.), Principles and Practice of Constraint Programming - CP98. Proceedings, 1998. XI, 482 pages. 1998.

Vol. 1521: B. Rovan (Ed.), SOFSEM'98: Theory and Practice of Informatics. Proceedings, 1998. XI, 453 pages. 1998.

Vol. 1522: G. Gopalakrishnan, P. Windley (Eds.), Formal Methods in Computer-Aided Design. Proceedings, 1998. IX, 529 pages. 1998.

Vol. 1524: G.B. Orr, K.-R. Müller (Eds.), Neural Networks: Tricks of the Trade. VI, 432 pages. 1998.

Vol. 1525: D. Aucsmith (Ed.), Information Hiding. Proceedings, 1998. IX, 369 pages. 1998.

Vol. 1526: M. Broy, B. Rumpe (Eds.), Requirements Targeting Software and Systems Engineering. Proceedings, 1997. VIII, 357 pages. 1998.

Vol. 1527: P. Baumgartner, Theory Reasoning in Connection Calculi. IX, 283. 1999. (Subseries LNAI).

Vol. 1528: B. Preneel, V. Rijmen (Eds.), State of the Art in Applied Cryptography. Revised Lectures, 1997. VIII, 395 pages. 1998.

Vol. 1529: D. Farwell, L. Gerber, E. Hovy (Eds.), Machine Translation and the Information Soup. Proceedings, 1998. XIX, 532 pages. 1998. (Subseries LNAI).

Vol. 1530: V. Arvind, R. Ramanujam (Eds.), Foundations of Software Technology and Theoretical Computer Science. XII, 369 pages. 1998.

Vol. 1531: H.-Y. Lee, H. Motoda (Eds.), PRICAI'98: Topics in Artificial Intelligence. XIX, 646 pages. 1998. (Subseries LNAI).

Vol. 1096: T. Schael, Workflow Management Systems for Process Organisations. Second Edition. XII, 229 pages. 1998.

Vol. 1532: S. Arikawa, H. Motoda (Eds.), Discovery Science. Proceedings, 1998. XI, 456 pages. 1998. (Subseries LNAI).

Vol. 1533: K.-Y. Chwa, O.H. Ibarra (Eds.), Algorithms and Computation. Proceedings, 1998. XIII, 478 pages. 1998.

Vol. 1534: J.S. Sichman, R. Conte, N. Gilbert (Eds.), Multi-Agent Systems and Agent-Based Simulation. Proceedings, 1998. VIII, 237 pages. 1998. (Subseries LNAI).

Vol. 1535: S. Ossowski, Co-ordination in Artificial Agent Societies. XV; 221 pages. 1999. (Subseries LNAI).

Vol. 1536: W.-P. de Roever, H. Langmaack, A. Pnueli (Eds.), Compositionality: The Significant Difference. Proceedings, 1997. VIII, 647 pages. 1998.

Vol. 1537: N. Magnenat-Thalmann, D. Thalmann (Eds.), Modelling and Motion Capture Techniques for Virtual Environments. Proceedings, 1998. IX, 273 pages. 1998. (Subseries LNAI).

Vol. 1538: J. Hsiang, A. Ohori (Eds.), Advances in Computing Science – ASIAN'98. Proceedings, 1998. X, 305 pages. 1998.

Vol. 1539: O. Rüthing, Interacting Code Motion Transformations: Their Impact and Their Complexity. XXI,225 pages. 1998.

Vol. 1540: C. Beeri, P. Buneman (Eds.), Database Theory – ICDT'99. Proceedings, 1999. XI, 489 pages. 1999.

Vol. 1541: B. Kågström, J. Dongarra, E. Elmroth, J. Waśniewski (Eds.), Applied Parallel Computing. Proceedings, 1998. XIV, 586 pages. 1998.

Vol. 1542: H.I. Christensen (Ed.), Computer Vision Systems. Proceedings, 1999. XI, 554 pages. 1999.

Vol. 1543: S. Demeyer, J. Bosch (Eds.), Object-Oriented Technology ECOOP'98 Workshop Reader. 1998. XXII, 573 pages. 1998.

Vol. 1544: C. Zhang, D. Lukose (Eds.), Multi-Agent Systems. Proceedings, 1998. VII, 195 pages. 1998. (Subseries LNAI).

Vol. 1545: A. Birk, J. Demiris (Eds.), Learning Robots. Proceedings, 1996. IX, 188 pages. 1998. (Subseries LNAI).

Vol. 1546: B. Möller, J.V. Tucker (Eds.), Prospects for Hardware Foundations. Survey Chapters, 1998. X, 468 pages. 1998.

Vol. 1547: S.H. Whitesides (Ed.), Graph Drawing. Proceedings 1998. XII, 468 pages. 1998.

Vol. 1548: A.M. Haeberer (Ed.), Algebraic Methodology and Software Technology. Proceedings, 1999. XI, 531 pages. 1999.

Vol. 1550: B. Christianson, B. Crispo, W.S. Harbison, M. Roe (Eds.), Security Protocols. Proceedings, 1998. VIII, 241 pages. 1999.

Vol. 1551: G. Gupta (Ed.), Practical Aspects of Declarative Languages. Proceedings, 1999. VIII, 367 pgages. 1999.

Vol. 1552: Y. Kambayashi, D.L. Lee, E.-P. Lim, M.K. Mohania, Y. Masunaga (Eds.), Advances in Database Technologies. Proceedings, 1998. XIX, 592 pages. 1999.

Vol. 1553: S.F. Andler, J. Hansson (Eds.), Active, Real-Time, and Temporal Database Systems. Proceedings, 1997. VIII, 245 pages. 1998.

Vol. 1554: S. Nishio, F. Kishino (Eds.), Advanced Multimedia Content Processing. Proceedings, 1998. XIV, 454 pages. 1999.

Vol. 1555: J.P. Müller, M.P. Singh, A.S. Rao (Eds.), Intelligent Agents V. Proceedings, 1998. XXIV, 455 pages. 1999. (Subseries LNAI).

Vol. 1557: P. Zinterhof, M. Vajteršic, A. Uhl (Eds.), Parallel Computation. Proceedings, 1999. XV, 604 pages. 1999.

Vol. 1558: H. J.v.d. Herik, H. Iida (Eds.), Computers and Games. Proceedings, 1998. XVIII, 337 pages. 1999.

Vol. 1559: P. Flener (Ed.), Logic-Based Program Synthesis and Transformation. Proceedings, 1998. X, 331 pages. 1999.

Vol. 1560: K. Imai, Y. Zheng (Eds.), Public Key Cryptography. Proceedings, 1999. IX, 327 pages. 1999.

Vol. 1561: I. Damgård (Ed.), Lectures on Data Security. VII, 250 pages. 1999.

Vol. 1563: Ch. Meinel, S. Tison (Eds.), STACS 99. Proceedings, 1999. XIV, 582 pages. 1999.

Vol. 1567: P. Antsaklis, W. Kohn, M. Lemmon, A. Nerode, S. Sastry (Eds.), Hybrid Systems V. X, 445 pages. 1999.

Vol. 1568: G. Bertrand, M. Couprie, L. Perroton (Eds.), Discrete Geometry for Computer Imagery. Proceedings, 1999. XI, 459 pages. 1999.

Vol. 1569: F.W. Vaandrager, J.H. van Schuppen (Eds.), Hybrid Systems: Computation and Control. Proceedings, 1999. X, 271 pages. 1999.

Vol. 1570: F. Puppe (Ed.), XPS-99: Knowledge-Based Systems. VIII, 227 pages. 1999. (Subseries LNAI).

Vol. 1572: P. Fischer, H.U. Simon (Eds.), Computational Learning Theory. Proceedings, 1999. X, 301 pages. 1999. (Subseries LNAI).

Vol. 1575: S. Jähnichen (Ed.), Compiler Construction. Proceedings, 1999. X, 301 pages. 1999.

Vol. 1576: S.D. Swierstra (Ed.), Programming Languages and Systems. Proceedings, 1999. X, 307 pages. 1999.

Vol. 1577: J.-P. Finance (Ed.), Fundamental Approaches to Software Engineering. Proceedings, 1999. X, 245 pages. 1999.

Vol. 1578: W. Thomas (Ed.), Foundations of Software Science and Computation Structures. Proceedings, 1999. X, 323 pages. 1999.

Vol. 1579: W.R. Cleaveland (Ed.), Tools and Algorithms for the Construction and Analysis of Systems. Proceedings, 1999. XI, 445 pages. 1999.

Vol. 1580: A. Včkovski, K.E. Brassel, H.-J. Schek (Eds.), Interoperating Geographic Information Systems. Proceedings, 1999. XI, 329 pages. 1999.

Vol. 1581: J.-Y. Girard (Ed.), Typed Lambda Calculi and Applications. Proceedings, 1999. VIII, 397 pages. 1999.

Vol. 1582: A. Lecomte, F. Lamarche, G. Perrier (Eds.), Logical Aspects of Computational Linguistics. Proceedings, 1997. XI, 251 pages. 1999. (Subseries LNAI).

Vol. 1586: J. Rolim et al. (Eds.), Parallel and Distributed Processing. Proceedings, 1999. XVII, 1443 pages. 1999.

Vol. 1587: J. Pieprzyk, R. Safavi-Naini, J. Seberry (Eds.), Information Security and Privacy. Proceedings, 1999. XI, 327 pages. 1999.

Vol. 1593: P. Sloot, M. Bubak, A. Hoekstra, B. Hertzberger (Eds.), High-Performance Computing and Networking. Proceedings, 1999. XXIII, 1318 pages. 1999.